D0305832

Practical
Multivariate Analysis
Fifth Edition

CHAPMAN & HALL/CRC
Texts in Statistical Science Series

Series Editors

Francesca Dominici, *Harvard School of Public Health, USA*
Julian J. Faraway, *University of Bath, UK*
Martin Tanner, *Northwestern University, USA*
Jim Zidek, *University of British Columbia, Canada*

Texts in Statistical Science

Practical
Multivariate Analysis
Fifth Edition

Abdelmonem Afifi
Susanne May
Virginia A. Clark

CRC Press
Taylor & Francis Group
Boca Raton London New York

CRC Press is an imprint of the
Taylor & Francis Group an **informa** business

A CHAPMAN & HALL BOOK

The previous edition of this book was entitled *Computer-Aided Multivariate Analysis, Fourth Edition*.

CRC Press
Taylor & Francis Group
6000 Broken Sound Parkway NW, Suite 300
Boca Raton, FL 33487-2742

© 2012 by Taylor & Francis Group, LLC
CRC Press is an imprint of Taylor & Francis Group, an Informa business

No claim to original U.S. Government works

Printed in the United States of America on acid-free paper
Version Date: 20110510

International Standard Book Number: 978-1-4398-1680-6 (Hardback)

Visit the Taylor & Francis Web site at
http://www.taylorandfrancis.com

and the CRC Press Web site at
http://www.crcpress.com

Contents

CONTENTS

Preface

In 1984, when the first edition of this book appeared, the title "Computer Aided Multivariate Analysis" distinguished it from other books that were more theoretically oriented. Today, it is impossible to think of a book on multivariate analysis for scientists and applied researchers that is not computer oriented. We therefore decided to change the title to "Practical Multivariate Analysis" to better characterize the nature of the book.

We wrote this book for investigators, specifically behavioral scientists, biomedical scientists, and industrial or academic researchers, who wish to perform **multivariate statistical analyses** and understand the results. We expect readers to be able to perform and understand the results, but also expect them to know when to ask for help from an expert on the subject. It can either be used as a self-guided textbook or as a text in an applied course in multivariate analysis. In addition, we believe that the book can be helpful to many statisticians who have been trained in conventional mathematical statistics who are now working as statistical consultants and need to explain multivariate statistical concepts to clients with a limited background in mathematics.

We do not present mathematical derivations of the techniques; rather we rely on geometric and graphical arguments and on examples to illustrate them. The mathematical level has been deliberately kept low. While the derivations of the techniques are referenced, we concentrate on applications to real-life problems, which we feel are the "fun" part of multivariate analysis. To this end, we assume that the reader will use a packaged software program to perform the analysis. We discuss specifically how each of six popular and comprehensive software packages can be used for this purpose. These packages are R, S-PLUS, SAS, SPSS, Stata, and STATISTICA. The book can be used, however, in conjunction with all other software packages since our presentation explains the output of most standard statistical programs.

We assume that the reader has taken a basic course in statistics that includes tests of hypotheses and covers one-way analysis of variance.

Approach of this book

The book has been written in a modular fashion. Part One, consisting of five chapters, provides examples of studies requiring multivariate analysis techniques, discusses characterizing data for analysis, computer programs, data entry, data management, data clean-up, missing values, and transformations; and presents a rough guide to assist in the choice of an appropriate multivariate analysis. We included these topics since many investigators have more difficulty with these preliminary steps than with running the multivariate analyses themselves. Also, if these steps are not done with care, the results of the statistical analysis can be faulty.

In the rest of the chapters, we follow a standard format. The first four sections of each chapter include a discussion of when the technique is used, a data example, and the basic assumptions and concepts of the technique. In subsequent sections, we present more detailed aspects of the analysis. At the end of each chapter, we give a summary table showing which features are available in the six software packages. We also include a section entitled "What to watch out for" to warn the reader about common problems related to data analysis. In those sections, we rely on our own experiences in consulting and those detailed in the literature to supplement the formal treatment of the subject.

Part Two covers regression analysis. Chapter 6 deals with simple linear regression and is included for review purposes to introduce our notation and to provide a more complete discussion of outliers and diagnostics than is found in some elementary texts. Chapters 7–9 are concerned with multiple linear regression. Multiple linear regression is used very heavily in practice and provides the foundation for understanding many concepts relating to residual analysis, transformations, choice of variables, missing values, dummy variables, and multicollinearity. Since these concepts are essential to a good grasp of multivariate analysis, we thought it useful to include these chapters in the book.

Part Three is what might be considered the heart of multivariate analysis. It includes chapters on canonical correlation analysis, discriminant analysis, logistic regression analysis, survival analysis, principal components analysis, factor analysis, cluster analysis, and log-linear analysis. The multivariate analyses have been discussed more as separate techniques than as special cases of some general framework. The advantage of this approach is that it allows us to concentrate on explaining how to analyze a certain type of data from readily available computer programs to answer realistic questions. It also enables the reader to approach each chapter independently. We did include interspersed discussions of how the different analyses relate to each other in an effort to describe the "big picture" of multivariate analysis. In Part Three, we include a chapter on regression of correlated data resulting from clustered or longitudinal samples. Although this chapter could have been included in Part Two, the fact that outcome data of such samples are correlated makes it appropriate to think of this as a multivariate analysis technique.

How to use the book

We have received many helpful suggestions from instructors and reviewers on how to order these chapters for reading or teaching purposes. For example, one instructor uses the following order in teaching: principal components, factor analysis, cluster analysis, and then canonical correlation. Another prefers presenting a detailed treatment of multiple regression followed by logistic regression and survival analysis. Instructors and self-learning readers have a wide choice of other orderings of the material because the chapters are largely self contained.

What's new in the Fifth Edition

During the nearly thirty years since we wrote the first edition of this book, tremendous advances have taken place in the fields of computing and software development. These advances have made it possible to quickly perform any of the multivariate analyses that were available only in theory at that time. They also spurred the invention of new multivariate analyses as well as new options for many of the standard methods. In this edition, we have taken advantage of these developments and made many changes, as will be described below.

For each of the techniques discussed, we used the most recent software versions available and discussed the most modern ways of performing the analysis. In each chapter, we updated the references to today's literature (while still including the fundamental original references) and, in order to eliminate duplication, we combined all the references into one section at the end of the book. We also added a number of problems to the many already included at the ends of the chapters. In terms of statistical software, we added the package R, a popular freeware that is being used increasingly by statisticians. In addition to the above-described modifications, we reorganized Chapter 3 to reflect the current software packages we used. In Chapter 5, we modified Table 5.2 to include the analyses added to this edition.

As mentioned above, we added a chapter on regression of correlated outcomes. Finally, in each chapter we updated and /or expanded the summary table of the options available in the six statistical packages to make it consistent with the most recent software versions.

We added three data sets to the six previously used in the book for examples and problems. These are described throughout the book as needed and summarized in Appendix A. Two web sites are available. The first one is the CRC web site:

    http://www.crcpress.com/product/isbn/9781439816806

From this site, you can download all the data sets used in the book. The other web site that is available to all readers is:

    http://statistics.ats.ucla.edu/stat/examples/pma5/default.htm

This site, developed by the UCLA Academic Technology Services, includes the data sets in the formats of various statistical software packages as well as illustrations of the examples in all chapters, complete with code from the software packages used in the book. We encourage readers to obtain data from either web site and frequently refer to the solutions given in the UCLA web site for practice.

As we might discover typographical errors in this book at a later time, we will include an *errata* document at the CRC web site which can be downloaded together with other data sets. Other updates may also be uploaded as needed.

Acknowledgments

We would like to express our appreciation to our colleagues and former students and staff who helped us over the years, both in the planning and preparation of the various editions. These include our colleagues Drs. Carol Aneshensel, Roger Detels, Robert Elashoff, Ralph Frerichs, Mary Ann Hill, and Roberta Madison. Our former students include Drs. Stella Grosser, Luohua Jiang, Jack Lee, Steven Lewis, Tim Morgan, Leanne Streja and David Zhang. Our former staff includes Ms. Dorothy Breininger, Jackie Champion, and Anne Eiseman. In addition, we would like to thank Meike Jantzen for her help with the references.

We also thank Rob Calver and Sarah Morris from CRC Press for their very capable assistance in the preparation of the fifth edition.

We especially appreciate the efforts of the staff of UCLA Academic Technology Services in putting together the UCLA web site of examples from the book (referenced above).

Our deep gratitude goes to our spouses Marianne Afifi, Bruce Jacobson, and Welden Clark for their patience and encouragement throughout the stages of conception, writing, and production of the book. Special thanks go to Welden Clark for his expert assistance and troubleshooting of the electronic versions of the manuscript.

A. Afifi
S. May
V.A. Clark

Authors' Biographies

Abdelmonem Afifi, Ph.D., has been Professor of Biostatistics in the School of Public Health, University of California, Los Angeles (UCLA) since 1965, and served as the Dean of the School from 1985 until June 2000. Currently, he is the senior statistician on several research projects. His research includes multivariate and multilevel data analysis, handling missing observations in regression and discriminant analyses, meta analysis, and variable selection. He teaches well-attended courses in biostatistics for Public Health students and clinical research physicians, and doctoral-level courses in multivariate statistics and multilevel modeling. He has authored many publications, including two widely used books (with multiple editions) on multivariate analysis.

Susanne May, Ph.D., is an Associate Professor in the Department of Biostatistics at the University of Washington in Seattle. Her areas of expertise and interest include survival analysis, longitudinal data analysis, and clinical trials methodology. She has more than 15 years of experience as a statistical collaborator and consultant on health related research projects. In addition to a number of methodological and applied publications, she is a coauthor (with Drs. Hosmer and Lemeshow) of *Applied Survival Analysis: Regression Modeling of Time-to-Event Data.* Dr. May has taught courses on introductory statistics, survival analysis, and multivariate analysis.

Virginia A. Clark, Ph. D., is professor emerita of Biostatistics and Biomathematics at UCLA. For 27 years, she taught courses in multivariate analysis and survival analysis, among others. In addition to this book, she is coauthor of four books on survival analysis, linear models and analysis of variance, and survey research as well as an introductory book on biostatistics. She has published extensively in statistical and health science journals. Currently, she is a statistical consultant to community-based organizations and statistical reviewer for medical journals.

Part I

Preparation for Analysis

Chapter 1

What is multivariate analysis?

1.1 Defining multivariate analysis

The expression **multivariate analysis** is used to describe analyses of data that are multivariate in the sense that numerous observations or variables are obtained for each individual or unit studied. In a typical survey 30 to 100 questions are asked of each respondent. In describing the financial status of a company, an investor may wish to examine five to ten measures of the company's performance. Commonly, the answers to some of these measures are interrelated. The challenge of disentangling complicated interrelationships among various measures on the same individual or unit and of interpreting these results is what makes multivariate analysis a rewarding activity for the investigator. Often results are obtained that could not be attained without multivariate analysis.

In the next section of this chapter several studies are described in which the use of multivariate analysis is essential to understanding the underlying problem. Section 1.3 gives a listing and a very brief description of the multivariate analysis techniques discussed in this book. Section 1.4 then outlines the organization of the book.

1.2 Examples of multivariate analyses

The studies described in the following subsections illustrate various multivariate analysis techniques. Some are used later in the book as examples.

Depression study example

The data for the depression study have been obtained from a complex, random, multiethnic sample of 1000 adult residents of Los Angeles County. The study was a **panel** or **longitudinal** design where the same respondents were interviewed four times between May 1979 and July 1980. About three-fourths of the respondents were re-interviewed for all four interviews. The field work for the survey was conducted by professional interviewers from the Institute for Social Science Research at UCLA.

This research is an epidemiological study of depression and help-seeking behavior among free-living (noninstitutionalized) adults. The major objectives are to provide estimates of the prevalence and incidence of depression and to identify causal factors and outcomes associated with this condition. The factors examined include demographic variables, life events stressors, physical health status, health care use, medication use, lifestyle, and social support networks. The major instrument used for classifying depression is the Depression Index (CESD) of the National Institute of Mental Health, Center of Epidemiological Studies. A discussion of this index and the resulting prevalence of depression in this sample is given in Frerichs, Aneshensel and Clark (1981).

The longitudinal design of the study offers advantages for assessing causal priorities since the time sequence allows us to rule out certain potential causal links. Nonexperimental data of this type cannot directly be used to establish causal relationships, but models based on an explicit theoretical framework can be tested to determine if they are consistent with the data. An example of such model testing is given in Aneshensel and Frerichs (1982).

Data from the first time period of the depression study are described in Chapter 3. Only a subset of the factors measured on a subsample of the respondents is included in this book's web site in order to keep the data set easily comprehensible. These data are used several times in subsequent chapters to illustrate some of the multivariate techniques presented in this book.

Parental HIV study

The data from the parental HIV study have been obtained from a clinical trial to evaluate an intervention given to increase coping skills (Rotheram-Borus *et al.*, 2001). The purpose of the intervention was to improve behavioral, social, and health outcomes for parents with HIV/AIDS and their children. Parents and their adolescent children were recruited from the New York City Division of Aids Services (DAS). Adolescents were eligible for the study if they were between the ages of 11 and 18 and if the parents and adolescents had given informed consent. Individual interviews were conducted every three months for the first two years and every six months thereafter. Information obtained in the interviews included background characteristics, sexual behavior, alcohol and drug use, medical and reproductive history, and a number of psychological scales.

A subset of the data from the study is available on this book's web site. To protect the identity of the participating adolescents we used the following procedures. We randomly chose one adolescent per family. In addition, we reduced the sample further by choosing a random subset of the original sample.

Adolescent case numbers were assigned randomly without regard to the original order or any other numbers in the original data set.

Data from the baseline assessment will be used throughout the book for problems as well as to illustrate various multivariate analysis techniques. A subset of the data which includes information from the baseline assessment is used for problems and for illustration of analysis techniques.

Northridge earthquake study

On the morning of January 17, 1994 a magnitude 6.7 earthquake centered in Northridge, CA awoke Los Angeles and Ventura County residents. Between August 1994 and May 1996, 1830 residents were interviewed about what happened to them in the earthquake. The study uses a telephone survey lasting approximately 48 minutes to assess the residents' experiences in and responses to the Northridge earthquake. Data from 506 residents are included in the data set posted on the book web site, and described in Appendix A.

Subjects were asked where they were, how they reacted, where they obtained information, whether their property was damaged or whether they experienced injury, and what agencies they were in contact with. The questionnaire included the Brief Symptom Inventory (BSI), a measure of psychological functioning used in community studies, and questions on emotional distress. Subjects were also asked about the impact of the damage to the transportation system as a result of the earthquake. Investigators not only wanted to learn about the experiences of the Southern California residents in the Northridge earthquake, but also wished to compare their findings to similar studies of the Los Angeles residents surveyed after the Whittier Narrows earthquake on October 1, 1987, and Bay Area residents interviewed after the Loma Prieta earthquake on October 17, 1989.

The Northridge earthquake data set is used in problems at the end of several chapters of the book to illustrate a number of multivariate techniques. Multivariate analyses of these data include, for example, exploring pre- and post-earthquake preparedness activities as taking into account several factors relating to the subject and the property (Nguyen *et al.*, 2006).

Bank loan study

The managers of a bank need some way to improve their prediction of which borrowers will successfully pay back a type of bank loan. They have data from the past on the characteristics of persons to whom the bank has lent money and the subsequent record of how well the person has repaid the loan. Loan payers can be classified into several types: those who met all of the terms of the loan, those who eventually repaid the loan but often did not meet deadlines, and those who simply defaulted. They also have information on age, sex, income, other indebtedness, length of residence, type of residence, family size, occupation, and the reason for the loan. The question is, can a simple rating system be devised that will help the bank personnel improve their prediction rate and lessen the time it takes to approve loans? The methods described in Chapters 11 and 12 can be used to answer this question.

Lung function study

The purpose of this lung function study of chronic respiratory disease is to determine the effects of various types of smog on lung function of children and adults in the Los Angeles area. Because they could not randomly assign people to live in areas that had different levels of pollutants, the investigators were very concerned about the interaction that might exist between the locations where persons chose to live and their values on various lung function tests. The investigators picked four areas of quite different types of air pollution and measured various demographic and other responses on all persons over seven years old who live there. These areas were chosen so that they are close to an air-monitoring station.

The researchers took measurements at two points in time and used the change in lung function over time as well as the levels at the two periods as outcome measures to assess the effects of air pollution. The investigators have had to do the lung function tests by using a mobile unit in the field, and much effort has gone into problems of validating the accuracy of the field observations. A discussion of the particular lung function measurements used for one of the four areas can be found in Detels *et al.* (1975). In the analysis of the data, adjustments must be made for sex, age, height, and smoking status of each person.

Over 15,000 respondents have been examined and interviewed in this study. The original data analyses were restricted to the first collection period, but now analyses include both time periods. The data set is being used to answer numerous questions concerning effects of air pollution, smoking, occupation, etc. on different lung function measurements. For example, since the investigators obtained measurements on all family members seven years old and older, it is possible to assess the effects of having parents who smoke on the lung function of their children (Tashkin *et al.*, 1984). Studies of this type require multivariate analyses so that investigators can arrive at plausible scientific conclusions that could explain the resulting lung function levels.

This data set is described in Appendix A. Lung function and associated data for nonsmoking families for the father, mother, and up to three children ages 7–17 are available from the book's web site.

Assessor office example

Local civil laws often require that the amount of property tax a homeowner pays be a percentage of the current value of the property. Local assessor's offices are charged with the function of estimating current value. Current value can be estimated by finding comparable homes that have been recently sold and using some sort of an average selling price as an estimate of the price of those properties not sold.

Alternatively, the sample of sold homes can indicate certain relationships between selling price and several other characteristics such as the size of the lot, the size of the livable area, the number of bathrooms, the location, etc. These relationships can then be incorporated into a mathematical equation used to estimate the current selling price from those other characteristics. Multiple regression analysis methods discussed in Chapters 7–9 can be used by many assessor's offices for this purpose (Tchira, 1973).

School data set

The school data set is a publicly available data set that is provided by the National Center for Educational Statistics. The data come from the National Education Longitudinal Study of 1988 (called NELS:88). The study collected data beginning with 8th graders and conducted initial interviews and four follow-up interviews which were performed every other year. The data used here contain only initial interview data. They represent a random subsample of 23 schools with 519 students out of more than a thousand schools with almost twenty five thousand students. Extensive documentation of all aspects of the study is available at the following web site: `http://nces.ed.gov/surveys/NELS88/`. The longitudinal component of NELS:88 has been used to investigate change in students' lives and school-related and other outcomes. The focus on the initial interview data provides the opportunity to examine associations between school and student-related factors and students' academic performance in a cross-sectional manner. This type of analysis will be illustrated in Chapter 18.

1.3 Multivariate analyses discussed in this book

In this section a brief description of the major multivariate techniques covered in this book is presented. To keep the statistical vocabulary to a minimum, we illustrate the descriptions by examples.

Simple linear regression

A nutritionist wishes to study the effects of early calcium intake on the bone density of postmenopausal women. She can measure the bone density of the arm (radial bone), in grams per square centimeter, by using a noninvasive device. Women who are at risk of hip fractures because of too low a bone density will tend to show low arm bone density also. The nutritionist intends to sample a group of elderly churchgoing women. For women over 65 years of age, she will plot calcium intake as a teenager (obtained by asking the women about their consumption of high-calcium foods during their teens) on the horizontal axis and arm bone density (measured) on the vertical axis. She expects the radial bone density to be lower in women who had a lower calcium intake. The nutritionist plans to fit a simple linear regression equation and test whether the slope of the regression line is zero. In this example a single outcome factor is being predicted by a single predictor factor.

Simple linear regression as used in this case would not be considered multivariate by some statisticians, but it is included in this book to introduce the topic of multiple regression.

Multiple linear regression

A manager is interested in determining which factors predict the dollar value of sales of the firm's personal computers. Aggregate data on population size, income, educational level, proportion of population living in metropolitan areas, etc. have been collected for 30 areas. As a first step, a multiple linear regression equation is computed, where dollar sales is the outcome variable and the other factors are considered as candidates for predictor variables. A linear combination of the predictors is used to predict the outcome or response variable.

Canonical correlation

A psychiatrist wishes to correlate levels of both depression and physical well-being from data on age, sex, income, number of contacts per month with family and friends, and marital status. This problem is different from the one posed in the multiple linear regression example because more than one outcome variable is being predicted. The investigator wishes to determine the linear function of age, sex, income, contacts per month, and marital status that is most highly correlated with a linear function of depression and physical well-being. After these two linear functions, called canonical variables, are determined, the investigator will test to see whether there is a statistically significant (canonical) correlation between scores from the two linear functions and whether a reasonable interpretation can be made of the two sets of coefficients from the functions.

Discriminant function analysis

A large sample of initially disease-free men over 50 years of age from a community has been followed to see who subsequently has a diagnosed heart attack. At the initial visit, blood was drawn from each man, and numerous other determinations were made, including body mass index, serum cholesterol, phospholipids, and blood glucose. The investigator would like to determine a linear function of these and possibly other measurements that would be useful in predicting who would and who would not get a heart attack within ten years. That is, the investigator wishes to derive a classification (discriminant) function that would help determine whether or not a middle-aged man is likely to have a heart attack.

Logistic regression

A television station staff has classified movies according to whether they have a high or low proportion of the viewing audience when shown. The staff has also measured factors such as the length and the type of story and the characteristics of the actors. Many of the characteristics are discrete yes–no or categorical types of data. The investigator will use logistic regression because some of the data do not meet the assumptions for statistical inference used in discriminant function analysis, but they do meet the assumptions for logistic regression. From logistic regression we derive an equation to estimate the probability of capturing a high proportion of the audience.

Poisson regression

In a health survey, middle school students were asked how many visits they made to the dentist in the last year. The investigators are concerned that many students in this community are not receiving adequate dental care. They want to determine what characterizes how frequently students go to the dentist so that they can design a program to improve utilization of dental care. Visits per year are counted data and Poisson regression analysis provides a good tool for analyzing this type of data. Poisson regression is covered in the logistics regression chapter.

Survival analysis

An administrator of a large health maintenance organization (HMO) has collected data since 1990 on length of employment in years for their physicians who are either family practitioners or internists. Some of the physicians are still employed, but many have left. For those still employed, the administrator can only know that their ultimate length of employment will be greater than their current length of employment. The administrator wishes to describe the distribution of length of employment for each type of physician, determine the possible effects of factors such as gender and location of work, and test whether or not the length of employment is the same for the two specialties. Survival analysis, or event history analysis (as it is often called by behavioral scientists), can be used to analyze the distribution of time to an event such as quitting work, having a relapse of a disease, or dying of cancer.

Principal components analysis

An investigator has made a number of measurements of lung function on a sample of adult males who do not smoke. In these tests each man is told to inhale deeply and then blow out as fast and as much as possible into a spirometer, which makes a trace of the volume of air expired over time. The maximum or forced vital capacity (FVC) is measured as the difference between maximum inspiration and maximum expiration. Also, the amount of air expired in the first second (FEV1), the forced mid-expiratory flow rate (FEF 25–75), the maximal expiratory flow rate at 50% of forced vital capacity (V50), and other measures of lung function are calculated from this trace. Since all these measures are made from the same flow–volume curve for each man, they are highly interrelated. From past experience it is known that some of these measures are more interrelated than others and that they measure airway resistance in different sections of the airway.

The investigator performs a principal components analysis to determine whether a new set of measurements called principal components can be obtained. These principal components will be linear functions of the original lung function measurements and will be uncorrelated with each other. It is hoped that the first two or three principal components will explain most of the variation in the original lung function measurements among the men. Also, it is anticipated that some operational meaning can be attached to these linear functions that will aid in their interpretation. The investigator may decide to do future analyses on these uncorrelated principal components rather than on the original data. One advantage of this method is that often fewer principal components are needed than original variables. Also, since the principal components are uncorrelated, future computations and explanations can be simplified.

Factor analysis

An investigator has asked each respondent in a survey whether he or she strongly agrees, agrees, is undecided, disagrees, or strongly disagrees with 15 statements concerning attitudes toward inflation. As a first step, the investigator will do a factor analysis on the resulting data to determine which statements belong together in sets that are uncorrelated with other sets. The particular statements that form a single set will be examined to obtain a better understanding of attitudes toward inflation. Scores derived from each set or factor will be used in subsequent analysis to predict consumer spending.

Cluster analysis

Investigators have made numerous measurements on a sample of patients who have been classified as being depressed. They wish to determine, on the basis of their measurements, whether these patients can be classified by type of depression. That is, is it possible to determine distinct types of depressed patients by performing a cluster analysis on patient scores on various tests?

Unlike the investigator of men who do or do not get heart attacks, these investigators do not possess a set of individuals whose type of depression can be known before the analysis is performed (see Andreasen and Grove, 1982, for an example). Nevertheless, the investigators want to separate the patients into separate groups and to examine the resulting groups to see whether distinct types do exist and, if so, what their characteristics are.

Log-linear analysis

An epidemiologist in a medical study wishes to examine the interrelationships among the use of substances that are thought to be risk factors for disease. These include four risk factors where the answers have been summarized into categories. The risk factors are smoking tobacco (yes at present, former smoker, never smoked), drinking (yes, no), marijuana use (yes, no), and other illicit drug use (yes, no). Previous studies have shown that people who drink are more apt than nondrinkers to smoke cigarettes, but the investigator wants to study the associations among the use of these four substances simultaneously.

Correlated outcomes regression

A health services researcher is interested in determining the hospital-related costs of appendectomy, the surgical removal of the appendix. Data are available for a number of patients in each of several hospitals. Such a sample is called a clustered sample since patients are clustered within hospitals. For each operation, the information includes the costs as well as the patient's age, gender, health status and other characteristics. Information is also available on the hospital, such as its number of beds, location and staff size. A multiple linear regression equation is computed, where cost is the outcome variable and the other factors are considered as candidates for predictor variables. As in multiple linear regression, a linear combination of the predictors is used to predict the outcome or response variable. However, adjustments to the analysis must be made to account for the clustered nature of the sample, namely the possibility that patients within any one hospital may be more similar to each other than to patients in other hospitals. Since the outcomes within a given hospital are correlated, the researcher plans to use correlated outcomes regression to analyze the data.

1.4   Organization and content of the book

This book is organized into three major parts. Part One (Chapters 1–5) deals with data preparation, entry, screening, missing values, transformations, and decisions about likely choices for analysis. Part 2 (Chapters 6–9) deals with regression analysis. Part Three (Chapters 10–18) deals with a number of multivariate analyses. Statisticians disagree on whether or not regression is properly considered as part of multivariate analysis. We have tried to avoid this argument by including regression in the book, but as a separate part. Statisticians certainly agree that regression is an important technique for dealing with problems having multiple variables. In Part Two on regression analysis we have included various topics, such as dummy variables, variable selection methods, and material on missing values that can be used in Part Three.

Chapters 2–5 are concerned with data preparation and the choice of what analysis to use. First, **variables** and how they are classified are discussed in Chapter 2. The next two chapters concentrate on the practical problems of getting data into the computer, handling nonresponse, data management, getting rid of erroneous values, checking assumptions of normality and independence, creating new variables, and preparing a useful codebook. The features of computer software packages used in this book are discussed. The choice of appropriate statistical analyses is discussed in Chapter 5.

Readers who are familiar with handling data sets on computers could skip these initial chapters and go directly to Chapter 6. However, formal course work in statistics often leaves an investigator unprepared for the complications and difficulties involved in real data sets. The material in Chapters 2–5 was deliberately included to fill this gap in preparing investigators for real world data problems.

For a course limited to multivariate analysis, Chapters 2–5 can be omitted if a carefully prepared data set is used for analysis. The depression data set, presented in Section 3.6, has been modified to make it directly usable for multivariate data analysis, but the user may wish to subtract one from the variables 2, 31, 33, and 34 to change the results to zeros and ones. Also, the lung function data, the lung cancer data, and the parental HIV data are briefly described in Appendix A. These data, along with the data in Table 8.1 and Table 16.1, are available on the web from the publisher. See Appendix A or the preface for the exact web site address.

In Chapters 6–18 we follow a standard format. The topics discussed in each chapter are given, followed by a discussion of when the techniques are used.

Then the basic concepts and formulas are explained. Further interpretation, and data examples with topics chosen that relate directly to the techniques, follow. Finally, a summary of the available computer output that may be obtained from six statistical software packages is presented. We conclude each chapter with a discussion of pitfalls to avoid and alternatives to consider when performing the analyses described.

As much as possible, we have tried to make each chapter self-contained. However, Chapters 11 and 12, on discriminant analysis and logistic regression, are somewhat interrelated, as are Chapters 14 and 15, covering principal components and factor analysis.

References for further information on each topic are given in each chapter. Most of the references do require more mathematics than this book, but special emphasis can be placed on references that include examples. If you wish primarily to learn the concepts involved in multivariate techniques and are not as interested in performing the analysis, then a conceptual introduction to multivariate analysis can be found in Kachigan (1991). Everitt and Dunn (2001) provide a highly readable introduction also. For a concise description of multivariate analysis see Manly (2004).

We believe that the best way to learn multivariate analysis is to do it on data that you are familiar with. No book can illustrate all the features found in computer output for a real-life data set. Learning multivariate analysis is similar to learning to swim: you can go to lectures, but the real learning occurs when you get into the water.

Chapter 2

Characterizing data for analysis

2.1 Variables: their definition, classification, and use

In performing multivariate analysis, the investigator deals with numerous variables. In this chapter, we define what a variable is in Section 2.2. Section 2.3 presents a method of classifying variables that is sometimes useful in multivariate analysis since it allows one to check that a commonly used analysis has not been missed. Section 2.4 explains how variables are used in analysis and gives the common terminology for distinguishing between the two major uses of variables. Section 2.5 includes some examples of classifying variables and Section 2.6 discusses other characteristics of data and references exploratory data analysis.

2.2 Defining statistical variables

The word **variable** is used in statistically oriented literature to indicate a characteristic or property that is possible to measure. When we measure something, we make a numerical model of the thing being measured. We follow some rule for assigning a number to each level of the particular characteristic being measured. For example, the height of a person is a variable. We assign a numerical value to correspond to each person's height. Two people who are equally tall are assigned the same numeric value. On the other hand, two people of different heights are assigned two different values. Measurements of a variable gain their meaning from the fact that there exists unique correspondence between the assigned numbers and the levels of the property being measured. Thus two people with different assigned heights are not equally tall. Conversely, if a variable has the same assigned value for all individuals in a group, then this variable does not convey useful information to differentiate individuals in the group.

Physical measurements, such as height and weight, can be measured directly by using physical instruments. On the other hand, properties such as reasoning ability or the state of depression of a person must be measured indirectly. We might choose a particular intelligence test and define the variable "intelligence" to be the score achieved on this test. Similarly, we may define the variable "depression" as the number of positive responses to a series of questions. Although what we wish to measure is the degree of depression, we end up with a count of yes answers to some questions. These examples point out a fundamental difference between direct physical measurements and abstract variables.

Often the question of how to measure a certain property can be perplexing. For example, if the property we wish to measure is the cost of keeping the air clean in a particular area, we may be able to come up with a reasonable estimate, although different analysts may produce different estimates. The problem becomes much more difficult if we wish to estimate the benefits of clean air.

On any given individual or thing we may measure several different characteristics. We would then be dealing with several variables, such as age, height, annual income, race, sex, and level of depression of a certain individual. Similarly, we can measure characteristics of a corporation, such as various financial measures. In this book we are concerned with analyzing data sets consisting of measurements on several variables for each individual in a given sample. We use the symbol P to denote the number of variables and the symbol N to denote the number of **individuals, observations, cases**, or **sampling units**.

2.3 Stevens's classification of variables

In the determination of the appropriate statistical analysis for a given set of data, it is useful to classify variables by type. One method for classifying variables is by the degree of sophistication evident in the way they are measured. For example, we can measure the height of people according to whether the top of their head exceeds a mark on the wall; if yes, they are tall; and if no, they are short. On the other hand, we can also measure height in centimeters or inches. The latter technique is a more sophisticated way of measuring height. As a scientific discipline advances, the measurement of the variables used in it tends to become more sophisticated.

Various attempts have been made to formalize variable classification. A commonly accepted system is that proposed by Stevens (1966). In this system, measurements are classified as **nominal, ordinal, interval,** or **ratio**. In deriving his classification, Stevens characterized each of the four types by a transformation that would not change a measurement's classification. In the subsections that follow, rather than discuss the mathematical details of these transformations, we present the practical implications for data analysis.

As with many classification schemes, Stevens's system is useful for some purposes but not for others. It should be used as a general guide to assist in characterizing the data and to make sure that a useful analysis is not overlooked. However, it should not be used as a rigid rule that ignores the purpose of the analysis or limits its scope (Velleman and Wilkinson, 1993).

Nominal variables

With **nominal variables** each observation belongs to one of several distinct categories. The categories are not necessarily numerical, although numbers may be used to represent them. For example, "sex" is a nominal variable. An individual's gender is either male or female. We may use any two symbols, such as M and F, to represent the two categories. In data analysis, numbers are used as the symbols since many computer programs are designed to handle only numerical symbols. Since the categories may be arranged in any desired order, any set of numbers can be used to represent them. For example, we may use 0 and 1 to represent males and females, respectively. We may also use 1 and 2 to avoid confusing zeros with blanks. Any two other numbers can be used as long as they are used consistently.

An investigator may rename the categories, thus performing a numerical operation. In doing so, the investigator must preserve the uniqueness of each category. Stevens expressed this last idea as a "basic empirical operation" that preserves the category to which the observation belongs. For example, two males must have the same value on the variable "sex," regardless of the two numbers chosen for the categories. Table 2.1 summarizes these ideas and presents further examples. Nominal variables with more than two categories, such as race or religion, may present special challenges to the multivariate data analyst. Some ways of dealing with these variables are presented in Chapter 9.

Ordinal variables

Categories are used for **ordinal variables** as well, but there also exists a known order among them. For example, in the Mohs Hardness Scale, minerals and rocks are classified according to ten levels of hardness. The hardest mineral is diamond and the softest is talc (Pough, 1996). Any ten numbers can be used to represent the categories, as long as they are ordered in magnitude. For instance, the integers 1–10 would be natural to use. On the other hand, any sequence of increasing numbers may also be used. Thus, the basic empirical operation defining ordinal variables is whether one observation is greater than another. For example, we must be able to determine whether one mineral is harder than another. Hardness can be tested easily by noting which mineral can scratch the other. Note that for most ordinal variables there is an underlying continuum being approximated by artificial categories. For example, in the above hardness scale fluorite is defined as having a hardness of 4, and calcite, 3. However, there is a range of hardness between these two numbers not accounted for by the scale.

Table 2.1: *Stevens's measurement system*

Type of measurement	Basic empirical operation	Examples
Nominal	Determine equality of categories	Company names Race Religion Soccer players' numbers
Ordinal	Determine greater than or less than (ranking)	Hardness of minerals Socioeconomic status Rankings of wines
Interval	Determine equality of differences between levels	Temperature in degrees Fahrenheit Calendar dates
Ratio	Determine equality of ratios of levels	Height Weight Density Difference in time

Often investigators classify people, or ask them to classify themselves, along some continuum (see Luce and Narens, 1987). For example, a physician may classify a patient's disease status as none = 1, mild = 2, moderate = 3, and severe = 4. Clearly, increasing numbers indicate increasing severity, but it is not certain that the difference between not having an illness and having a mild case is the same as between having a mild case and a moderate case. Hence, according to Stevens's classification system, this is an ordinal variable.

Interval variables

An **interval variable** is a variable in which the differences between successive values are always the same. For example, the variable "temperature," in degrees Fahrenheit, is measured on the interval scale since the difference between $12°$ and $13°$ is the same as the difference between $13°$ and $14°$ or the difference between any two successive temperatures. In contrast, the Mohs Hardness Scale does not satisfy this condition since the intervals between successive categories are not necessarily the same. The scale must satisfy the basic empirical operation of preserving the equality of intervals.

Ratio variables

Ratio variables are interval variables with a natural point representing the origin of measurement, i.e., a natural zero point. For instance, height is a ratio variable since zero height is a naturally defined point on the scale. We may change the unit of measurement (e.g., centimeters to inches), but we would still preserve the zero point and also the ratio of any two values of height. Temperature is not a ratio variable since we may choose the zero point arbitrarily, thus not preserving ratios.

There is an interesting relationship between interval and ratio variables. The difference between two interval variables is a ratio variable. For example, although time of day is measured on the interval scale, the length of a time period is a ratio variable since it has a natural zero point.

Other classifications

Other methods of classifying variables have also been proposed. Many authors use the term **categorical** to refer to nominal and ordinal variables where categories are used.

We mention, in addition, that variables may be classified as discrete or continuous. A variable is called **continuous** if it can take on any value in a specified range. Thus the height of an individual may be 70 or 70.4539 inches. Any numerical value in a certain range is a conceivable height.

A variable that is not continuous is called **discrete**. A discrete variable may take on only certain specified values. For example, counts are discrete variables since only zero or positive integers are allowed. In fact, all nominal and ordinal variables are discrete. Interval and ratio variables can be continuous or discrete. This latter classification carries over to the possible distributions assumed in the analysis. For instance, the normal distribution is often used to describe the distribution of continuous variables.

Statistical analyses have been developed for various types of variables. In Chapter 5 a guide to selecting the appropriate descriptive measures and multivariate analyses will be presented. The choice depends on how the variables are used in the analysis, a topic that is discussed next.

2.4   How variables are used in data analysis

The type of data analysis required in a specific situation is also related to the way in which each variable in the data set is used. Variables may be used to measure outcomes or to explain why a particular outcome resulted. For example, in the treatment of a given disease a specific drug may be used. The **outcome variable** may be a discrete variable classified as "cured" or "not cured." The outcome variable may depend on several characteristics of the patient such as age, genetic background, and severity of the disease. These characteristics are sometimes called **explanatory** or **predictor variables**. Equivalently, we may call the outcome the **dependent variable** and the characteristics the **independent variable**. The latter terminology is very common in statistical literature. This choice of terminology is unfortunate in that the "independent" variables do not have to be statistically independent of each other. Indeed, these independent variables are usually interrelated in a complex way. Another disadvantage of this terminology is that the common connotation of the words implies a causal model, an assumption not needed for the multivariate analyses described in this book. In spite of these drawbacks, the widespread use of these terms forces us to adopt them.

In other situations the dependent or outcome variable may be treated as a continuous variable. For example, in household survey data we may wish to relate monthly expenditure on cosmetics per household to several explanatory or independent variables such as the number of individuals in the household, their gender, and the household income.

In some situations the roles that the various variables play are not obvious and may also change, depending on the question being addressed. Thus a data set for a certain group of people may contain observations on their sex, age, diet, weight, and blood pressure. In one analysis, we may use weight as a dependent or outcome variable with height, sex, age, and diet as the independent or predictor variables. In another analysis, blood pressure might be the dependent or outcome variable, with weight and other variables considered as independent or predictor variables.

In certain exploratory analyses all the variables may be used as one set with no regard to whether they are dependent or independent. For example, in the social sciences a large number of variables may be defined initially, followed by attempts to combine them into a smaller number of summary variables. In such an analysis the original variables are not classified as dependent or independent. The summary variables may later be used either as outcome or predictor variables. In Chapter 5 multivariate analyses described in this book will be characterized by the situations in which they apply according to the types of variables analyzed and the roles they play in the analysis.

2.5   Examples of classifying variables

In the depression data example several variables are measured on the nominal scale: sex, marital status, employment, and religion. The general health scale is an example of an ordinal variable. Income and age are both ratio variables. No interval variable is included in the data set. A partial listing and a codebook for this data set are given in Chapter 3.

One of the questions that may be addressed in analyzing these data is "Which factors are related to the degree of psychological depression of a person?" The variable "cases" may be used as the dependent or outcome variable since an individual is considered a case if his or her score on the depression scale exceeds a certain level. "Cases" is an ordinal variable, although it can be considered nominal because it has only two categories. The independent or predictor variable could be any or all of the other variables (except ID and measures of depression). Examples of analyses without regard to variable roles are given in Chapters 14 and 15 using the variables C_1 to C_{20} in an attempt to summarize them into a small number of components or factors.

Sometimes, Stevens's classification system is difficult to apply, and two investigators could disagree on a given variable. For example, there may be disagreement about the ordering of the categories of a socioeconomic status variable. Thus the status of blue-collar occupations with respect to the status of certain white-collar occupations might change over time or from culture to culture. So such a variable might be difficult to justify as an ordinal variable, but we would be throwing away valuable information if we used it as a nominal variable. Despite these difficulties, Stevens's system is useful in making decisions on appropriate statistical analysis, as will be discussed in Chapter 5.

2.6   Other characteristics of data

Data are often characterized by whether the measurements are accurately taken and are relatively error free, and by whether they meet the assumptions that were used in deriving statistical tests and confidence intervals. Often, an investigator knows that some of the variables are likely to have observations that have errors. If the effect of an error causes the numerical value of an observation to not be in line with the numerical values of most of the other observations, these extreme values may be called **outliers** and should be considered for removal from the analysis. But other observations may not be accurate and still be within the range of most of the observations. Data sets that contain a sizeable portion of inaccurate data or errors are called "dirty" data sets.

Special statistical methods have been developed that are resistant to the effects of dirty data. Other statistical methods, called robust methods, are insensitive to departures from underlying model assumptions. In this book, we do not present these methods but discuss finding outliers and give methods of determining if the data meet the assumptions. For further information on statistical methods that are well suited for dirty data or require few assumptions, see Hoaglin *et al.* (1985), Schwaiger and Opitz (2003), or Fox and Long (1990).

2.7   Summary

In this chapter statistical variables were defined. Their types and the roles they play in data analysis were discussed. Stevens's classification system was described.

These concepts can affect the choice of analyses to be performed, as will be discussed in Chapter 5.

2.8 Problems

2.1 Classify the following types of data by using Stevens's measurement system: decibels of noise level, father's occupation, parts per million of an impurity in water, density of a piece of bone, rating of a wine by one judge, net profit of a firm, and score on an aptitude test.

2.2 In a survey of users of a walk-in mental health clinic, data have been obtained on sex, age, household roster, race, education level (number of years in school), family income, reason for coming to the clinic, symptoms, and scores on screening examination. The investigator wishes to determine what variables affect whether or not coercion by the family, friends, or a governmental agency was used to get the patient to the clinic. Classify the data according to Stevens's measurement system. What would you consider to be possible independent variables? Dependent variables? Do you expect the dependent variables to be independent of each other?

2.3 For the chronic respiratory study data described in Appendix A, classify each variable according to Stevens's scale and according to whether it is discrete or continuous. Pose two possible research questions and decide on the appropriate dependent and independent variables.

2.4 Repeat problem 2.3 for the lung cancer data set described in Table 13.1.

2.5 From a field of statistical application (perhaps your own field of specialty), describe a data set and repeat the procedures described in Problem 2.3.

2.6 If the RELIG variable described in Table 3.4 of this text was recoded 1 = Catholic, 2 = Protestant, 3 = Jewish, 4 = none, and 5 = other, would this meet the basic empirical operation as defined by Stevens for an ordinal variable?

2.7 Give an example of nominal, ordinal, interval, and ratio variables from a field of application you are familiar with.

2.8 Data that are ordinal are often analyzed by methods that Stevens reserved for interval data. Give reasons why thoughtful investigators often do this.

2.9 The Parental HIV data set described in Appendix A includes the following variables: job status of mother (JOBMO, 1=employed, 2=unemployed, and 3=retired/disabled) and mother's education (EDUMO, 1=did not complete high school, 2=high school diploma/GED, and 3=more than high school). Classify these two variables using Stevens's measurement system.

2.10 Give an example from a field that you are familiar with of an increased sophistication of measuring that has resulted in a measurement that used to be ordinal now being interval.

Chapter 3

Preparing for data analysis

3.1 Processing data so they can be analyzed

Once the data are available from a study there are still a number of steps that must be undertaken to get them into shape for analysis. This is particularly true when multivariate analyses are planned since these analyses are often done on large data sets. In this chapter we provide information on topics related to data processing.

Section 3.2 describes the statistical software packages used in this book. Note that several other statistical packages offer an extensive selection of multivariate analyses. In addition, almost all statistical packages and even some of the spreadsheet programs include at least multiple regression as an option.

The next topic discussed is data entry (Section 3.3). Survey data collection is performed more and more using computers directly via Computer Assisted Personal Interviewing (CAPI), Audio Computer Assisted Self Interviewing (ACASI), or via the Internet. For example, SurveyMonkey is a commercially available program that facilitates sending and collecting surveys via the Internet. Nonetheless, paper and pencil interviews or mailed questionnaires are still a major form of data collection. The methods that need to be used to enter the information obtained from paper and pencil interviews into a computer depend on the size of the data set. For a small data set there are a variety of options since cost and efficiency are not important factors. Also, in that case the data can be easily screened for errors simply by visual inspection. But for large data sets, careful planning of data entry is necessary since costs are an important consideration along with getting an error-free data set available for analysis. Here we summarize the data input options available in the statistical software packages used in this book and discuss the useful options.

Section 3.4 covers combining and updating data sets. The operations used and the options available in the various packages are described. Initial discussion of missing values, outliers, and transformations is given and the need to save results is stressed. Finally, in Section 3.5 we introduce a multivariate data set that will be widely used in this book and summarize the data in a codebook.

We want to stress that the procedures discussed in this chapter can be time consuming and frustrating to perform when large data sets are involved. Often the amount of time used for data entry, editing, and screening can far exceed that used on statistical analysis. It is very helpful to either have computer expertise yourself or have access to someone you can get advice from occasionally.

3.2 Choice of a statistical package

There is a wide choice of statistical software packages available. Many of these packages, however, are quite specialized and do not include many of the multivariate analyses given in this book. For example, there are statistical packages that are aimed at particular areas of application or give tests for exact statistics that are more useful for other types of work. In choosing a package for multivariate analysis, we recommend that you consider the statistical analyses listed in Table 5.2 and check whether the package includes them.

In some cases the statistical package is sold as a single unit and in others you purchase a basic package, but you have a choice of additional programs so you can buy what you need. Some programs require yearly license fees.

Ease of use

Some packages are easier to use than others, although many of us find this difficult to judge–we like what we are familiar with. In general, the packages that are simplest to use have two characteristics. First, they have fewer options to choose from and these options are provided automatically by the program with little need for programming by the user. Second, they use the "point and click" method of choosing what is done rather than writing out statements. The point and click method is even simpler to learn if the package uses options similar to ones found in word processing, spreadsheet, or database management packages. But many current point and click programs do not leave the user with an audit trail of what choices have been made.

On the other hand, software programs with extensive options have obvious advantages. Also, the use of written statements (or *commands*) allows you to have a written record of what you have done. This record can be particularly useful in large-scale data analyses that extend over a considerable period of time and involve numerous investigators. Still other programs provide the user with a programming language that allows the users great freedom in what output they can obtain.

Packages used in this book

In this book, we make specific reference to six general-purpose statistical software packages: R, S-PLUS, SAS, SPSS, Stata, and STATISTICA, listed in alphabetical order. Although S-PLUS has been renamed TIBCO Spotfire S+, we continue to use the name S-PLUS since it is more familiar to many readers. For this edition, we also include R, a popular package that is being increasingly used by statisticians. It is a freeware, i.e., it consists of software that is available free of charge. It offers a large number of multivariate analyses, including most of the ones discussed in this book. Many of the commands in R are similar to those in S-PLUS and therefore we discuss them together. The software versions we use in this book are those available at the end of December 2010.

S-PLUS and R can be used on quite different levels. The simplest is to access it through Microsoft Excel and then run the programs using the usual point and click operations. The most commonly used analyses can be performed this way. Alternatively, S-PLUS and R can be used as a language created for performing statistical analyses. The user writes the language expressions that are read and immediately executed by the program. This process allows the user to write a function, run it, see what happens, and then use the result of the function in a second function. Effort is required to learn the language but, similar to SAS and Stata, it provides the user with a highly versatile programming tool for statistical computing. There are numerous books written on writing programs in S-PLUS and R for different areas of application; for example, see Braun and Murdoch (2007), Crawley (2002), Dalgaard (2008), Everitt (2007), Hamilton (2009) or Heiberger and Neuwirth (2009).

The SAS philosophy is that the user should string together a sequence of procedures to perform the desired analysis. SAS provides a lot of versatility to the user since the software provides numerous possible procedures. It also provides extensive capability for data manipulations. Effort is required to learn the language, but it provides the user with a highly versatile programming tool for statistical computing. Some data management and analysis features are available via point and click operations (SAS/LAB and SAS/ASSIST). Numerous texts have been written on using SAS; for example, see Khattree and Naik (2003), Der and Everitt (2008), or Freund and Littell (2006).

SPSS was originally written for survey applications. The manuals are easy to read and the available options are well chosen. It offers a number of comprehensive programs and users can choose specific options that they desire. It provides excellent data entry programs and data manipulation procedures. It can be used either with point and click or command modes. In addition to the manuals, books such as the ones by Abu-Bader (2010) or Gerber and Finn (2005) are available.

Stata is similar to SAS in that an analysis consists of a sequence of commands with their own options. Analyses can also be performed via a point and click environment. Features are available to easily log and rerun an analysis. Stata is relatively inexpensive and contains features that are not found in other programs. It also includes numerous data management features and a very rich set of graphics options. Several books are available which discuss statistical analysis using Stata; see Rabe-Hesketh and Everitt (2007), Hills and De Stavola (2009), or Cleves, Gould and Gutierrez (2008).

STATISTICA is a very easy software to use. It runs using the usual Windows point and click operations. It has a comprehensive list of options for multivariate analysis. STATISTICA 6 was used in this text. It allows the user to easily record logs of analyses so that routine jobs can be run the same way in the future. The user can also develop custom applications. See the statsoft.com web site for a wide listing of books on using STATISTICA.

When you are learning to use a package for the first time, there is no substitute for reading the on-line HELP, manuals, or texts that present examples. However, at times the sheer number of options presented in these programs may seem confusing, and advice from an experienced user may save you time. Many programs offer default options, and it often helps to use these when you run a program for the first time. In this book, we frequently recommend which options to use. On-line HELP is especially useful when it is programmed to offer information needed for the part of the program you are currently using (context sensitive). Links to the preceding software programs can be found in the UCLA web site cited in the preface.

There are numerous statistical packages that are not included in this book. We have tried to choose those that offer a wide range of multivariate techniques. This is a volatile area with new packages being offered by numerous software firms.

For information on other packages, you can refer to the statistical computing software review sections of *The American Statistician, PC Magazine,* or journals in your own field of interest.

3.3   Techniques for data entry

Appropriate techniques for entering data for analysis depend mainly on the size of the data set and the form in which the data set is stored. As discussed below, all statistical packages use data in a spreadsheet (or rectangular) format. Each column represents a specific variable and each row has the data record for a case or observation. The variables are in the same order for each case. For example, for the depression data set given later in this chapter, looking only at the first three variables and four cases, we have

ID	Sex	Age
1	2	68
2	1	58
3	2	45
4	2	50

where for the variable "sex," 1 = male and 2 = female, and "age" is given in years.

Normally each row represents an individual case. What is needed in each row depends on the unit of analysis for the study. By unit of analysis, we mean what is being studied in the analysis. If the individual is the unit of analysis, as it usually is, then the data set just given is in a form suitable for analysis. Another situation is when the individuals belong to one household, and the unit of analysis is the household but data have been obtained from several individuals in the household. Alternatively, for a company, the unit of analysis may be a sales district and sales made by different salespersons in each district are recorded. Data sets given in the last two examples are called hierarchical data sets and their form can get to be quite complex. Some statistical packages have limited capacity to handle hierarchical data sets. In other cases, the investigator may have to use a relational database package such as Access to first get the data set into the rectangular or spreadsheet form used in the statistical package.

As discussed below, either one or two steps are involved in data entry. The first one is actually entering the data into the computer. If the data are not entered directly into the statistical package being used, a second step of transferring the data to the desired statistical package must be performed.

Data entry

Before entering the actual data in most statistical, spreadsheet, or database management packages, the investigator first names the file where the data are stored, states how many variables will be entered, names the variables, and provides information on these variables. Note that in the example just given we listed three variables which were named for easy use later. The file could be called "depress." Statistical packages commonly allow the user to designate the format and type of variable, e.g., numeric or alphabetic, calendar date, or categorical. They allow you to specify missing value codes, the length of each variable, and the placement of the decimal points. Each program has slightly different features so it is critical to read the appropriate online HELP statements or manual, particularly if a large data set is being entered.

The two commonly used formats for data entry are the **spreadsheet** and the **form**. By spreadsheet, we mean the format given previously where the columns are the variables and the rows the cases. This method of entry allows the user to see the input from previous records, which often gives useful clues if an error in entry is made. The spreadsheet method is very commonly used, particularly when all the variables can be seen on the screen without scrolling. Many persons doing data entry are familiar with this method due to their experience with spreadsheet programs, so they prefer it.

With the form method, only one record, the one being currently entered, is on view on the screen. There are several reasons for using the form method. An entry form can be made to look like the original data collection form so that the data entry person sees data in the same place on the screen as it is in the collection form. A large number of variables for each case can be seen on a computer monitor screen and they can be arranged in a two-dimensional array, instead of just the one-dimensional array available for each case in the spreadsheet format. Flipping pages (screens) in a display may be simpler than scrolling left or right for data entry. Short coding comments can be included on the screen to assist in data entry. Also, if the data set includes alphabetical information such as short answers to open-ended questions, then the form method is preferred.

The choice between these two formats is largely a matter of personal preference, but in general the spreadsheet is used for data sets with a small or medium number of variables and the form is used for a larger number of variables. In some cases a scanner can be used to enter the data and then an optical character recognition program converts the image to the desired text and numbers.

To make the discussion more concrete, we present the features given in a specific data entry package. The SPSS data entry program provides a good mix of features that are useful in entering large data sets. It allows either spreadsheet or form entry and switching back and forth between the two modes. In addition to the features already mentioned, SPSS provides what is called "skip and fill." In medical studies and surveys, it is common that if the answer to a certain question is no, a series of additional questions can then be skipped. For example, subjects might be asked if they ever smoked, and if the answer is yes they are asked a series of questions on smoking history. But if they answer no, these questions are not asked and the interviewer skips to the next section of the interview. The skip-and-fill option allows the investigator to specify that if a person answers no, the smoking history questions are automatically filled in with specified values and the entry cursor moves to the start of the next section. This saves a lot of entry time and possible errors.

Another feature available in many packages is range checking. Here the investigator can enter upper and lower values for each variable. If the data entry person enters a value that is either lower than the low value or higher than the high value, the data entry program provides a warning. For example, for the variable "sex," if an investigator specifies 1 and 2 as possible values and the data entry person hits a 3 by mistake, the program issues a warning. This feature, along with input by forms or spreadsheet, is available also in SAS.

Each software has its own set of features and the reader is encouraged to examine them before entering medium or large data sets, to take advantage of them.

Mechanisms of entering data

Data can be entered for statistical computation from different sources. We will discuss four of them.

1. entering the data along with the program or procedure statements for a batch-process run;

2. using the data entry features of the statistical package you intend to use;

3. entering the data from an outside file which is constructed without the use of the statistical package;

4. importing the data from another package using the operating system such as Windows or MAC OS.

The first method can only be used with a limited number of programs which use program or procedure statements, for example R, S-PLUS or SAS. It is only recommended for very small data sets that are not going to be used very many times. For example, a SAS data set called "depress" could be made by stating:

```
data depress;
    input id sex age;
    cards;
1    2    68
2    1    58
3    2    45
4    2    50
:
run;
```

Similar types of statements can be used for the other programs which use the spreadsheet type of format.

The disadvantage of this type of data entry is that there are only limited editing features available to the person entering the data. No checks are made as to whether or not the data are within reasonable ranges for this data set. For example, all respondents were supposed to be 18 years old or older, but there is no automatic check to verify that the age of the third person, who was 45 years old, was not erroneously entered as 15 years. Another disadvantage is that the data set disappears after the program is run unless additional statements are made. In small data sets, the ability to save the data set, edit typing, and have range checks performed is not as important as in larger data sets.

The second strategy is to use the data entry package or system provided by the statistical program you wish to use. This is always a safe choice as it means that the data set is in the form required by the program and no data transfer problems will arise. Table 3.1 summarizes the built-in data entry features of the six statistical packages used in this book. Note that for SAS, Proc COMPARE can be used to verify the data after they are entered. In general, as can be seen in Table 3.1, SPSS and SAS have extensive data entry features.

The third method is to use another statistical software package, data entry package, word processor, spreadsheet, or data management program to enter

Table 3.1: *Built-in data entry features of the statistical packages*

	S-PLUS/R	SAS	SPSS	Stata	STATISTICA
Spreadsheet entry	Yes	Yes	Yes	Yes	Yes
Form entry	No	Yes	Yes	No	No
Range check	User	Yes	Yes	No	No
Logical check	User	Yes	Yes	No	No
Skip and fill	User	Use SCL	Yes	No	No
Verify mode	No	No	Yes	No	No

the data into a spreadsheet format. The advantage of this method is that an available program that you are familiar with can be used to enter the data. Two commonly used programs for data entry are Excel and Access. Excel provides entry in the form of a spreadsheet and is widely available. Access allows entry using forms and provides the ability to combine different data sets. Once the data sets are combined in Access, it is straightforward to transfer them to Excel. Many of the statistical software packages import Excel files. In addition, many of the statistical packages allow the user to import data from other statistical packages. For example, R and S-PLUS will import SAS, SPSS, and Stata data files. One suggestion is to first check the manual or HELP for the statistical package you wish to use to see which types of data files it can import.

A widely used transfer method is to create an ASCII file from the data set. ASCII (American Standard Code for Information Interchange) files can be created by almost any spreadsheet, data management, or word processing program. Instructions for reading ASCII files are given in the statistical packages. The disadvantage of transferring ASCII files is that typically only the data are transferred, and variable names and information concerning the variables have to be reentered into the statistical package. This is a minor problem if there are not too many variables. If this process appears to be difficult, or if the investigator wishes to retain the variable names, then they can run a special-purpose program such as STAT/TRANSFER, DBMS/COPY (available from Data Flux Corporation) or DATA JUNCTION that will copy data files created by a wide range of programs and put them into the right format for access by any of a wide range of statistical packages.

Finally, if the data entry program and the statistical package both use the Windows operating system, then three methods of transferring data may be considered depending on what is implemented in the programs. First, the data in the data entry program may be highlighted and moved to the statistical package using the usual copy and paste options. Second, dynamic data exchange

Table 3.2: *Data management features of the statistical packages*

	S-PLUS/R	SAS	SPSS	Stata	STATISTICA
Merging data sets	merge	MERGE statement	MATCH FILES	merge	merge
Adding data sets	rbind cbind	PROC APPEND or set statement	ADD FILES	append	add cases add variables
Hierarchical data sets	User written functions	Write multiple output statements: RETAIN	CASESTOVARS	reshape	stacking
Importing data (types)	ASCII, spreadsheets, databases stat-packages	ASCII, ACCESS: spreadsheets, databases	ASCII, spreadsheets, databases	ASCII, spreadsheets, databases	ASCII, spreadsheets, databases
Exporting data (types)	ASCII, spreadsheet, databases	ASCII, ACCESS: spreadsheets, databases	ASCII, spreadsheets, databases	ASCII, spreadsheets, databases	ASCII, spreadsheets, databases
Calender dates	Yes	Yes	Yes	Yes	Yes
Transpose data	t	PROC TRANSPOSE	FLIP	xpose	Transpose
Range limit checks	Yes	Yes	Yes	Yes	Yes
Missing value imputation	Yes	MI and MIANALYZE	MULTIPLE IMPUTATION	mi	Mean substitution

(DDE) can be used to transfer data. Here the data set in the statistical package is dynamically linked to the data set in the entry program. If you correct a variable for a particular case in the entry program, the identical change is made in the data set in the statistical package, Third, object linking and embedding (OLE) can be used to share data between a program used for data entry and statistical analysis. Here also the data entry program can be used to edit the data in the statistical program. The investigator can activate the data entry program from within the statistical package program. MAC users often find it simplest to enter their data in Excel and then transfer their Excel file to Windows-based computers for analysis. Transfers can be made by disk or by a third party software package called DAVE.

If you have a very large data set to enter, it is often sensible to use a professional data entering service. A good service can be very fast and can offer different levels of data checking and advice on which data entry method to use. But whether or not a professional service is used, the following suggestions may be helpful for data entry.

1. Whenever possible, code information in numbers not letters.

2. Code information in the most detailed form you will ever need. You can use the statistical program to aggregate the data into coarser groupings later. For example, it is better to record age as the exact age at the last birthday rather than to record the ten-year age interval into which it falls.

3. The use of range checks or maximum and minimum values can eliminate the entry of extreme values but they do not guard against an entry error that falls within the range. If minimizing errors is crucial then the data can be entered twice into separate data files. One data file can be subtracted from the other and the resulting nonzeros examined. Alternatively, some data entry programs have a verify mode where the user is warned if the first entry does not agree with the second one (SPSS).

4. If the data are stored on a personal computer, then backup copies should be made on an external storage device such as a CD or DVD. Backups should be updated regularly as changes are made in the data set. Particularly when using Windows programs, if dynamic linking is possible between analysis output and the data set, it is critical to keep an unaltered data set.

5. For each variable, use a code to indicate missing values. The various programs each have their own way to indicate missing values. The manuals or HELP statements should be consulted so that you can match what they require with what you do.

To summarize, there are three important considerations in data entry: accuracy, cost, and ease of use of the data file. Whichever system is used, the investigator should ensure that the data file is free of typing errors, that time and money are not wasted, and that the data file is readily available for future data management and statistical analysis.

3.4   Organizing the data

Prior to statistical analysis, it is often necessary to make some changes in the data set. Table 3.2 summarizes the common options in the programs described in this book.

Combining data sets

Combining data sets is an operation that is commonly performed. For example, in biomedical studies, data may be taken from medical history forms, a questionnaire, and laboratory results for each patient. These data for a group of patients need to be combined into a single rectangular data set where the rows are the different patients and the columns are the combined history, questionnaire, and laboratory variables. In longitudinal studies of voting intentions, the questionnaire results for each respondent must be combined across time periods in order to analyze change in voting intentions of an individual over time. There are essentially two steps in this operation. The first is sorting on some key variable (given different names in different packages) which must be included in the separate data sets to be merged. Usually this key variable is an identification or ID variable (case number). The second step is combining the separate data sets side-by-side, matching the correct records with the correct person using the key variable. Sometimes one or more of the data items are missing for an individual. For example, in a longitudinal study it may not be possible to locate a respondent for one or more of the interviews. In such a case, a symbol or symbols indicating missing values will be inserted into the spaces for the missing data items by the program. This is done so that you will end up with a rectangular data set or file, in which information for an individual is put into the proper row, and missing data are so identified.

    Data sets can be combined in the manner described above in SAS by using the MERGE statement followed by a BY statement and the variable(s) to be used to match the records. (The data must first be sorted by the values of the matching variable, say ID.) An UPDATE statement can also be used to add variables to a master file. In SPSS, you simply use the JOIN MATCH command followed by the data files to be merged if you are certain that the cases are already listed in precisely the same order and each case is present in all the data files. Otherwise, you first sort the separate data files on the key variable and use the JOIN MATCH command followed by the BY key variable. In Stata, you USE the first data file and then use a MERGE key variable USING the second data file statement. STATISTICA has a merge function and S-PLUS also has the CBIND function or a merge BY.X or BY.Y argument can be used for more complex situations (see their help file). This step can also be done using a cut and paste operation in many programs.

In any case, it is highly desirable to list the data set to determine that the merging was done in the manner that you intended. If the data set is large, then only the first and last 25 or so cases need to be listed to see that the results are correct. If the separate data sets are expected to have missing values, you need to list sufficient cases so you can see that missing records are correctly handled.

Another common way of combining data sets is to put one data set at the end of another data set or to interleave the cases together based on some key variable. For example, an investigator may have data sets that are collected at different places and then combined together. In an education study, student records could be combined from two high schools, with one simply placed at the bottom of the other set. This is done by using the Proc APPEND in SAS. In SPSS the JOIN command with the keyword ADD can be used to combine cases from two to five data files or with specification of a key variable, to interleave. In Stata the APPEND command is used and in S-PLUS the RBIND function. In STATISTICA, the MERGE procedure is used. This step can also be done as a cut and paste operation in many programs.

It is also possible to update the data files with later information using the editing functions of the package. Thus a single data file can be obtained that contains the latest information, if this is desired for analysis. This option can also be used to replace data that were originally entered incorrectly.

When using a statistical package that does not have provision for merging data sets, it is recommended that a spreadsheet program be used to perform the merging and then, after a rectangular data file is obtained, the resulting data file can be transferred to the desired statistical package. In general, the newer spreadsheet programs have excellent facilities for combining data sets side-by-side or for adding new cases.

Missing values

There are two types of missing data. The first type occurs when no information is obtained from a case, individual, or sampling unit. This type is called **unit nonresponse**. For example, in a survey it may be impossible to reach a potential respondent or the subject may refuse to answer. In a biomedical study, records may be lost or a laboratory animal may die of unrelated causes prior to measuring the outcome. The second type of nonresponse occurs when the case, individual, or sampling unit is available but yields incomplete information. For example, in a survey the respondent may refuse to answer questions on income or only fill out the first page of a questionnaire. Busy physicians may not completely fill in a medical record. This type of nonresponse is called **item nonresponse**. In general, the more control the investigator has of the sampling units, the less apt unit nonresponse or item nonresponse is to occur. In surveys the investigator often has little or no control over the respondent, so both types of nonresponse are apt to happen. For this reason, much of the research on handling nonresponse has been done in the survey field and the terminology used reflects this emphasis.

The seriousness of either unit nonresponse or item nonresponse depends mainly on the magnitude of the nonresponse and on the characteristics of the nonresponders. If the proportion of nonresponse is very small, it is seldom a problem and if the nonresponders can be considered to be a random sample of the population then it can be ignored (see Section 9.2 for a more complete classification of nonresponse). Also, if the units sampled are highly homogeneous then most statisticians would not be too concerned. For example, some laboratory animals have been bred for decades to be quite similar in their genetic background. In contrast, people in most major countries have very different backgrounds and their opinions and genetic makeup can vary greatly.

When only unit nonresponse occurs, the data gathered will look complete in that information is available on all the variables for each case. Suppose in a survey of students 80% of the females respond and 60% of the males respond and the investigator expects males and females to respond differently to a question (X). If in the population 55% are males and 45% are females, then instead of simply getting an overall average of responses for all the students, a weighted average could be reported. For males $w_1 = .55$ and for females $w_2 = .45$. If $\overline{X}_1$ is the mean for males and $\overline{X}_2$ is the mean for females, then a weighted average could be computed as

$$\overline{X} = \frac{\sum w_i \overline{X}_i}{\sum w_i} = \frac{w_1 \overline{X}_1 + w_2 \overline{X}_2}{w_1 + w_2}$$

Another common technique is to assign each observation a weight and the weight is entered into the data set as if it were a variable. Observations are weighted more if they come from subgroups that have a low response rate. This weight may be adjusted so that the sum of the weights equals the sample size. When weighting data, the investigator is assuming that the responders and nonresponders in a subgroup are similar.

In this book, we do not discuss such weighted analyses in detail. A more complete discussion of using weights for adjustment of unit nonresponse can be found in Groves et al. (2002) or Little and Rubin (2002). Several types of weights can be used and it is recommended that the reader consider the various options before proceeding. The investigator would need to obtain information on the units in the population to check whether the units in the sample are proportional to the units in the population. For example, in a survey of professionals taken from a listing of society members if the sex, years since graduation, and current employment information is available from both the listing of the members and the results of the survey, these variables could be used to compute subgroup weights.

The data set should also be screened for item nonresponse. As will be discussed in Section 9.2, most multivariate analyses require complete data on all the variables used in the analysis. If even one variable has a missing value for a case, that case will not be used. Most statistical packages provide programs that indicate how many cases were used in computing common univariate statistics such as means and standard deviations (or report how many cases were missing). Thus it is simple to find which variables have few or numerous missing values.

Some programs can also indicate how many missing values there are for each case. Other programs allow you to transpose or flip your data file so the rows become the columns and the columns become the rows (Table 3.2). Thus the cases and variables are switched as far as the statistical package is concerned. The number of missing values by case can then be found by computing the univariate statistics on the transposed data. Examination of the pattern of missing values is important since it allows the investigator to see if it appears to be distributed randomly or only occurs in some variables. Also, it may have occurred only at the start of the study or close to the end.

Once the pattern of missing data is determined, a decision must be made on how to obtain a complete data set for multivariate analysis. For a first step, most statisticians agree on the following guidelines.

1. If a variable is missing in a very high proportion of cases, then that variable could be deleted.

2. If a case is missing many variables that are crucial to your analysis, then that case could be deleted.

You should also carefully check if there is anything special about the cases that have numerous missing data as this might give you insight into problems in data collection. It might also give some insight into the population to which the results actually apply. Likewise, a variable that is missing in a high proportion of the respondents may be an indication of a special problem. Following the guidelines listed previously can reduce the problems in data analysis but it will not eliminate the problems of reduced efficiency due to discarded data or potential bias due to differences between the data that are complete and the grossly incomplete data. For example, this process may result in a data set that is too small or that is not representative of the total data set. That is, the missing data may not be missing completely at random (see Section 9.2). In such cases, you should consider methods of imputing (or filling-in) the missing data (see Section 9.2 and the books by Rubin, 2004, Little and Rubin, 2002, Schafer, 1997, or Molenberghs and Kenward, 2007).

Item nonresponse can occur in two ways. First, the data may be missing from the start. In this case, the investigator enters a code for missing values at the time the data are entered into the computer. One option is to enter a symbol that the statistical package being used will automatically recognize as a missing value. For example, a period, an asterisk (*), or a blank space may be recognized as a missing value by some programs. Commonly, a numerical value is used that is outside the range of possible values. For example, for the variable "sex" (with 1 = male and 2 = female) a missing code could be 9. A string of 9s is often used; thus, for the weight of a person 999 could be used as a missing code. Then that value is declared to be missing. For example, for SAS, one could state

if sex = 9, then sex = . ;

Similar statements are used for the other programs. The reader should check the manual for the precise statement.

If the data have been entered into a spreadsheet program, then commonly blanks are used for missing values. In this case, most spreadsheet (and word processor) programs have search-and-replace features that allow you to replace all the blanks with the missing value symbol that your statistical package automatically recognizes. This replacement should be done before the data file is transferred to the statistical package.

The second way in which values can be considered missing is if the data values are beyond the range of the stated maximum or minimum values. For example, if the age of a respondent is entered as 167 and it is not possible to determine the correct value, then the 167 should be replaced with a missing value code so an obviously incorrect value is not used.

Further discussion of the types of missing values and of ways of handling item nonresponse in multivariate data analysis is given in Section 9.2. Here, we will briefly mention one simple method.

The replacement of missing values with the mean value of that variable is a common option in statistical software packages and is the simplest method of imputation. We do not recommend the use of this method when using the multivariate methods given in later chapters. This method results in underestimation of the variances and covariances that are subsequently used in many multivariate analyses.

Detection of outliers

Outliers are observations that appear inconsistent with the remainder of the data set (Barnett and Lewis, 2000). One method for determining outliers has already been discussed, namely, setting minimum and maximum values. By applying these limits, extreme or unreasonable outliers are prevented from entering the data set.

Often, observations are obtained that seem quite high or low but are not impossible. These values are the most difficult ones to cope with. Should they be removed or not? Statisticians differ in their opinions, from "if in doubt, throw it out" to the point of view that it is unethical to remove an outlier for fear of biasing the results. The investigator may wish to eliminate these outliers from the analyses but report them along with the statistical analysis. Another possibility is to run the analyses twice, both with the outliers and without them, to see if they make an appreciable difference in the results. Most investigators would hesitate, for example, to report rejecting a null hypothesis if the removal of an outlier would result in the hypothesis not being rejected.

A review of formal tests for detection of outliers is given in Barnett and Lewis (2000). To make the formal tests you usually must assume normality of the data. Some of the formal tests are known to be quite sensitive to nonnormality and should only be used when you are convinced that this assumption is reasonable. Often an alpha level of 0.10 or 0.15 is used for testing if it is suspected that outliers are not extremely unusual. Smaller values of alpha can be used if outliers are thought to be rare.

The data can be examined one variable at a time by using histograms and box plots if the variable is measured on the interval or ratio scale. A questionable value would be one that is separated from the remaining observations. For nominal or ordinal data, the frequency of each outcome can be noted. If a recorded outcome is impossible, it can be declared missing. If a particular outcome occurs only once or twice, the investigator may wish to consolidate that outcome with a similar one. We will return to the subject of outliers in connection with the statistical analyses starting in Chapter 6, but mainly the discussion in this book is not based on formal tests.

Transformations of the data

Transformations are commonly made either to create new variables with a form more suitable for analysis or to achieve an approximate normal distribution. Here we discuss the first possibility. Transformations to achieve approximate normality are discussed in Chapter 4.

Transformations to create new variables can either be performed as a step in organizing the data or can be included later when the analyses are being performed. It is recommended that they be done as a part of organizing the data. The advantage of this is that the new variables are created once and for all, and sets of instructions for running data analysis from then on do not have to include the data transformation statements. This results in shorter sets of instructions with less repetition and chance for errors when the data are being analyzed. This is almost essential if several investigators are analyzing the same data set.

One common use of transformations occurs in the analysis of questionnaire data. Often the results from several questions are combined to form a new variable. For example, in studying the effects of smoking on lung function it is common to ask first a question such as:

Have you ever smoked cigarettes?   yes___

or   no___

If the subjects answer no, they skip a set of questions and go on to another topic. If they answer yes, they are questioned further about the amount in terms of packs per day and length of time they smoked (in years). From this information, a new pack–year variable is created that is the number of years times the average number of packs. For the person who has never smoked, the answer is zero. Transformation statements are used to create the new variable.

Each package offers a slightly different set of transformation statements, but some general options exist. The programs allow you to select cases that meet certain specifications using IF statements. Here for instance, if the response is no to whether the person has ever smoked, the new variable should be set to zero. If the response is yes, then pack–years is computed by multiplying the average amount smoked by the length of time smoked. This sort of arithmetic operation is provided for and the new variable is added to the end of the data set.

Additional options include taking means of a set of variables or the maximum value of a set of variables. Another common arithmetic transformation involves simply changing the numerical values coded for a nominal or ordinal variable. For example, for the depression data set, sex was coded male = 1 and female = 2. In some of the analyses used in this book, we have recoded that to male = 0 and female = 1 by simply subtracting one from the given value.

Saving the results

After the data have been screened for missing values and outliers, and transformations made to form new variables, the results are saved in a master file that can be used for analysis. We recommend that a copy or copies of this master file be made on an external storage device such as a CD or DVD so that it can be stored outside the computer. A summary of decisions made in data screening and transformations used should be stored with the master file. Enough information should be stored so that the investigator can later describe what steps were taken in organizing the data.

If the steps taken in organizing the data were performed by typing in control language, it is recommended that a copy of this control language be stored along with the data sets. Then, should the need arise, the manipulation can be redone by simply editing the control language instructions rather than completely recreating them.

If results are saved interactively (point and click), then it is recommended that multiple copies be saved along the way until you are perfectly satisfied with the results and that the Windows notepad or some similar memo facility be used to document your steps. Figure 3.1 summarizes the steps taken in data entry and data management.

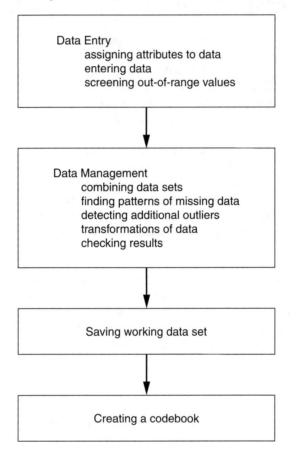

Figure 3.1: *Preparing Data for Statistical Analysis*

3.5   Example: depression study

In this section we discuss a data set that will be used in several succeeding chapters to illustrate multivariate analyses. The depression study itself is described in Chapter 1.

The data given here are from a subset of 294 observations randomly chosen from the original 1000 respondents sampled in Los Angeles. This subset of observations is large enough to provide a good illustration of the statistical techniques but small enough to be manageable. Only data from the first time period are included. Variables are chosen so that they would be easily understood and would be sensible to use in the multivariate statistical analyses described in Chapters 6–17.

The codebook, the variables used, and the data set are described below.

Codebook

In multivariate analysis, the investigator often works with a data set that has numerous variables, perhaps hundreds of them. An important step in making the data set understandable is to create a written codebook that can be given to all the users. The codebook should contain a description of each variable and the variable name given to each variable for use in the statistical package. Some statistical packages have limits on the length of the variable names so that abbreviations are used. Often blank spaces are not allowed, so dashes or underscores are included. Some statistical packages reserve certain words that may not be used as variable names. The variables should be listed in the same order as they are in the data file. The codebook serves as a guide and record for all users of the data set and for future documentation of the results.

Table 3.3 contains a codebook for the depression data set. In the first column the variable number is listed, since that is often the simplest way to refer to the variables in the computer. A variable name is given next, and this name is used in later data analysis. These names were chosen to be eight characters or less so that they could be used by all the package programs. It is helpful to choose variable names that are easy to remember and are descriptive of the variables, but short to reduce space in the display.

Finally a description of each variable is given in the last column of Table 3.3. For nominal or ordinal data, the numbers used to code each answer are listed. For interval or ratio data, the units used are included. Note that income is given in thousands of dollars per year for the household; thus an income of 15 would be $15,000 per year. Additional information that is sometimes given includes the number of cases that have missing values, how missing values are coded, the largest and smallest value for that variable, simple descriptive statistics such as frequencies for each answer for nominal or ordinal data, and means and standard deviations for interval or ratio data. We note that one package (Stata) can produce a codebook for its users that includes much of the information just described.

Depression variables

The 20 items used in the depression scale are variables 9–28 and are named C1, C2,..., C20. (The wording of each item is given later in the text, in Table 14.2.) Each item was written on a card and the respondent was asked to tell the

Table 3.3: *Codebook for depression data*

Variable number	Variable name	Description
1	ID	Identification number from 1 to 294
2	SEX	1 = male; 2 = female
3	AGE	Age in years at last birthday
4	MARITAL	1 = never married; 2 = married; 3 = divorced; 4 = separated; 5 = widowed
5	EDUCAT	1 = less than high school; 2 = some high school; 3 = finished high school; 4 = some college; 5 = finished bachelor's degree; 6 = finished master's degree; 7 = finished doctorate
6	EMPLOY	1 = full time; 2 = part time; 3 = unemployed; 4 = retired; 5 = houseperson; 6 = in school; 7 = other
7	INCOME	Thousands of dollars per year
8	RELIG	1 = Protestant; 2 = Catholic; 3 = Jewish; 4 = none; 5 = other
9–28	C1–C20	"Please look at this card and tell me the number that best describes how often you felt or behaved this way during the past week." 20 items from depression scale (already reflected; see text) 0 = rarely or none of the time (less than 1 day); 1 = some or a little of the time (1–2 days); 2 = occasionally or a moderate amount of the time (3–4 days); 3 = most or all of the time (5–7 days)
29	CESD	Sum of C1–20; 0 = lowest level possible; 60 = highest level possible
30	CASES	0 = normal; 1 = depressed, where depressed is CESD$\geq$16
31	DRINK	Regular drinker? 1 = yes; 2 = no
32	HEALTH	General health? 1 = excellent; 2 = good; 3 = fair; 4 = poor
33	REGDOC	Have a regular doctor? 1 = yes; 2 = no
34	TREAT	Has a doctor prescribed or recommended that you take medicine, medical treatments, or change your way of living in such areas as smoking, special diet, exercise, or drinking? 1 = yes; 2 = no
35	BEDDAYS	Spent entire day(s) in bed in last two months? 0 = no; 1 = yes
36	ACUTEILL	Any acute illness in last two months? 0 = no; 1 = yes
37	CHRONILL	Any chronic illness in last year? 0 = no; 1 = yes

interviewer the number that best describes how often he or she felt or behaved this way during the past week. Thus respondents who answered item C2, "I felt depressed," could respond 0–3, depending on whether this particular item applied to them rarely or none of the time (less than 1 day: 0), some or little of the time (1–2 days: 1), occasionally or a moderate amount of the time (3–4 days: 2), or most or all of the time (5–7 days: 3).

Most of the items are worded in a negative fashion, but items C8–C11 are positively worded. For example, C8 is "I felt that I was as good as other people." For positively worded items the scores are **reflected**: that is, a score of 3 is changed to be 0, 2 is changed to 1, 1 is changed to 2, and 0 is changed to 3. In this way, when the total score of all 20 items is obtained by summation of variables C1–C20, a large score indicates a person who is depressed. This sum is the 29th variable, named CESD.

Persons whose CESD score is greater than or equal to 16 are classified as depressed since this value is the common cutoff point used in the literature (Frerichs, Aneshensel and Clark, 1981). These persons are given a score of 1 in variable 30, the CASES variable. The particular depression scale employed here was developed for use in community surveys of noninstitutionalized respondents (Comstock and Helsing, 1976; Radloff, 1977).

Data set

As can be seen by examining the codebook given in Table 3.3 demographic data (variables 2–8), depression data (variables 9–30), and general health data (variables 32–37) are included in this data set. Variable 31, DRINK, was included so that it would be possible to determine if an association exists between drinking and depression. Frerichs *et al.* (1981) have already noted a lack of association between smoking and scores on the depression scale.

The actual data for the first 30 of the 294 respondents are listed in Table 3.4. The rest of the data set, along with the other data sets used in this book, are available on the CRC Press and UCLA web sites (see Appendix A).

3.6 Summary

In this chapter we discussed the steps necessary before statistical analysis can begin. The first of these is the decision of what computer and software packages to use. Once this decision is made, data entry and organizing the data can be started.

Note that investigators often alter the order of these operations. For example, some prefer to check for missing data and outliers and to make transformations prior to combining the data sets. This is particularly true in analyzing longitudinal data when the first data set may be available well before the others. This may also be an iterative process in that finding errors may lead to entering

Table 3.4: Depression data

OBS	SEX	AGE	MARITAL	EDUCAT	EMPLOY	INCOME	RELIG	C1	C2	C3	C4	C5	C6	C7	C8	C9	C10	C11	C12	C13	C14	C15	C16	C17	C18	C19	C20	CESD	CASES	DRINK	HEALTH	REGDOC	TREAT	BEDDAYS	ACUTEILL	CHRONILL
1	2	68	5	2	4	4	1	0	0	0	0	0	0	0	0	0	0	0	0	0	0	0	0	0	0	0	0	4	0	2	1	1	1	0	0	1
2	1	58	5	4	1	15	1	0	0	1	0	0	0	0	0	0	0	0	1	0	0	1	0	1	0	0	0	4	0	1	1	1	1	0	0	0
3	2	45	2	3	1	28	1	0	0	0	0	1	0	0	0	0	0	0	0	0	1	0	1	0	0	0	0	5	0	1	1	1	1	0	0	1
4	2	50	3	3	3	9	1	0	0	0	0	1	1	0	3	3	0	0	0	0	0	0	0	0	0	0	0	6	0	2	1	1	2	0	0	0
5	2	33	4	3	1	35	1	0	0	0	0	0	0	0	0	1	0	0	0	0	0	0	2	0	0	0	0	7	0	1	1	1	1	1	1	1
6	1	24	2	3	1	11	1	2	0	0	2	0	0	0	3	2	0	2	1	2	0	3	0	0	0	0	0	15	0	1	3	1	1	0	1	0
7	2	58	2	2	5	9	1	0	1	1	0	1	0	0	0	0	0	0	0	0	0	0	1	1	0	0	1	10	0	1	1	2	1	0	1	1
8	1	22	1	3	1	23	2	0	1	2	0	2	3	0	2	0	0	0	0	0	0	2	3	0	1	1	0	16	1	1	4	1	2	0	1	1
9	2	47	2	2	4	35	4	0	0	1	0	0	0	0	0	0	0	0	3	0	3	0	0	0	0	0	0	0	1	1	1	2	1	1	0	0
10	1	30	2	2	1	25	4	0	0	0	0	0	2	3	0	0	0	0	0	0	0	1	1	2	0	0	0	18	1	1	2	1	2	0	0	1
11	2	20	1	3	3	24	4	0	1	1	0	1	0	0	0	0	1	0	0	2	2	2	0	0	3	0	0	4	0	1	1	1	1	1	0	0
12	2	57	2	2	2	28	1	1	0	0	0	0	0	0	0	0	0	0	1	0	0	0	2	0	0	0	0	8	0	1	3	1	1	0	1	0
13	1	39	2	3	1	13	1	0	0	0	0	0	3	0	0	0	0	2	1	0	2	0	1	0	0	0	1	4	0	1	1	2	1	0	1	1
14	2	61	5	3	2	15	2	0	1	0	0	1	0	0	0	0	0	2	0	0	0	1	0	1	0	2	1	8	0	1	3	1	2	0	1	0
15	2	23	2	3	4	6	1	1	0	0	3	0	0	2	1	2	2	0	0	0	2	3	3	1	0	0	1	21	1	2	1	2	1	1	0	0
16	2	21	1	2	1	8	1	3	0	0	1	3	0	0	0	3	2	2	2	1	0	0	0	0	2	0	0	42	1	1	3	1	2	1	0	0
17	2	23	1	4	1	19	1	1	0	2	0	1	0	0	1	0	0	0	0	0	0	0	0	0	1	0	3	6	0	1	2	1	2	0	1	1
18	2	55	4	2	3	15	4	0	1	2	0	0	0	0	2	0	0	0	0	0	0	0	0	0	0	0	0	0	0	2	2	2	2	1	1	0
19	1	26	1	6	1	9	2	0	0	1	0	0	0	0	0	0	0	0	0	0	0	1	0	0	0	0	0	3	0	1	1	1	1	1	1	0
20	2	64	5	3	4	6	1	0	0	0	1	0	0	0	0	3	0	0	0	0	0	0	2	0	0	0	0	3	0	1	2	1	1	1	1	0
21	2	44	1	3	1	35	2	0	0	0	0	0	0	1	3	0	2	0	0	1	0	2	2	0	0	0	0	4	0	1	3	1	1	0	0	1
22	2	25	2	3	1	7	1	0	0	0	0	0	0	2	0	0	0	0	0	2	2	1	2	1	0	0	0	2	0	1	3	1	2	0	0	0
23	2	72	5	3	4	19	2	0	0	0	0	1	0	0	0	0	0	0	0	2	0	3	0	0	0	0	0	4	0	1	3	2	1	0	1	0
24	2	61	2	3	1	6	1	0	0	0	0	1	2	0	0	2	0	0	1	0	1	0	0	0	0	0	0	10	0	1	3	1	2	0	1	1
25	2	43	3	2	1	19	2	0	2	1	0	0	0	0	2	1	0	0	1	1	0	0	2	0	0	0	0	12	0	2	2	1	1	0	0	1
26	2	52	2	3	5	13	1	1	0	0	0	2	0	0	0	0	0	2	0	1	1	3	0	0	1	0	0	6	0	1	2	1	1	0	1	1
27	1	23	2	4	5	5	2	0	1	2	0	1	2	0	3	0	2	1	0	0	0	0	0	2	0	0	0	9	0	1	1	1	1	0	1	1
28	2	73	4	2	4	19	2	0	2	2	0	0	0	2	0	0	1	0	2	3	2	0	3	0	0	0	2	28	1	1	2	1	2	0	0	0
29	1	34	2	3	2	2	2	0	0	0	0	0	0	0	1	1	0	0	1	0	0	0	0	2	0	0	0	5	0	1	1	1	1	1	1	0
30	2	34	2	3	1	20	1	0	0	0	0	0	0	0	0	0	0	0	0	0	0	0	0	0	0	0	0	1	0	1	2	1	2	0	0	1

new data to replace erroneous values. Again, we stress saving the results on CDs or DVDs or some other external storage device after each set of changes.

Six statistical packages — R, S-PLUS, SAS, SPSS, Stata, and STATIS-TICA — were noted as the packages used in this book. In evaluating a package it is often helpful to examine the data entry and data manipulation features they offer. The tasks performed in data entry and organization are often much more difficult and time consuming than running the statistical analyses, so a package that is easy and intuitive to use for these operations is a real help. If the package available to you lacks needed features, then you may wish to perform these operations in one of the spreadsheet or relational database packages and then transfer the results to your statistical package.

3.7 Problems

3.1 Enter the data set given in Table 8.1, Chemical companies' financial performance (Section 8.3), using a data entry program of your choice. Make a codebook for this data set.

3.2 Using the data set entered in Problem 3.1 delete the P/E variable for the Dow Chemical company and D/E for Stauffer Chemical and Nalco Chemical in a way appropriate for the statistical package you are using. Then, use the missing value features in your statistical package to find the missing values and replace them with an imputed value.

3.3 Using your statistical package, compute a scatter diagram of income versus employment status from the depression data set. From the data in this table, decide if there are any adults whose income is unusual considering their employment status. Are there any adults in the data set whom you think are unusual?

3.4 Transfer a data set from a spreadsheet program into your statistical software package.

3.5 Describe the person in the depression data set who has the highest total CESD score.

3.6 Using a statistical software program, compute histograms for mothers' and fathers' heights and weights from the lung function data set described in Section 1.2 and in Section A.5 of Appendix A. The data and codebook can be obtained from the web site listed in Section A.7 of Appendix A. Describe cases that you consider to be outliers.

3.7 From the lung function data set (Problem 3.6), determine how many families have one child, two children, and three children between the ages of 7 and 18.

3.8 For the lung function data set, produce a two-way table of gender of child 1 versus gender of child 2 (for families with at least two children). Comment.

3.9 For the lung cancer data set (see the codebook in Table 13.1 of Section 13.3 and Section A.7 of Appendix A for how to obtain the data) use a statistical package of your choice to

(a) compute a histogram of the variable Days, and

(b) for every other variable produce a frequency table of all possible values.

3.10 For the raw data in Problem 3.9

(a) produce a separate histogram of the variable Days for small and large tumor sizes (0 and 1 values of the variable Staget),

(b) compute a two-way frequency table of the variable Staget versus the variable Death, and

(c) comment on the results of (a) and (b).

3.11 For the depression data set, determine if any of the variables have observations that do not fall within the ranges given in Table 3.4, codebook for depression data.

3.12 For the statistical package you intend to use, describe how you would add data from three more time periods for the same subjects to the depression data set.

3.13 For the lung function data set, create a new variable called AGEDIFF = (age of child 1) − (age of child 2) for families with at least two children. Produce a frequency count of this variable. Are there any negative values? Comment.

3.14 Combine the results from the following two questions into a single variable:

a. Have you been sick during the last two weeks?

    Yes, go to b.   ——

    No   ——

b. How many days were you sick? ——

3.15 Consistency checks are sometimes performed to detect possible errors in the data. If a data set included information on sex, age, and use of contraceptive pill, describe a consistency check that could be used for this data set.

3.16 In the Parental HIV data set, the variable LIVWITH (who the adolescent was living with) was coded 1=both parents, 2=one parent, and 3=other. Transform the data so it is coded 1=one parent, 2=two parents, and 3=other using the features available in the statistical package you are using or Excel.

Chapter 4

Data screening and transformations

4.1 Transformations, assessing normality and independence

In Section 3.5 we discussed the use of transformations to create new variables. In this chapter we discuss transforming the data to obtain a distribution that is approximately normal. Section 4.2 shows how transformations change the shape of distributions. Section 4.3 discusses several methods for deciding when a transformation should be made and how to find a suitable transformation. An iterative scheme is proposed that helps to zero in on a good transformation. Statistical tests for normality are evaluated. Section 4.4 presents simple graphical methods for determining if the data are independent. In this chapter, we rely heavily on graphical methods: see Cook and Weisberg (1994) and Tufte (1997, 2001).

Each computer package offers the users information to help decide if their data are normally distributed. The packages provide convenient methods for transforming the data to achieve approximate normality. They also include output for checking the independence of the observations. Hence the assumption of independent, normally distributed data that is made in many statistical tests can be assessed, at least approximately. Note that most investigators will try to discard the most obvious outliers prior to assessing normality because such outliers can grossly distort the distribution.

4.2 Common transformations

In analysis of data it is often useful to transform certain variables before performing the analyses. Examples are found in the next section and in Chapter 6. In this section we present some common transformations. If you are familiar with this subject, you may wish to skip to the next section.

To develop a feel for transformations, let us examine a plot of transformed values versus the original values of the variable. To begin with, a plot of values of a variable X against itself produces a 45° diagonal line going through the origin, as shown in Figure 4.1.

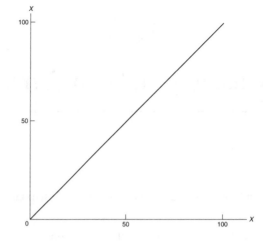

Figure 4.1: *Plot of Variable X versus Variable X*

One of the most commonly performed transformations is taking the **logarithm** (log) to base 10. Recall that the logarithm is the number that satisfies the relationship $X = 10^Y$. That is, the logarithm of X is the power Y to which 10 must be raised in order to produce X. As shown in Figure 4.2 in plot **a**, the logarithm of 10 is 1 since $10 = 10^1$. Similarly, the logarithm of 1 is 0 since $1 = 10^0$, and the logarithm of 100 is 2 since $100 = 10^2$. Other values of logarithms can be obtained from tables of common logarithms, from a hand calculator with a log function, or from statistical packages by using the transformation options. All statistical packages discussed in this book allow the user to make this transformation as well as others.

Note that an increase in X from 1 to 10 increases the logarithm from 0 to 1, that is, an increase of one unit. Similarly, an increase in X from 10 to 100 increases the logarithm also by one unit. For larger numbers it takes a great increase in X to produce a small increase in log X. Thus the logarithmic transformation has the effect of stretching small values of X and condensing large values of X. Note also that the logarithm of any number less than 1 is negative, and the logarithm of a value of X that is less than or equal to 0 is not defined. In practice, if negative or zero values of X are possible, the investigator may first add an appropriate constant to each value of X, thus making them all positive prior to taking the logarithms. The choice of the additive constant can have an important effect on the statistical properties of the transformed variable, as will be seen in the next section. The value added must be larger than the magnitude of the minimum value of X.

Logarithms can be taken to any base. A familiar base is the number e = 2.7183 ... . Logarithms taken to base e are called **natural logarithms** and are

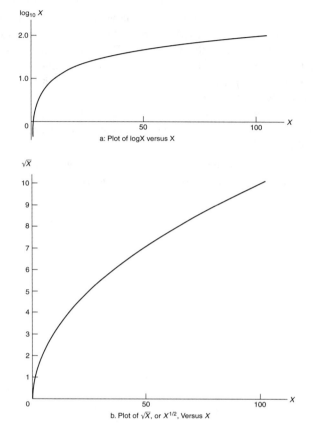

Figure 4.2: *Plots of Log and Square Root of X versus Variable X*

denoted by $\log_e$ or ln. The natural logarithm of X is the power to which e must be raised to produce X. There is a simple relationship between the natural and common logarithms, namely,

$$\log_e X \;=\; 2.3026 \log_{10} X$$
$$\log_{10} X \;=\; 0.4343 \log_e X$$

If we graph $\log_e X$ versus X, we would get a figure with the same shape as $\log_{10} X$, with the only difference being in the vertical scale; i.e., $\log_e X$ is larger than $\log_{10} X$ for $X > 1$ and smaller than $\log_{10} X$ for $X < 1$. The natural logarithm is used frequently in theoretical studies because of certain appealing mathematical properties.

Another class of transformations is known as **power transformations**. For example, the transformation X^2 (X raised to the power of 2) is used frequently in statistical formulas such as computing the variance. The most commonly used power transformations are the square root (X raised to the power $\frac{1}{2}$) and the inverse $1/X$ (X raised to the power -1). Figure 4.2b shows a plot of the square root of X versus X. Note that this function is also not defined for negative values of X. Compared with taking the logarithm, taking the square root also progressively condenses the values of X as X increases. However, the degree of condensation is not as severe as in the case of logarithms. That is, the square root of X tapers off slower than $\log X$, as can be seen by comparing plots a and b in Figure 4.2.

Unlike $X^{\frac{1}{2}}$ and $\log X$, the function $1/X$ decreases with increasing X (see Figure 4.3a). To obtain an increasing function, you may use $-1/X$.

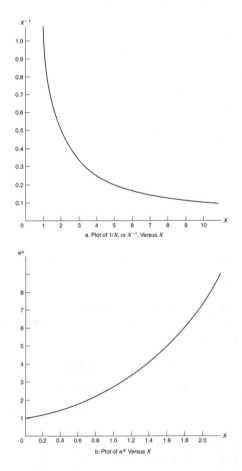

a. Plot of 1/X, or X^{-1}, Versus X

b: Plot of e^X Versus X

Figure 4.3: *Plots of $1/X$ and e^X versus the Variable X*

One way to characterize transformations is to note the power to which X is raised. We denote that power by p. For the square root transformation, $p = \frac{1}{2}$ and for the inverse transformation, $p = -1$. The logarithmic transformation can be thought of as corresponding to a value of $p = 0$ (Tukey, 1977). We can think of these three transformations as ranging from p values of -1 to 0 to $\frac{1}{2}$. The effects of these transformations in reducing a long-tailed distribution to the right are greater as the value of p decreases from 1 (the p value for no transformation) to $\frac{1}{2}$ to 0 to -1. Smaller values of p result in a greater degree of transformation for $p < 1$. Thus, p of -2 would result in a greater degree of transformation than a p of -1. The logarithmic transformation reduces a long tail more than a square root transformation (Figure 4.2a versus 4.2b) but not as much as the inverse transformation given in Figure 4.3a. The changes in the amount of transformation depending on the value of p provide the background for one method of choosing an appropriate transformation. In the next section, after a discussion of how to assess whether you have a normal distribution, we give several strategies to assist in choosing appropriate p values.

Finally, exponential functions are also sometimes used in data analysis. An **exponential function** of X may be thought of as the antilogarithm of X: for example, the antilogarithm of X to base 10 is 10 raised to the power X; similarly, the antilogarithm of X to base e is e raised to the power X. The exponential function e^x is illustrated in Figure 4.3b. The function 10^x has the same shape but increases faster than e^x. Both have the opposite effect of taking logarithms; i.e., they increasingly stretch the larger values of X. Additional discussion of these interrelationships can be found in Tukey (1977).

4.3 Selecting appropriate transformations

In the theoretical development of statistical methods some assumptions are usually made regarding the distribution of the variables being analyzed. Often the form of the distribution is specified. The most commonly assumed distribution for continuous observations is the **normal**, or **Gaussian**, distribution. Although the assumption is sometimes not crucial to the validity of the results, some check of normality is advisable in the early stages of analysis. For a review of the role of the normality assumption in the validity of statistical analyses, see Lumley *et al.* (2002). In this section, methods for assessing normality and for choosing a transformation to induce normality are presented.

Assessing normality using histograms

The left graph of Figure 4.4a illustrates the appearance of an ideal histogram, or density function, of normally distributed data. The values of the variable X are plotted on the horizontal axis. The range of X is partitioned into numerous intervals of equal length and the proportion of observations in each interval is plotted on the vertical axis. The mean is in the center of the distribution and is equal to zero in this hypothetical histogram. The distribution is symmetric about the mean, that is, intervals equidistant from the mean have equal proportions of observations (or the same height in the histogram). If you place a mirror vertically at the mean of a symmetric histogram, the right side should be a mirror image of the left side. A distribution may be symmetric and still not be normal, but often distributions that are symmetric are close to normal. If the population distribution is normal, the sample histogram should resemble the famous symmetric bell-shaped Gaussian curve. For small sample sizes, the sample histogram may be irregular in appearance and the assessment of normality difficult.

Plots of various histograms and the appearance of their normal probability plot are given in Figure 4.4 and Figure 4.5.

All statistical packages described in Chapter 3 plot histograms for specific variables in the data set. The best fitting normal density function can also be superimposed on the histogram. Such graphs, produced by some packages, can enable you to make a crude judgment as to how well the normal distribution approximates the histogram.

Assessing symmetry using box plots

Symmetry of the empirical distribution can be assessed by examining certain percentiles (or quantiles). Quantiles are obtained by first ordering the N observations from smallest to largest. Using these ordered observations, the common rule for computing the quantile of the ith ordered observation is to compute $Q(i) = (i - 0.5)/N$. If N was five the third quantile would be $(3 - 0.5)/5 = .5$ or the median (see Cleveland, 1993). Note that there is a quantile for each observation. Percentiles are similar but they are computed to divide the sample into 100 equal parts. Quantiles or percentiles can be estimated from all six statistical packages. Usually plots are done of quantiles since in that case there is a point at each observation.

Suppose the observations for a variable such as income are ordered from smallest to largest. The person with an income at the 25th percentile would have an income such that 25% of the individuals have an income less than or equal to that person. This income is denoted as $P(25)$. Equivalently, the 0.25 quantile, often written as $Q(0.25)$, is the income that divides the cases into two groups where a fraction 0.25 of the cases has an income less than or equal to $Q(0.25)$ and a fraction 0.75 greater. Numerically $P(25) = Q(0.25)$. For further discussion on quantiles and how they are computed in statistical packages, see Frigge, Hoaglin and Iglewicz (1989).

Some quantiles are widely used. One such quantile is the sample median $Q(0.5)$. The median divides the observations into two equal parts. The median of ordered variables is either the middle observation if the number of observa-

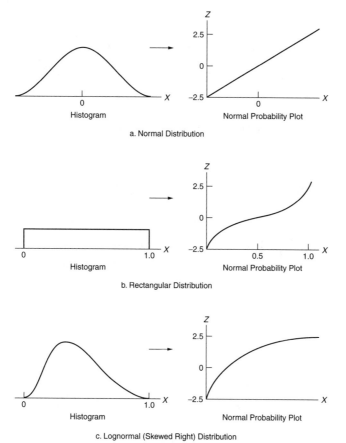

Figure 4.4: *Plots (a,b,c) of Common Histograms and the Resulting Normal Probability Plots from those Distributions*

tions N is odd, or the average of the middle two if N is even. Theoretically, for the normal distribution, the median equals the mean. In samples from normal distributions, they may not be precisely equal but they should not be far different. Two other widely used quantiles are $Q(0.25)$ and $Q(0.75)$. These are called "quartiles" since, along with the median, they divide the values of a variable into four equal parts.

If the distribution is symmetric, then the difference between the median and $Q(0.25)$ would equal the difference between $Q(0.75)$ and the median. This is displayed graphically in **box plots**. $Q(0.75)$ and $Q(0.25)$ are plotted at the top and the bottom of the box (or rectangle) and the median is denoted either by a line or a dot within the box. Lines (called **whiskers**) extend from the ends

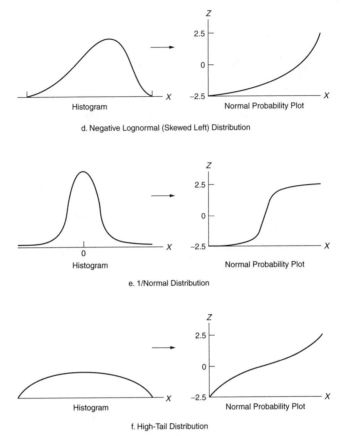

Figure 4.5: *Plots (d,e,f) of Other Histograms and the Resulting Normal Probability Plots from those Distributions*

of the box out to what are called **adjacent values**. The numerical values of the adjacent values and quantiles are not precisely the same in all statistical packages (Frigge, Hoaglin and Iglewicz, 1989).

Usually the top adjacent value is the largest observation that is less than or equal to $Q(0.75)$ plus $1.5 \ (Q(0.75) - Q(0.25))$. The bottom adjacent value is the smallest observation that is greater than or equal to $Q(0.25)$ minus $1.5(Q(0.75) - Q(0.25))$. Values that are beyond the adjacent value are sometimes examined to see if they are outliers. Symmetry can be assessed from box plots primarily by looking at the distances from the median to the top of the box and from the median to the bottom of the box. If these two lengths are decidedly unequal, then symmetry would be questionable. If the distribution is not symmetric, then it is not normal.

All six statistical packages can produce box plots.

Assessing normality using normal probability or normal quantile plots

Normal probability plots present an appealing option for checking for normality. One axis of the probability plot shows the values of X and the other shows expected values, Z, of X if its distribution were exactly normal. The computation of these expected Z values is discussed in Johnson and Wichern (2007). Equivalently, this graph is a plot of the **cumulative distribution** found in the data set, with the vertical axis adjusted to produce a straight line if the data followed an exact normal distribution. Thus if the data were from a normal distribution, the normal probability plot should approximate a straight line, as shown in the right-hand graph of Figure 4.4a. In this graph, values of the variable X are shown on the horizontal axis and values of the expected normal are shown on the vertical axis. Even when the data are normally distributed, if the sample size is small, the normal probability plot may not be perfectly straight, especially at the extremes. You should concentrate your attention on the middle 80% or 90% of the graph and see if that is approximately a straight line. Most investigators can visually assess whether or not a set of connected points follows a straight line, so this method of determining whether or not data are normally distributed is highly recommended.

In some packages the X and Z axes are interchanged. For your package, you should examine the axes to see which represents the data (X) and which represents the expected values (Z).

The remaining plots of Figure 4.4 and 4.5 illustrate other distributions with their corresponding normal probability plots. The purpose of these plots is to help you associate the appearance of the probability plot with the shape of the histogram or frequency distribution. One especially common departure from normality is illustrated in Figure 4.4c where the distribution is skewed to the right (has a longer tail on the right side). In this case, the data actually follow a **log-normal distribution**. Note that the curvature resembles a quarter of a circle (or an upside-down bowl) with the ends pointing downward. In this case we can create a new variable by taking the logarithm of X to either base 10 or e. A normal probability plot of log X will follow a straight line.

If the axes in Figure 4.4c were interchanged so X was on the vertical axis, then the bowl would be right side up with the ends pointing upward. It would look like Figure 4.5d.

Figures 4.4b and 4.5f show distributions that have higher tails than a normal distribution (more extreme observations). The corresponding normal probability plot is an inverted S. Figure 4.5e illustrates a distribution where the tails have fewer observations than a normal does. It was in fact obtained by taking the inverse of normally distributed data. The resulting normal probability plot is S-shaped. If the inverse of the data $(1/X)$ were plotted, then a straight line normal probability plot would be obtained. If X were plotted on the vertical axis then Figure 4.5e would illustrate a heavy tail distribution and Figures 4.4b and 4.5f would illustrate distributions that have fewer expected observations than in the tails of a normal distribution.

The advantage of using the normal probability plot is that it not only tells you if the data are approximately normally distributed, but it also offers insight into how the data are distributed if they are not normally distributed. This insight is helpful in deciding what transformation to use to induce normality.

Many software programs also produce theoretical quantile–quantile plots. Here a theoretical distribution such as a normal is plotted against the empirical quantiles of the data. When a normal distribution is used, they are often called **normal quantile** plots. These plots are interpreted similarly to normal probability plots with the difference being that quantile plots emphasize the tails more than the center of the distribution. Normal quantile plots may have the variable X plotted on the vertical axis in some packages (see S-PLUS). For further information on their interpretation, see Gan, Koehler and Thompson (1991).

Normal probability or normal quantile plots are available also in all six packages.

Selecting a transformation to induce normality

As we noted in the previous discussion, transforming the data can sometimes produce an approximate normal distribution. In some cases the appropriate transformation is known. Thus, for example, the square root transformation is used with Poisson-distributed variables. Such variables represent counts of events occurring randomly in space or time with a small probability such that the occurrence of one event does not affect another. Examples include the number of cells per unit area on a slide of biological material, the number of incoming phone calls per second in a telephone exchange, and counts of radioactive material per unit of time. Similarly, the logarithmic transformation has been used on variables such as household income, the time elapsing before a specified number of Poisson events have occurred, systolic blood pressure of older individuals, and many other variables with long tails to the right. Finally, the inverse transformation has been used frequently on the length of time it takes an individual or an animal to complete a certain task. Tukey (1977) gives numerous examples of variables and appropriate transformations on them to produce approximate normality. Tukey (1977), Box and Cox (1964), Draper and Hunter (1969), and Bickel and Doksum (1981) discuss some systematic ways to find an appropriate transformation. A table of common transformations is included in van Belle *et al.* (2004).

Another strategy for deciding what transformation to use is to progress up or down the values of p depending upon the shape of the normal probability plot. If the plot looks like Figure 4.4c, then a value of p less than 1 is tried. Suppose $p = \frac{1}{2}$ is tried. If this is not sufficient and the normal probability plot is still curved downward at the ends, then $p = 0$, i.e., the logarithmic transformation, can be tried. If the plot appears to be almost correct but a little more transformation is needed, a positive constant can be subtracted from each X prior to the transformation. If the transformation appears to be a little too much, a positive constant can be added. Thus, various transformations of the form $(X + C)^p$ are tried until the normal probability plot is as straight as possible. An example of the use of this type of transformation occurs when considering systolic blood pressure (SBP) for older adult males. It has been found that a transformation such as $\log(\text{SBP} - 75)$ results in data that appear to be normally distributed.

The investigator decreases or increases the value of p until the resulting observations appear to have close to a normal distribution. Note that this does not mean that theoretically you have a normal distribution since you are only working with a single sample. Particularly if the sample is small, the transformation that is best for your sample may not be the one that is best for the entire population. For this reason, most investigators tend to round off the numerical values of p. For example, if they found that $p = 0.46$ worked best with their data, they might actually use $p = 0.5$ or the square root transformation.

All the statistical packages discussed in Chapter 3 allow the user to perform a large variety of transformations. Stata has an option, called ladder of powers, that automatically produces power transformations for various values of p.

Hines and O'Hara Hines (1987) developed a method for reducing the number of iterations needed by providing a graph which produces a suggested value of p from information available in most statistical packages. Using their graph one can directly estimate a reasonable value of p.

Their method is based on the use of quantiles. What are needed for their method are the values of the median and a pair of symmetric quantiles. As noted earlier, the normal distribution is symmetric about the median. The difference between the median and a lower quantile (say $Q(0.2)$) should equal the difference between the upper symmetric quantile (say $Q(0.8)$) and the median if the data are normally distributed. By choosing a value of p that results in those differences being equal, one is choosing a transformation that tends to make the data approximately normally distributed. Using Figure 4.6 (kindly supplied by W. G. S. Hines and R. J. O'Hara Hines), one plots the ratio of the lower quantile of X to the median on the vertical axis and the ratio of the median to the upper quantile on the horizontal axis for at least one set of symmetric quantiles. The resulting p is read off the closest curve.

For example, in the depression data set given in Table 3.4 a variable called INCOME is listed. This is family income in thousands of dollars. The median is 15, $Q(0.25) = 9$, and $Q(0.75) = 28$. Hence the median is not halfway between the two quantiles (which in this case are the quartiles), indicating a nonsymmetric distribution with the long tail to the right. If we plot $9/15 = 0.60$ on the vertical axis and $15/28 = 0.54$ on the horizontal axis of Figure 4.3, we get a value of p of approximately $-\frac{1}{3}$. Since $-\frac{1}{3}$ lies between 0 and $-\frac{1}{2}$, trying first the log transformation seems reasonable. The median for log(INCOME) is 1.18, $Q(0.25) = 0.95$, and $Q(0.75) = 1.45$. The median $- 0.95 = 0.23$ and $1.45 -$ median $= 0.27$, so from the quartiles it appears that the data are still slightly skewed to the right. Additional quantiles could be estimated to better approximate p from Figure 4.6. When we obtained an estimate of the skewness of log(INCOME), it was negative, indicating a long tail to the left. The reason for the contradiction is that seven respondents had an income of only 2 ($2000 per year) and these low incomes had a large effect on the estimate of the skewness but less effect on the quartiles. The skewness statistic is sensitive to extreme values.

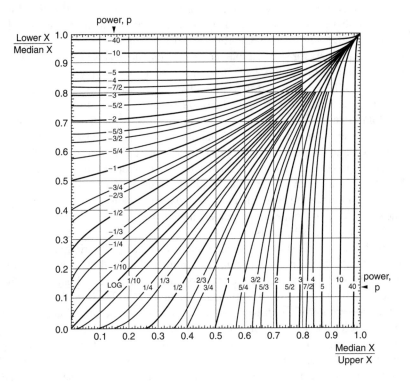

Figure 4.6: *Determination of the Value of p in the Power Transformation to Produce Approximate Normality*

Stata, through the command "lnskew0," considers all transformations of the form $\ln(X - k)$ and finds the value of k which makes the skewness approximately equal to 0. For the variable INCOME in the depression data set, the value of k chosen by the program is $k = -2.01415$. For the transformed variable $\ln(\text{INCOME} + 2.01415)$ the value of the skewness is -0.00016 compared with 1.2168 for INCOME. Note that adding a constant reduces the effect of taking the log. Examining the histograms with the normal density imposed upon them shows that the transformed variable fits a normal distribution much better than the original data. There does not appear to be much difference between the log transformation suggested by Figure 4.3 and the transformation obtained from Stata when the histograms are examined.

It should be noted that not every distribution can be transformed to a normal distribution. For example, variable 29 in the depression data set is the sum of the results from the 20-item depression scale. The mode (the most commonly occurring score) is at zero, thus making it virtually impossible to transform these CESD scores to a normal distribution. However, if a distribution has a single mode in the interior of the distribution, then it is usually not too difficult to find a distribution that appears at least closer to normal than the original one. Ideally, you should search for a transformation that has the proper scientific meaning in the context of your data.

Statistical tests for normality

It is also possible to do formal tests for normality. These tests can be done to see if the null hypothesis of a normal distribution is rejected and hence whether a transformation should be considered. They can also be used after transformations to assist in assessing the effect of the transformation. A commonly used procedure is the Shapiro–Wilk W statistic (Mickey, Dunn and Clark, 2009). This test has been shown to have good power against a wide range of non-normal distributions (see D'Agostino, Belanger and D'Agostino (1990) for a discussion of this test and others). Several of the statistical programs give the value of W and the associated p value for testing that the data came from a normal distribution. If the value of p is small, then the data may not be considered normally distributed. For example, we used the command "swilk" in Stata to perform the Shapiro–Wilk test on INCOME and $\ln(\text{INCOME} + 2.01415)$ from the depression data set. The p values for the two variables were 0.00000 and 0.21347, respectively. These results indicate that INCOME has a distribution that is significantly different from normal, while the transformed data fit a normal distribution reasonably well. A more practical transformation for data analysis is $\ln(\text{INCOME} + 2)$. This transformed variable is also fitted well with a normal distribution $(p = 0.217)$.

Some programs also compute the Kolmogorov–Smirnov D statistic and approximate p values. Here the null hypothesis is rejected for large D values and small p values. van Belle *et al.* (2004) discuss the characteristics of this test. It is also possible to perform a chi-square goodness of fit test. Note that both the Kolmogorov–Smirnov D test and the chi-square test have poor power properties. In effect, if you use these tests you will tend to reject the null hypothesis when your sample size is large and accept it when your sample size is small. As one statistician put it, they are to some extent a complicated way of determining your sample size.

Another way of testing for normality is to examine the **skewness** of your data. Skewness is a measure of how nonsymmetric a distribution is. If the data are symmetrically or normally distributed, the computed skewness will be close to zero. The numerical value of the ratio of the skewness to its standard error can be compared with normal Z tables, and symmetry, and hence normality, would be rejected if the absolute value of the ratio is large. Positive skewness indicates that the distribution has a long tail to the right and probably looks like Figure 4.4c. (Note that it is important to remove outliers or erroneous observations prior to performing this test.) D'Agostino, Belanger and D'Agostino (1990) give an approximate formula for the test of skewness for sample sizes > 8. Skewness is computed in all six packages. STATISTICA provides estimates of the standard error that can be used with large sample sizes. Stata computes the test of D'Agostino, Belanger and D'Agostino (1990) via the sktest command. In general, a distribution with skewness greater than one will appear to be noticeably skewed unless the sample size is very small.

Very little is known about what significance levels α should be chosen to compare with the p value obtained from formal tests of normality. So the sense of increased preciseness gained by performing a formal test over examining a plot is somewhat of an illusion. From the normal probability plot you can both decide whether the data are normally distributed and get a suggestion about what transformation to use.

Assessing the need for a transformation

In general, transformations are more effective in inducing normality when the standard deviation of the untransformed variable is large relative to the mean. If the standard deviation divided by the mean is less than $\frac{1}{4}$, then the transformation may not be necessary. Alternatively, if the largest observation divided by the smallest observation is less than 2, then the data are likely not to be sufficiently variable for the transformation to have a decisive effect on the results (Hoaglin, Mosteller and Tukey, 1985). These rules are not meaningful for data without a natural zero (interval data but not ratio data by Stevens's classification). For example, the above rules will result in different values if temperature is measured in Fahrenheit or Celsius. For variables without a natural zero, it is often possible to estimate a natural minimum value below which observations are seldom if ever found. For such data, the rule of thumb is if

$$\frac{\text{Largest value} - \text{natural minimum}}{\text{smallest observation} - \text{natural minimum}} < 2$$

then a transformation is not likely to be useful.

Other investigators examine how far the normal probability plot is from a straight line. If the amount of curvature is slight, they would not bother to transform the data.

Note that the usefulness of transformations is difficult to evaluate when the sample sizes are small or if numerous outliers are present. In deciding whether to make a transformation, you may wish to perform the analysis with and without the proposed transformation. Examining the results will frequently convince you that the conclusions are not altered after making the transformation. In this case it is preferable to present the results in terms of the most easily interpretable units. And it is often helpful to conform to the customs of the particular field of investigation.

Sometimes, transformations are made to simplify later analyses rather than to approximate normal distributions. For example, it is known that FEV1 (forced expiratory volume in 1 second) and FVC (forced vital capacity) decrease in adults as they grow older. (See Section 1.3 for a discussion of these variables.) Some researchers will take the ratio FEV1/FVC and work with it because this ratio is less dependent on age. Using a variable that is independent of age can make analyses of a sample including adults of varying ages much simpler. In a practical sense, then, the researcher can use the transformation capabilities of the computer program packages to create new variables to be added to the set of initial variables rather than only modify and replace them.

If transformations alter the results, then you should select the transformation that makes the data conform as much as possible to the assumptions. If a particular transformation is selected, then all analyses should be performed on the transformed data, and the results should be presented in terms of the transformed values. Inferences and statements of results should reflect this fact.

4.4 Assessing independence

Measurements on two or more variables collected from the **same** individual are not expected to be, nor are they assumed to be, independent of each other. On the other hand, independence of observations collected from **different** individuals or items is an assumption made in the theoretical derivation of most multivariate statistical analyses. This assumption is crucial to the validity of the results, and violating it may result in erroneous conclusions. Unfortunately, little is published about the quantitative effects of various degrees of nonindependence. Also, few practical methods exist for checking whether the assumption is valid.

In situations where the observations are collected from people, it is frequently safe to assume independence of observations collected from different individuals. Dependence could exist if a factor or factors exist to affect all of the individuals in a similar manner with respect to the variables being measured. For example, political attitudes of adult members of the same household cannot be expected to be independent. Inferences that assume independence are not valid when drawn from such responses. Similarly, biological data on twins or siblings are usually not independent.

Data collected in the form of a sequence either in time or space can also be dependent. For example, observations of temperature on successive days are likely to be dependent. In those cases it is useful to plot the data in the appropriate sequence and search for trends or changes over time. Some programs allow you to plot an outcome variable against the program's own ID variable. Other programs do not; so the safe procedure is to always type in a variable or variables that represent the order in which the observations are taken and any other factor that you think might result in lack of independence, such as location or time. Also, if more complex sampling methods are used, then information on clusters or strata should be entered.

Figure 4.7 presents a series of plots that illustrate typical outcomes. Values of the outcome variable being considered appear on the vertical axis and values of order of time, location, or observation identification number (ID) appear on the horizontal axis. If the plot resembles that shown in Figure 4.7a, little reason exists for suspecting lack of independence or nonrandomness. The data in that plot were, in fact, obtained as a sequence of random standard normal numbers. In contrast, Figure 4.7b shows data that exhibit a positive trend. Figure 4.7c is an example of a temporary shift in the level of the observations, followed by a return to the initial level. This result may occur, for example, in laboratory data when equipment is temporarily out of calibration or when a substitute technician, following different procedures, temporarily performs the work. Finally, a common situation occurring in some business or industrial data is the existence of seasonal cycles, as shown in Figure 4.7d.

Such plots or scatter diagrams are available in all six statistical packages mentioned in this book and are widely available elsewhere. If a two-dimensional display of location is desired with the outcome variable under consideration on the vertical axis, then this can be displayed in three dimensions in some packages.

Some formal tests for the randomness of a sequence of observations are given in Brownlee (1965). One such test (the Durbin-Watson test) is presented in Chapter 6 of this book in the context of regression and correlation analysis. If you are dealing with series of observations you may also wish to study the area of forecasting and time series analysis. Some books on this subject are referenced in Chapter 6.

4.5   Summary

In this chapter we emphasized methods for determining if the data were normally distributed and for finding suitable transformations to make the data

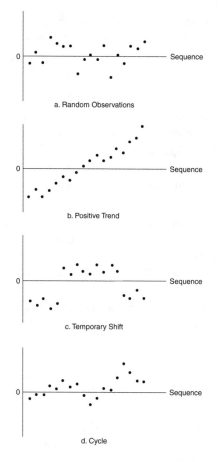

Figure 4.7: *Graphs of Hypothetical Data Sequences Showing Lack of Independence*

closer to a normal distribution. We also discussed methods for checking whether the data were independent.

No special order is best for all situations in data screening, However, most investigators would delete the obvious outliers first. They may then check the assumption of normality, attempt some transformations, and recheck the data for outliers. Other combinations of data-screening activities are usually dictated by the problem.

We wish to emphasize that data screening can be the most time-consuming and costly portion of data analysis. Investigators should not underestimate this aspect. If the data are not screened properly, much of the analysis may have to be repeated, resulting in an unnecessary waste of time and resources. Once the data have been carefully screened, the investigator will then be in a position to select the appropriate analysis to answer specific questions. In Chapter 5 we present a guide to the selection of the appropriate data analysis.

4.6   Problems

4.1 Using the depression data set described in Tables 3.3 and 3.4, create a variable equal to the negative of one divided by the cubic root of income. Display a normal probability plot of the new variable.

4.2 Take the logarithm to base 10 of the income variable in the depression data set. Compare the histogram of income with the histogram of log(INCOME). Also, compare the normal probability plots of income and log(INCOME).

4.3 Repeat Problems 4.1 and 4.2, taking the square root of income.

4.4 Generate a set of 100 random normal deviates by using a computer program package. Display a histogram and normal probability plot of these values. Square these numbers by using transformations. Compare histograms and normal probability plots of the logarithms and the square roots of the set of squared normal deviates.

4.5 Use the two sets of 100 random numbers from Problem 4.4. Display boxplots of these two sets of values and state which of the three graphical methods (histograms, normal probability plots, and boxplots) are most useful in assessing normality.

4.6 Take the logarithm of the CESD score plus 1 and compare the histograms of CESD and log(CESD + 1). (A small constant must be added to CESD because CESD can be zero.)

4.7 Obtain normal probability plots of mothers' and fathers' weights from the lung function data set described in Appendix A. Discuss whether or not you consider weight to be normally distributed in the population from which this sample is taken.

4.8 The accompanying data are from the New York Stock Exchange Composite Index for the period August 9 through September 17, 1982. Run a program to plot these data in order to assess the lack of independence of successive observations. (Note that the data cover five days per week, except for a holiday on Monday, September 6). The daily volume is the number of transactions given in millions of shares. This time period saw an unusually rapid rise in stock prices (especially for August), coming after a protracted falling market. Compare this time series with prices for the current year.

Day	Month	Index	Volume	Day	Month	Index	Volume
9	Aug.	59.3	63	1	Sept.	67.9	98
10	Aug.	59.1	63	2	Sept.	69.0	87
11	Aug.	60.0	59	3	Sept.	70.3	150
12	Aug.	58.8	59	6	Sept.	–	–
13	Aug.	59.5	53	7	Sept.	69.6	81
16	Aug.	59.8	66	8	Sept.	70.0	91
17	Aug.	62.4	106	9	Sept.	70.0	87
18	Aug.	62.3	150	10	Sept.	69.4	82
19	Aug.	62.6	93	13	Sept.	70.0	71
20	Aug.	64.6	113	14	Sept.	70.6	98
23	Aug.	66.4	129	15	Sept.	71.2	83
24	Aug.	66.1	143	16	Sept.	71.0	93
25	Aug.	67.4	123	17	Sept.	70.4	77
26	Aug.	68.0	160				
27	Aug.	67.2	87				
30	Aug.	67.5	70				
31	Aug.	68.5	100				

4.9 Generate ten random normal deviates. Display a probability plot of these data. Suppose you didn't know the origin of these data. Would you conclude they were normally distributed? What is your conclusion based on the Shapiro–Wilk test? Do ten observations provide sufficient information to check normality?

4.10 Obtain a normal probability plot of the index given in Problem 4.8. Suppose that you had been ignorant of the lack of independence of these data and had treated them as if they were independent samples. Assess whether they are normally distributed.

4.11 Repeat problem 4.7 with weights expressed in ounces instead of pounds. How will your conclusions change? Obtain normal probability plots of the logarithm of mothers' weights expressed in pounds and then in ounces, and compare.

4.12 From the variables ACUTEILL and BEDDAYS described in Table 3.3, create a single variable that takes on the value 1 if the person has been both bedridden and acutely ill in the last two months and that takes on the value 0 otherwise.

4.13 Using the Parental HIV data set (see Appendix A), plot a histogram, boxplot, and a normal probability plot for the variable AGESMOKE. This variable is the age in years when the respondent started smoking. If the respondent did not start smoking, AGESMOKE was assigned a value of zero. Decide what to do about the zero values and if a transformation should be used for this variable if the assumption of normality is made when it is used in a statistical analysis.

4.14 Using the Parental HIV data calculate an overall Brief Symptom Inventory (BSI) score for each adolescent (see the codebook for details). Log-transform the BSI score. Obtain a normal probability plot for the log-transformed variables. Does the log-transformed variable seem to be normally distributed? As you might notice, the numbers of adolescents with a missing value on the overall BSI score and the log-transformed BSI score are different. Why is this the case? Could this influence our conclusions regarding the normality of the transformed variable? How could this be avoided?

4.15 Using the lung cancer data described in Appendix A, examine the distribution of the variable *days* separately for those who died (death=1) and for those who did not (death=0). Plot a normal probability plot, a histogram, and a boxplot for each. Use the methods described in this chapter to choose appropriate transformations to induce approximate normality. Are the chosen transformations the same for the two groups? Discuss the results.

Chapter 5

Selecting appropriate analyses

5.1 Which analyses to perform?

When you have a data set and wish to analyze it using the computer, obvious
questions that arise are "What descriptive measures should be used to examine the data?" and "What statistical analyses should be performed?" Section
5.2 explains why sometimes these are confusing questions, particularly to investigators who lack experience in real-life data analysis. Section 5.3 presents
measures that are useful in describing data. Usually, after the data have been
"cleaned" and any needed transformations made, the next step is to obtain a
set of descriptive statistics and graphs. It is also useful to obtain the number
of missing values for each variable. The suggested descriptive statistics and
graphs are guided by Stevens's classification system introduced in Chapter 2.
Section 5.4 presents a table that summarizes suggested multivariate analyses
and indicates in which chapters of this book they can be found. Readers with
experience in data analysis may be familiar with all or parts of this chapter.

5.2 Why selection is often difficult

There are two reasons why deciding what descriptive measures or analyses to
perform and report is often difficult for an investigator with real-life data. First,
in statistics textbooks, statistical methods are presented in a logical order from
the viewpoint of learning statistics but not from the viewpoint of doing data
analysis by using statistics. Most texts are either mathematical statistics texts,
or are imitations of them with the mathematics simplified or left out. Also,
when learning statistics for the first time, the student often finds mastering the
techniques themselves tough enough without worrying about how to use them
in the future. The second reason is that real-life data often contain mixtures
of types of data, which makes the choice of analysis somewhat arbitrary. Two
trained statisticians presented with the same set of data will often opt for different ways of analyzing the set, depending on what assumptions they are willing
to take into account in the interpretation of the analysis.

    Acquiring a sense of when it is safe to ignore assumptions is difficult both to learn and to teach. Here, for the most part, an empirical approach will be suggested. For example, it is often a good idea to perform several different analyses, one where all the assumptions are met and one where some are not, and compare the results. The idea is to use statistics to obtain insights into the data and to determine how the system under study works.

One point to keep in mind is that the examples presented in many statistics books are often ones the authors have selected after a long period of working with a particular technique. Thus they usually are "ideal" examples, designed to suit the technique being discussed. This feature makes learning the technique simpler but does not provide insight into its use in typical real-life situations. In this book we will attempt to be more flexible than standard textbooks so that you will gain experience with commonly encountered difficulties.

In the next section, suggested graphical and descriptive statistics are given for several types of data collected for analysis. Note, however, that these suggestions should not be applied rigidly in all situations. They are meant to be a framework for assisting the investigator in analyzing and reporting the data.

5.3   Appropriate statistical measures

In Chapter 2, Stevens's system of classifying variables into nominal (naming results), ordinal (determination of greater than or less than), interval (equal intervals between successive values of a variable), and ratio (equal intervals and a true zero point) was presented. In this section this system is used to obtain suggested descriptive measures. Table 5.1 shows appropriate graphical and computed measures for each type of variable. It is important to note that the descriptive measures are appropriate to that type of variable listed on the left **and to all below it**. Note also that $\sum$ signifies addition, $\prod$ multiplication, and P percentile in Table 5.1.

Measures of nominal data

For nominal data, the order of the numbers has no meaning. For example, in the depression data set the respondent's religion was coded 1 = Protestant, 2 = Catholic, 3 = Jewish, 4 = none, and 5 = other. Any other five distinct numbers could be chosen, and their order could be changed without changing the empirical operation of equality. The measures used to describe these data should **not** imply a sense of order.

Suitable graphical measures for nominal data are bar graphs and pie charts. These bar graphs and pie charts show the proportion of respondents who have each of the five responses for religion. The length of the bar represents the proportion for the bar graph and the size (or angle) of the piece represents the proportion for the pie chart. Fox and Long (1990) note that both of these graphical methods can be successfully used by observers to make estimates

Table 5.1: *Descriptive measures depending upon Stevens's scale*

Classification	Graphical measures	Measures of the center of a distribution	Measures of the variability of a distribution
Nominal	Bar graphs Pie charts	Mode	Binomial or multi-nomial variance
Ordinal	Bar graphs Histogram	Median	Range $P_{75} - P_{25}$
Interval	Histogram areas measurable	Mean $= \bar{X}$	Standard deviation $= S$
Ratio	Histogram areas measurable	Geometric mean $= (\prod_{i=1}^{N} X_i)^{1/N}$ Harmonic mean $= N / \sum_{i=1}^{N} 1/X_i$	Coefficient of variation $= S/\bar{X}$

of proportions or counts. Others are less impressed with the so-called stacked bar graphs (where each bar is divided into a number of subdistances based on another variable, say gender). It is difficult to compare the subdistances in stacked bar graphs since they all do not start at a common base. Bar graphs and pie charts are available in all six packages.

Note that there are two types of pie charts: the "value" and the "count" pie charts. Some packages only have the former. The count pie chart has a pie piece corresponding to each category of a given variable. On the other hand, the value pie charts represent the sum of values for each of a group of variables in the pie pieces. Counts can be produced by creating a dummy variable for each category: suppose you create a variable called "rel1" which equals 1 if religion = 1 and 0 otherwise. Similarly, you can create "rel2," "rel3," "rel4," and "rel5." The sum of rel1 over the cases will then equal the number of cases for which religion = 1; and so forth. The value pie chart with rel1 to rel5 will then have five pieces representing the five categories of religion.

The **mode**, or outcome, of a variable that occurs most frequently is the only appropriate measure of the center of the distribution. For a variable such as sex, where only two outcomes are available, the variability, or variance, of the proportion of cases who are male (female) can be measured by the **binomial variance**, which is

$$\text{Estimated variance of } p = \frac{p(1-p)}{N}$$

where p is the proportion of respondents who are males (females). If more than two outcomes are possible, then the variance of the ith proportion is given by

$$\text{Estimated variance of } p_i = \frac{p_i(1-p_i)}{N}$$

Measures for ordinal data

For ordinal variables, order or ranking does have relevance, and so more descriptive measures are available. In addition to the pie charts and bar graphs used for nominal data, histograms can now be used. The area under the histogram still has **no** meaning because the intervals between successive numbers are not necessarily equal. For example, in the depression data set a general health question was asked and later coded 1 = excellent, 2 = good, 3 = fair, and 4 = poor. The distance between 1 and 2 is not necessarily equal to the distance between 3 and 4 when these numbers are used to identify answers to the health question.

An appropriate measure of the center of the distribution is the median. Roughly speaking, the **median** is the value of the variable that half the respondents exceed and half do not. The **range**, or largest minus smallest observation, is a measure of how variable or disperse the distribution is. Another measure that is sometimes reported is the difference between two **percentiles**. For example, sometimes the fifth percentile, $P(5)$, is subtracted from the 95th percentile, $P(95)$. As explained in Section 4.3, $P(5)$ equals the quantile $Q(0.05)$. Some investigators prefer to report the **interquartile range**, $Q(0.75) - Q(0.25)$, or the **quartile deviation**, $(Q(0.75) - Q(0.25))/2$. Boxplots can also be used if they display the median and quartiles.

Histograms are available in all six statistical packages used in this book. Also, medians, $Q(0.25)$ and $Q(0.75)$, are widely available. Percentiles or quartiles for each distinct value of a variable are available in SAS and SPSS. Stata has values for a wide range of percentiles and in STATISTICA any quantile can be computed. Using the summary command S-PLUS will give the first and third quartile.

Measures for interval data

For interval data the full range of descriptive statistics generally used is available to the investigator. This set includes graphical measures such as histograms, with the area under the histogram now having meaning. Boxplots can now be used that display either medians and quartiles or means and standard deviations. Stem and leaf displays that are partly graphical and partly numerical are available in S-PLUS, e.g., see Selvin (1998). Additional graphical measures that are useful in visualizing the distribution of the data are given in Cleveland (1993).

The well-known **mean** and **standard deviation** can now be used for describing the data numerically. These descriptive statistics are part of the output in most programs and hence are easily obtainable.

Measures for ratio data

The additional measures available for ratio data are seldom used. The **geometric mean (GM)** is sometimes used when the log transformation is used, since

$$\log \text{GM} = \frac{\sum \log X}{N}$$

It is also used when computing the mean of a process where there is a constant rate of change. For example, suppose a rapidly growing city has a population of 2500 in 1990 and a population of 5000 according to the 2000 census. An estimate of the 1995 population (or halfway between 1990 and 2000) can be estimated as

$$\text{GM} = \left[\prod^{2} X_i \right]^{1/2} = (2500 \times 5000)^{1/2} = 3525$$

The **harmonic mean** (HM) is the reciprocal of the arithmetic mean of the reciprocals of the data. It is used for obtaining a mean of rates when the quantity in the numerator is fixed. For example, if an investigator wishes to analyze distance per unit of time that N cars require to run a fixed distance, then the harmonic mean should be used.

The **coefficient of variation** can be used to compare the variability of distributions that have different means. It is a unitless statistic. It is sometimes used as a measure of the accuracy of observations.

Note that the geometric and harmonic mean can both be computed with any package by first transforming the data. The coefficient of variation is widely available and is simple to compute in any case since it is the standard deviation divided by the mean.

Stretching assumptions

In the data analyses given in the next section, it is the ordinal variables that often cause confusion. Some statisticians treat them as if they were nominal data, often splitting them into two categories if they are dependent variables or using the dummy variables described in Section 9.3 if they are independent variables. Other statisticians treat them as if they were interval data. It is usually possible to assume that the underlying scale is continuous, and that because of a lack of a sophisticated measuring instrument, the investigator is not measuring with an interval scale.

One important question is how far off is the ordinal scale from an interval scale? If it is close, then using an interval scale makes sense; otherwise it is more questionable. There are many more statistical analyses available for interval data than for ordinal data so there are real incentives for ignoring the distinction between interval and ordinal data. If you are unwilling to stretch the assumptions, then you may wish to use nonparametric methods for your analysis. This subject is not covered in this book, but the interested reader may consult Conover (1999) or Sprent and Smeeton (2007).

Velleman and Wilkinson (1993) proposed that the data analysis used should be guided by what you hope to learn from the data, not by rigid adherence to Stevens's classification system. Further discussion of these issues can be found in Andrews *et al.* (1998).

Although assumptions are sometimes stretched so that ordinal data can be treated as interval, this stretching should not be done with nominal data, because complete nonsense is likely to result.

5.4   Selecting appropriate multivariate analyses

To decide on possible analyses, we can classify variables as follows:

1. independent versus dependent;
2. nominal or ordinal versus interval or ratio.

The classification of independent or dependent may differ from analysis to analysis, but the classification into Stevens's system usually remains constant throughout the analysis phase of the study. Once these classifications are determined, it is possible to refer to Table 5.2 and decide what analysis should be considered. This table should be used as a general guide to analysis selection rather than as a strict set of rules.

In Table 5.2 nominal and ordinal variables have been combined because this book does not cover analyses appropriate only to nominal or ordinal data separately. An extensive summary of measures and tests for these types of variables is given in Powers and Xie (2008) and Agresti (2002). Interval and ratio variables have also been combined because the same analyses are used for both types of variables. There are many measures of association and many statistical methods not listed in the table. For further information on choosing analyses appropriate to various data types, see Andrews *et al.* (1998).

The first row in the body of Table 5.2 includes analyses that can be done if there are no dependent variables. Note that if there is only one variable, it can be considered either dependent or independent. A single independent variable that is either interval or ratio can be screened by methods given in Chapters 3 and 4, and descriptive statistics can be obtained from many statistical programs. If there are several interval or ratio independent variables, then several techniques are listed in the table.

In Table 5.2 the numbers in the parentheses following some techniques refer to the chapters of this book where those techniques are described. For example, to determine the advantages of doing a principal components analysis and how to obtain results and interpret them, you would consult Chapter 11. A very brief description of this technique is also given in Chapter 1.

If no number is given in parentheses, then that technique is not discussed in this book. For example, in the lower right hand corner of Table 5.2, path analysis and structural models are listed. Those topics are not covered in this book. Structural models can be obtained from the Lisrel or Mplus program (see Section 15.10 for information on obtaining the programs). The Mplus program also provides a wide selection of methods for modeling cross-sectional or longitudinal data.

For interval or ratio dependent variables and nominal or ordinal independent variables, analysis of variance is the appropriate technique. Analysis of variance is not discussed in this book; for discussions of this topic, see Mickey, Dunn and Clark (2009) or Dean and Voss (1999). Multivariate analysis of variance and Hotelling's T^2 are discussed in Afifi and Azen (1979) and Timm (2002). Structural models are discussed in Pugesek *et al.* (2003). The term log-linear models used here applies to the analysis of nominal and ordinal data (Chapter 17). The same term is used in Chapter 13 in connection with one method of survival analysis.

Table 5.2 provides a general guide for what analyses should be **considered**. We do not mean that other analyses could not be done but simply that the usual analyses are the ones that are listed. For example, methods of performing discriminant function analyses have been studied for noninterval variables, but this technique was originally derived with interval or ratio data.

Judgment will be called for when the investigator has, for example, five independent variables, three of which are interval, while one is ordinal and one is nominal, with one dependent variable that is interval. Most investigators would use multiple regression, as indicated in Table 5.2. They might pretend that the one ordinal variable is interval and use dummy variables for the nominal variable (Chapter 9). Another possibility is to categorize all the independent variables and to perform an analysis of variance on the data. That is, analyses that require fewer assumptions in terms of types of variables can always be done. Sometimes, both analyses are done and the results are compared. Because the packaged programs are so simple to run, multiple analyses are a realistic option.

In the examples given in Chapters 6–18, the data used will often not be ideal. In some chapters a data set has been created that fits all the usual assumptions to explain the technique, but then a nonideal, real-life data set is also run and analyzed. It should be noted that when inappropriate variables are used in a statistical analysis, the association between the statistical results and the real-life situation is weakened. However, the statistical models do not have to fit perfectly in order for the investigator to obtain useful information from them.

5.5 Summary

The Stevens system of classification of variables can be a helpful tool in deciding on the choice of descriptive measures as well as in sophisticated data analyses. In this chapter we presented a table to assist the investigator in each of these two areas. A beginning data analyst may benefit from practicing the advice given in this chapter and from consulting more experienced researchers.

The recommended analyses are intended as general guidelines and are by no means exclusive. It is a good idea to try more than one way of analyzing the data whenever possible. Also, special situations may require specific analyses, perhaps ones not covered thoroughly in this book.

5.6 Problems

5.1 Compute an appropriate measure of the center of the distribution for the following variables from the depression data set: MARITAL, INCOME, AGE, and HEALTH.

5.2 An investigator is attempting to determine the health effects on families of living in crowded urban apartments. Several characteristics of the apartment have been measured, including square feet of living area per person, cleanliness, and age of the apartment. Several illness characteristics for the families have also been measured, such as number of infectious diseases and number of bed days per month for each child, and overall health rating for the mother. Suggest an analysis to use with these data.

5.3 A coach has made numerous measurements on successful basketball players, such as height, weight, and strength. He also knows which position each player is successful at. He would like to obtain a function from these data that would predict which position a new player would be best at. Suggest an analysis to use with these data.

5.4 A college admissions committee wishes to predict which prospective students will successfully graduate. To do so, the committee intends to obtain the college grade point averages for a sample of college seniors and compare these with their high school grade point averages and Scholastic Aptitude Test scores. Which analysis should the committee use?

5.5 Data on men and women who have died have been obtained from health maintenance organization records. These data include age at death, height and weight, and several physiological and lifestyle measurements such as blood pressure, smoking status, dietary intake, and usual amount of exercise. The immediate and underlying causes of death are also available. From

Table 5.2: Suggested data analysis under Stevens's classification

Dependent variable(s)	Independent variables			
	Nominal or ordinal		Interval or ratio	
	1 variable	>1 variable	1 variable	>1 variable
No dependent variables	χ^2 goodness of fit	Measures of association Log-linear model (17) χ^2 test of independence	Univariate statistics (e.g., one-sample t tests) Descriptive measures (5) Tests for normality (4)	Correlation matrix (7) Principal components (14) Factor analysis (15) Cluster analysis (16)
Nominal or ordinal 1 variable	χ^2 test Fisher's exact test	Log-linear model (17) Logistic regression (12) Poisson regression (12)	Discriminant function (11) Logistic regression (12) Univariate statistics (e.g., two-sample t tests)	Discriminant function (11) Logistic regression (12) Poisson regression (12)
>1 variable	Log-linear model (17)	Log-linear model (17)	Discriminant function (11)	Discriminant function (11)
Interval or ratio 1 variable	t-test Analysis of variance Survival analysis (13)	Analysis of variance Multiple-classification analysis Survival analysis (13)	Linear regression (6, 18) Correlation (6) Survival analysis (13)	Multiple regression (7–9, 18) Survival analysis (13)
>1 variable	Multivariate analysis of variance Analysis of variance on principal components Hotelling's T^2 Profile analysis (16)	Multivariate analysis of variance Analysis of variance on principal components	Canonical correlation (10)	Canonical correlation (10) Path analysis Structural models (LISREL, Mplus)

these data we would like to find out which variables predict death due to various underlying causes. (This procedure is known as **risk factor analysis**.) Suggest possible analyses.

5.6 Large amounts of data are available from the United Nations and other international organizations on each country and sovereign state of the world, including health, education, and commercial data. An economist would like to invent a descriptive system for the degree of development of each country on the basis of these data. Suggest possible analyses.

5.7 For the data described in Problem 5.6 we wish to put together similar countries into groups. Suggest possible analyses.

5.8 For the data described in Problem 5.6 we wish to relate health data such as infant mortality (the proportion of children dying before the age of one year) and life expectancy (the expected age at death of a person born today if the death rates remain unchanged) to other data such as gross national product per capita, percentage of people older than 15 who can read and write (literacy), average daily caloric intake per capita, average energy consumption per year per capita, and number of persons per practicing physician. Suggest possible analyses. What other variables would you include in your analysis?

5.9 A member of the admissions committee notices that there are several women with high grade point averages but low SAT scores. He wonders if this pattern holds for both men and women in general, only for women in general, or only in a few cases. Suggest ways to analyze this problem.

5.10 Two methods are currently used to treat a particular type of cancer. It is suspected that one of the treatments is twice as effective as the other in prolonging survival regardless of the severity of the disease at diagnosis. A study is carried out. After the data are collected, what analysis should the investigators use?

5.11 A psychologist would like to predict whether or not a respondent in the depression study described in Chapter 3 is depressed. To do this, she would like to use the information contained in the following variables: MARITAL, INCOME, and AGE. Suggest analyses.

5.12 Using the data described in Table A.1, an investigator would like to predict a child's lung function based on that of the parents and the area they live in. What analyses would be appropriate to use?

5.13 In the depression study, information was obtained on the respondent's religion (Chapter 3). Describe why you think it is incorrect to obtain an average score for religion across the 294 respondents.

5.14 Suppose you would like to analyze the relationship between the number of times an adolescent has been absent from school without a reason and how much the adolescent likes/liked going to school for the Parental HIV data (Appendix A). Suggest ways to analyze this relationship. Suggest other variables that might be related to the number of times an adolescent has been absent from school without a reason.

5.15 The Parental HIV data include information on the age at which adolescents started smoking. Where does this variable fit into Stevens's classification scheme? Particularly comment on the issue relating to adolescents who had not starting smoking by the time they were interviewed.

Part II

Applied Regression Analysis

Chapter 6

Simple regression and correlation

6.1 Chapter outline

Simple linear regression analysis is commonly performed when investigators wish to examine the relationship between two variables. Section 6.2 describes the two basic models used in linear regression analysis and a data example is given in Section 6.3. Section 6.4 presents the assumptions, methodology, and usual output from the first model while Section 6.5 does the same for the second model. Sections 6.6 and 6.7 discuss the interpretation of the results for the two models. In Section 6.8 a variety of useful output options that are available from statistical programs are described. These outputs include standardized regression coefficients, the regression analysis of variance table, determining whether or not the relationship is linear, and how to find outliers and influential observations. Section 6.9 defines robustness in statistical analysis and discusses how critical the various assumptions are. The use of transformations for simple linear regression is also described. In Section 6.10 regression through the origin and weighted regression are introduced. How to obtain a loess curve and when it is used are also given in this section. In Section 6.11 a variety of uses of linear regressions are presented. A brief discussion of how to obtain the computer output given in this chapter is found in Section 6.12. Finally, Section 6.13 describes what to watch out for in regression analysis.

If you are reading about simple linear regression for the first time, skip Sections 6.9, 6.10, and 6.11 in your first reading. If this chapter is a review for you, you can skim most of it, but read the above-mentioned sections in detail.

6.2 When are regression and correlation used?

The methods described in this chapter are appropriate for studying the relationship between two variables X and Y. By convention, X is called the **independent or predictor variable** and is plotted on the horizontal axis. The variable Y is called the **dependent or outcome variable** and is plotted on the vertical axis. The dependent variable is assumed to be continuous, while the independent variable may be continuous or discrete.

The data for regression analysis can arise in two forms.

1. **Fixed-X case.** The values of X are selected by the researchers or forced on them by the nature of the situation. For example, in the problem of predicting the sales for a company, the total sales are given for each year. Year is the fixed-X variable, and its values are imposed on the investigator by nature. In an experiment to determine the growth of a plant as a function of temperature, a researcher could randomly assign plants to three different preset temperatures that are maintained in three greenhouses. The three temperature values then become the fixed values for X.

2. **Variable-X case.** The values of X and Y are both random variables. In this situation, cases are selected randomly from the population, and both X and Y are measured. All survey data are of this type, whereby individuals are chosen and various characteristics are measured on each.

Regression and correlation analysis can be used for either of two main purposes.

1. **Descriptive.** The kind of relationship and its strength are examined.

2. **Predictive.** The equation relating Y and X can be used to predict the value of Y for a given value of X. Prediction intervals can also be used to indicate a likely range of the predicted value of Y.

6.3   Data example

In this section we present an example used in the remainder of the chapter to illustrate the methods of regression and correlation. Lung function data were obtained from an epidemiological study of households living in four areas with different amounts and types of air pollution. The data set used in this book is a subset of the total data. In this chapter we use only the data taken on the fathers, all of whom are nonsmokers (see Appendix A for more details).

One of the major early indicators of reduced respiratory function is FEV1 or forced expiratory volume in the first second (amount of air exhaled in 1 second). Since it is known that taller males tend to have higher FEV1, we wish to determine the relationship between height and FEV1. We exclude the data from the mothers as several studies have shown a different relationship for women. The sample size is 150. These data belong to the variable-X case, where X is height (in inches) and Y is FEV1 (in liters). Here we may be concerned with describing the relationship between FEV1 and height, a **descriptive** purpose. We may also use the resulting equation to determine expected or normal FEV1 for a given height, a **predictive** use.

In Figure 6.1 a scatter diagram of the data is produced by the STATISTICA package. In this graph, height is given on the horizontal axis since it is the independent or predictor variable and FEV1 is given on the vertical axis since it is the dependent or outcome variable. Note that heights have been rounded to

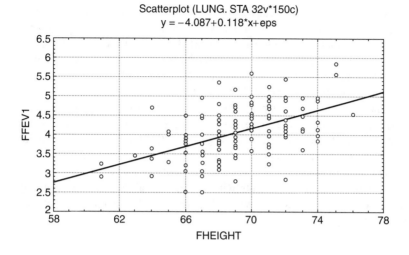

Figure 6.1: *Scatter Diagram and Regression Line of FEV1 versus Height for Fathers*

the nearest inch in the original data and the program marked every four inches on the horizontal axis. The circles in Figure 6.1 represent the location of the data. There does appear to be a tendency for taller men to have higher FEV1. The program also draws the **regression line** in the graph. The line is tilted upwards, indicating that we expect larger values of FEV1 with larger values of height. We will define the regression line in Section 6.4. The equation of the regression line is given under the title as

$$Y = -4.087 + 0.118X$$

The program also includes a "+ eps" which we will ignore until Section 6.6. The quantity 0.118 in front of X is greater than zero, indicating that as we increase X, Y will increase. For example, we would expect a father who is 70 inches tall to have an FEV1 value of

$$\text{FEV1} = -4.087 + (0.118)(70) = 4.173$$

or

$$\text{FEV1} = 4.17 \text{ (rounded off)}$$

If the height was 66 inches then we would expect an FEV1 value of only 3.70.

To take an extreme example, suppose a father was 2 feet tall. Then the equation would predict a negative value of FEV1 (-1.255). This example illustrates the danger of using the regression equation outside the appropriate range. A safe policy is to restrict the use of the equation to the range of the X observed in the sample.

In order to get more information about these men, we requested descriptive statistics. These included the sample size $N = 150$, the mean for each variable ($\bar{X} = 69.260$ inches for height and $\bar{Y} = 4.093$ liters for FEV1), and the standard deviations ($S_X = 2.779$ and $S_Y = 0.651$). Note that the mean height is approximately in the middle of the heights and the mean FEV1 is approximately in the middle of the FEV1 values in Figure 6.1.

The various statistical packages include a wide range of options for displaying scatter plots and regression lines. Different symbols and colors can often be used for the points. Usually, however, the default option is fine unless other options are desirable for inclusion in a report or an article.

One option that is useful for reports is controlling the numerical values to be displayed on the axes. Plots with numerical values not corresponding to what the reader expects or to what others have used are sometimes difficult to interpret.

6.4   Regression methods: fixed-X case

In this section we present the background assumptions, models, and formulas necessary for understanding simple linear regression. The theoretical background is simpler for the fixed-X case than for the variable-X case, so we will begin with it.

Assumptions and background

For each value of X we conceptualize a distribution of values of Y. This distribution is described partly by its mean and variance at each fixed X value:

$$\text{Mean of } Y \text{ values at a given } X = \alpha + \beta X$$

and

$$\text{Variance of } Y \text{ values at a given } X = \sigma^2$$

The basic idea of **simple linear regression** is that the means of Y lie on a straight line when plotted against X. Secondly, the variance of Y at a given X is assumed to be the same for all values of X. The latter assumption is called **homoscedasticity**, or homogeneity of variance. Figure 6.2 illustrates the distribution of Y at three values of X. Note from Figure 6.2 that σ^2 is not the variance of all the Y's from their mean but is, instead, the variance of Y at a given X. It is clear that the means lie on a straight line in the range of concern and that the variance or the degree of variation is the same at the different values of X. Outside the range of concern it is immaterial to our analysis what the curve looks like, and in most practical situations, linearity will hold over a limited range of X. This figure illustrates that extrapolation of a linear relationship beyond the range of concern can be dangerous.

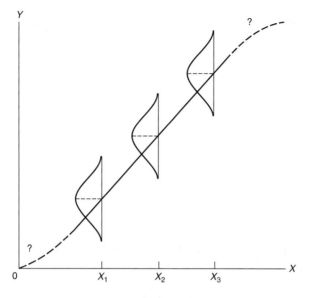

Figure 6.2: *Simple Linear Regression Model for Fixed X's*

The expression $\alpha + \beta X$, relating the mean of Y to X, is called the **population regression equation**. Figure 6.3 illustrates the meaning of the parameters α and β. The parameter α is the **intercept** of this line. That is, it is the mean of Y when $X = 0$. The **slope** β is the amount of change in the mean of Y when the value of X is increased by one unit. A negative value of β signifies that the mean of Y decreases as X increases.

Least squares method

The parameters α and β are estimated from a sample collected according to the fixed-X model. The **sample estimates** of α and β are denoted by A and B, respectively, and the resulting regression line is called the **sample least squares regression equation**.

To illustrate the method, we consider a hypothetical sample of four points, where X is fixed at $X_1 = 5, X_2 = 5, X_3 = 10$, and $X_4 = 10$. The sample values of Y are $Y_1 = 14, Y_2 = 17, Y_3 = 27$, and $Y_4 = 22$. These points are plotted in Figure 6.4.

The output from the computer would include the following information:

	MEAN	ST.DEV.	REGRESSION LINE	RES.MS.
X	7.5	2.8868		
Y	20.0	5.7155	$Y = 6.5 + 1.8X$	8.5

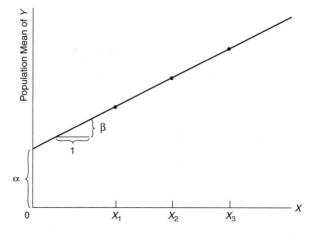

Figure 6.3: *Theoretical Regression Line Illustrating α and β*

The **least squares method** finds the line that minimizes the sum of squared vertical deviations from each point in the sample to the point **on** the line corresponding to the X-value. It can be shown mathematically that the least squares line is

$$\hat{Y} = A + BX$$

where

$$B = \frac{\Sigma(X - \bar{X})(Y - \bar{Y})}{\Sigma(X - \bar{X})^2}$$

and

$$A = \bar{Y} - B\bar{X}$$

Here $\bar{X}$ and $\bar{Y}$ denote the sample means of X and Y, and $\hat{Y}$ denotes the predicted value of Y for a given X.

The **deviation** (or **residual**) for the first point is computed as follows:

$$Y(1) - \hat{Y}(1) = 14 - [6.5 + 1.8(5)] = -1.5$$

Similarly, the other residuals are $+1.5, -2.5$, and $+2.5$, respectively. The sum of the squares of these deviations is 17.0. No other line can be fitted to produce a smaller sum of squared deviations than 17.0. The slope of the sample regression line is given by $B = 1.8$ and the intercept by $A = 6.5$.

The estimate of σ^2 is called the **residual mean square** (RES. MS.) or the **mean square error** (MSE) and is computed as

$$S^2 = \text{RES. MS.} = \frac{\Sigma(Y - \hat{Y})^2}{N - 2}$$

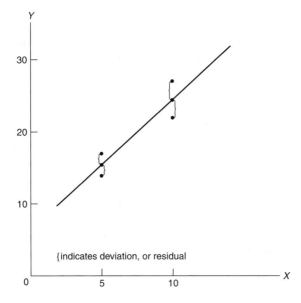

Figure 6.4: *Simple Data Example Illustrating Computations of Output Given in Figure 6.1*

The number $N - 2$, called the **residual degrees of freedom**, is the sample size minus the number of parameters in the line (in this case, α and β). Using $N - 2$ as a divisor in computing S^2 produces an **unbiased estimate** of σ^2. In the example,

$$S^2 = \text{RES. MS.} = \frac{17}{4 - 2} = 8.5$$

The square root of the residual mean square is called the **standard error of the estimate** and is denoted by S.

Software regression programs will also produce the **standard errors** of A and B. These statistics are computed as

$$\text{SE}(A) = S\left[\frac{1}{N} + \frac{\bar{X}^2}{\sum(X - \bar{X})^2}\right]^{1/2}$$

and

$$\text{SE}(B) = \frac{S}{[\sum(X - \bar{X})^2]^{1/2}}$$

Confidence and prediction intervals

For each value of X under consideration a population of Y-values is assumed to exist. Confidence intervals and tests of hypotheses concerning the intercept, slope, and line may be made with assurance when three assumptions hold.

1. The Y-values are assumed to be normally distributed.
2. Their means lie on a straight line.
3. Their variances are all equal.

For example, confidence for the slope B can be computed by using the standard error of B. The **confidence interval** (CI) for B is

$$CI = B \pm t[SE(B)]$$

where t is the $100(1 - \alpha/2)$ percentile of the t distribution with $N - 2$ df (**degrees of freedom**). Similarly, the confidence interval for A is

$$CI = A \pm t[SE(A)]$$

where the same degrees of freedom are used for t.

The value $\hat{Y}$ computed for a particular X can be interpreted in two ways:

1. $\hat{Y}$ is the **point estimate of the mean** of Y at that value of X.
2. $\hat{Y}$ is the **estimate of the Y value** for any individual with the given value of X.

The investigator may supplement these point estimates with interval estimates. The **confidence interval** (CI) **for the mean of Y** at a given value of X, say X^*, is

$$CI = \hat{Y} \pm tS \left[\frac{1}{N} + \frac{(X^* - \bar{X})^2}{\Sigma(X - \bar{X})^2} \right]^{1/2}$$

where t is the $100(1 - \alpha/2)$ percentile of the t distribution with $N - 2$ df.

For an individual Y-value the confidence interval is called the **prediction interval** (PI). The prediction interval (PI) for an individual Y at X^* is computed as

$$PI = Y \pm tS \left[1 + \frac{1}{N} + \frac{(X^* - \bar{X})^2}{\Sigma(X - \bar{X})^2} \right]^{1/2}$$

where t is the same as for the confidence interval for the mean of Y.

In summary, for the fixed-X case we have presented the model for simple regression analysis and methods for estimating the parameters of the model. Later in the chapter we will return to this model and present special cases and other uses. Next, we present the variable-X model.

6.5   Regression and correlation: variable-X case

In this section we present, for the variable-X case, material similar to that given in the previous section.

For this model both X and Y are random variables measured on cases that are randomly selected from a population. One example is the lung function data set, where FEV1 was predicted from height. The fixed-X regression model applies here when we treat the X-values as if they were preselected. (This technique is justifiable theoretically by **conditioning** on the X-values that happened to be obtained in the sample.) Therefore all the previous discussion and formulas are precisely the same for this case as for the fixed-X case. In addition, since both X and Y are considered random variables, other parameters can be useful for describing the model. These include the means and variances for X and Y over the entire population (μ_X, μ_Y, σ_X^2, and σ_Y^2). The sample estimates for these parameters are usually included in computer output. For example, in the analysis of the lung function data set these estimates were obtained by requesting descriptive statistics.

As a measure of how the variables X and Y vary together, a parameter called the **population covariance** is often estimated. The population covariance of X and Y is defined as the average of the product $(X - \mu_X)(Y - \mu_Y)$ over the entire population. This parameter is denoted by σ_{XY}. If X and Y tend to increase together, σ_{XY} will be positive. If, on the other hand, one tends to increase as the other decreases, σ_{XY} will be negative.

To standardize the magnitude of σ_{XY} we divide it by the product $\sigma_X \sigma_Y$. The resulting parameter, denoted by ρ, is called the **product moment correlation coefficient**, or simply the **correlation coefficient**. The value of

$$\rho = \frac{\sigma_{XY}}{\sigma_X \sigma_Y}$$

lies between -1 and $+1$, inclusive. The sample estimate for ρ is the **sample correlation coefficient** r, or

$$r = \frac{S_{XY}}{S_X S_Y}$$

where

$$S_{XY} = \frac{\sum (X - \bar{X})(Y - \bar{Y})}{N - 1}$$

The sample statistic r also lies between -1 and $+1$, inclusive. Further interpretation of the correlation coefficient is given in Cohen *et al.* (2002).

Tests of hypotheses and confidence intervals for the variable-X case require that X and Y be jointly normally distributed. Formally, this requirement is that X and Y follow a bivariate normal distribution. Examples of the appearance of bivariate normal distributions are given in Section 6.7. If this condition is true, it can be shown that Y also satisfies the three conditions for the fixed-X case.

6.6   Interpretation: fixed-X case

In this section we present methods for interpreting the results of a regression output.

First, the type of the sample must be determined. If it is a fixed-X sample, the statistics of interest are the intercept and slope of the line and the standard error of the estimate; point and interval estimates for α and β have already been discussed.

The investigator may also be interested in **testing hypotheses** concerning the parameters. A commonly used test is for the null hypothesis

$$H_0 : \beta = \beta_0$$

The test statistic is

$$t = \frac{(B - \beta_0)[\sum(X - \bar{X})^2]^{1/2}}{S}$$

where S is the square root of the residual mean square and the computed value of t is compared with the tabled t percentiles with $N - 2$ degrees of freedom to obtain the P value. Many computer programs will print the standard error of B. Then the t statistic is simply

$$t = \frac{B - \beta_0}{SE(B)}$$

A common value of β_0 is $\beta_0 = 0$, indicating independence of X and Y, i.e., the mean value of Y does not change as X changes.

A test concerning α can also be performed for the null hypothesis $H_0 : \alpha = \alpha_0$, using

$$t = \frac{A - \alpha_0}{S\{(1/N) + [\bar{X}^2/\sum(X - \bar{X})^2]\}^{1/2}}$$

Values of this statistic can also be compared with tables of t percentiles with $N - 2$ degrees of freedom to obtain the t value. If the standard error of A is printed by the program, the test statistic can be computed simply as

$$t = \frac{A - \alpha_0}{SE(A)}$$

For example, to test whether the line passes through the origin, the investigator would test the hypothesis $\alpha_0 = 0$.

It should be noted that rejecting the null hypothesis $H_0 : \beta = 0$ is itself no indication of the magnitude of the slope. An observed $B = 0.1$, for instance, might be found significantly different from zero, while a slope of $B = 1.0$ might be considered inconsequential in a particular application. The importance and strength of a relationship between Y and X is a separate question from the question of whether certain parameters are significantly different from zero. The test of the hypothesis $\beta = 0$ is a preliminary step to determine whether the magnitude of B should be further examined. If the null hypothesis is rejected, then the magnitude of the effect of X on Y should be investigated.

One way of investigating the **magnitude of the effect** of a typical X value on Y is to multiply B by $\bar{X}$ and to contrast this result with $\bar{Y}$. If $B\bar{X}$ is small relative to $\bar{Y}$, then the magnitude of the effect of B in predicting Y is small. Another interpretation of B can be obtained by first deciding on two typical values of X, say X_1 and X_2, and then calculating the difference $B(X_2 - X_1)$. This difference measures the change in Y when X goes from X_1 to X_2.

To infer **causality**, we must justify that all other factors possibly affecting Y have been controlled in the study. One way of accomplishing this control is to design an experiment in which such intervening factors are held fixed while only the variable X is set at various levels. Standard statistical wisdom also requires randomization in the assignment to the various X levels in the hope of controlling for any other factors not accounted for (Box, 1966).

6.7 Interpretation: variable-X case

In this section we present methods for interpreting the results of a regression and correlation output. In particular, we will look at the ellipse of concentration and the coefficient of correlation.

For the variable-X model the regression line and its interpretation remain valid. Strictly speaking, however, causality cannot be inferred from this model. Here we are concerned with the bivariate distribution of X and Y. We can safely estimate the means, variances, covariance, and correlation of X and Y, i.e., the distribution of pairs of values of X and Y measured on the same individual. (Although these parameter estimates are printed by the computer, they are meaningless in the fixed-X model.) For the variable-X model the interpretations of the means and variances of X and Y are the usual measures of location (center of the distribution) and variability. We will now concentrate on how the **correlation coefficient** should be interpreted.

Ellipse of concentration

The **bivariate distribution** of X and Y is best interpreted by a look at the scatter diagram from a random sample. If the sample comes from a bivariate normal distribution, the data will tend to cluster around the means of X and Y and will approximate an ellipse called the **ellipse of concentration**. Note in Figure 6.1 that the points representing the data could be enclosed by an ellipse of concentration.

An ellipse can be characterized by the following:

1. The center.
2. The major axis, i.e., the line going from the center to the farthest point on the ellipse.

3. The minor axis, i.e., the line going from the center to the nearest point on the ellipse (the minor axis is always perpendicular to the major axis).

4. The ratio of the length of the minor axis to the length of the major axis. If this ratio is small, the ellipse is thin and elongated; otherwise, the ellipse is fat and rounded.

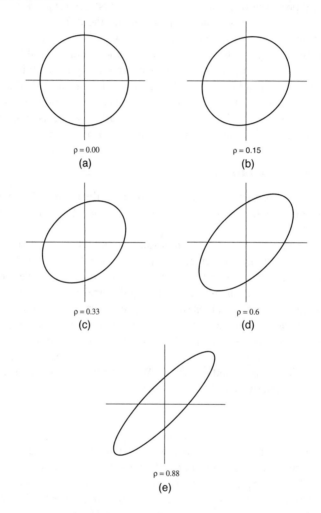

Figure 6.5: *Ellipses of Concentration for Various ρ Values*

Interpreting the correlation coefficient

For ellipses of concentration the center is at the point defined by the means of X and Y. The directions and lengths of the major and minor axes are determined by the two variances and the correlation coefficient. For fixed values of the variances the ratio of the length of the minor axis to that of the major axis, and hence the shape of the ellipse, is determined by the correlation coefficient ρ.

In Figure 6.5 we represent ellipses of concentration for various bivariate normal distributions in which the means of X and Y are both zero and the variances are both one (standardized X's and Y's). The case $\rho = 0$, Figure 6.5a, represents independence of X and Y. That is, the value of one variable has no effect on the value of the other, and the ellipse is a perfect circle. Higher values of ρ correspond to more elongated ellipses, as indicated in Figure 6.5b–e. Monette in Fox and Long (1990), Cohen *et al.* (2002), and Greene (2008) give further interpretation of the correlation coefficient in relation to the ellipse.

We see that for very high values of ρ one variable conveys a lot of information about the other. That is, if we are given a value of X, we can guess the corresponding Y value quite accurately. We can do so because the range of the possible values of Y for a given X is determined by the width of the ellipse at that value of X. This width is small for a large value of ρ. For negative values of ρ similar ellipses could be drawn where the major axis (long axis) has a negative slope, i.e., in the northwest/southeast direction.

Another interpretation of ρ stems from the concept of the **conditional distribution**. For a specific value of X the distribution of the Y value is called the conditional distribution of Y given that value of X. The word **given** is translated symbolically by a vertical line, so Y given X is written as $Y|X$. The variance of the conditional distribution of Y, variance $(Y|X)$, can be expressed as

$$\text{variance } (Y|X) = \sigma_Y^2 (1 - \rho^2)$$

or

$$\sigma_{Y|X}^2 = \sigma_Y^2 (1 - \rho^2)$$

Note that $\sigma_{Y|X} = \sigma$ as given in Section 6.4.

This equation can be written in another form:

$$\rho^2 = \frac{\sigma_Y^2 - \sigma_{Y|X}^2}{\sigma_Y^2}$$

The term $\sigma_{Y|X}^2$ measures the variance of Y when X has a specific fixed value. Therefore this equation states that ρ^2 is the proportion of variance of Y reduced because of knowledge of X. This result is often loosely expressed by saying that ρ^2 is the proportion of the variance of Y "explained" by X.

A better interpretation of ρ is to note that

$$\frac{\sigma_{Y|X}}{\sigma_Y} = (1 - \rho^2)^{1/2}$$

This value is a measure of the proportion of the standard deviation of Y not explained by X. For example, if $\rho = \pm 0.8$, then 64% of the variance of Y is explained by X. However, $(1 - 0.8^2)^{1/2} = 0.6$, saying that 60% of the standard deviation of Y is not explained by X. Since the standard deviation is a better measure of variability than the variance, it is seen that when $\rho = 0.8$, more than half of the variability of Y is still not explained by X. If instead of using ρ^2 from the population we use r^2 from a sample, then

$$S^2_{Y|X} = S^2 = \left(\frac{N-1}{N-2} \right) S^2_Y (1 - r^2)$$

and the results must be adjusted for sample size.

An important property of the correlation coefficient is that its value is not affected by the units of X or Y or any linear transformation of X or Y. For instance, X was measured in inches in the example shown in Figure 6.1, but the correlation between height and FEV1 is the same if we change the units of height to centimeters or the units of FEV1 to milliliters. In general, adding (subtracting) a constant to either variable or multiplying either variable by a constant will not alter the value of the correlation. Since $\hat{Y} = A + BX$, it follows that the correlation between Y and $\hat{Y}$ is the same as that between Y and X.

If we make the additional assumption that X and Y have a bivariate normal distribution, then it is possible to test the null hypothesis $H_0 : \rho = 0$ by computing the test statistic

$$t = \frac{r(N-2)^{1/2}}{(1 - r^2)^{1/2}}$$

with $N - 2$ degrees of freedom. For the fathers from the lung function data set given in Figure 6.1, $r = 0.504$. To test the hypothesis $H_0 : \rho = 0$ versus the alternative $H_1 : \rho \neq 0$, we compute

$$t = \frac{0.504(150 - 2)^{1/2}}{(1 - 0.504^2)^{1/2}} = 7.099$$

with 148 degrees of freedom. This statistic results in $P < 0.0001$, and the observed r is significantly different from zero.

Tests of null hypotheses other than $\rho = 0$ and confidence intervals for ρ can be found in many textbooks (Mickey, Dunn and Clark, 2009 or van Belle et al., 2004). As before, a test of $\rho = 0$ should be made before attempting to interpret the magnitude of the sample correlation coefficient. Note that the test of $\rho = 0$ is equivalent to the test of $\beta = 0$ given earlier.

All of the above interpretations were made with the assumption that the data follow a bivariate normal distribution, which implies that the mean of Y is related to X in a linear fashion. If the regression of Y on X is nonlinear, it is conceivable that the sample correlation coefficient is near zero when Y is, in fact, strongly related to X. For example, in Figure 6.6 we can quite accurately predict Y from X (in fact, the points fit a curve $Y = 100 - (X - 10)^2$ exactly). However, the sample correlation coefficient is $r = 0.0$. (An appropriate regression equation can be fitted by the techniques of polynomial regression — Section 7.8. Also, in Section 6.9 we discuss the role of transformations in reducing nonlinearities.)

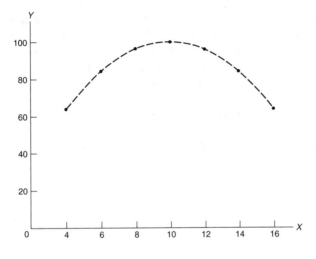

Figure 6.6: *Example of Nonlinear Regression*

6.8 Other available computer output

Most packaged regression programs include other useful statistics in their output. To obtain some of this output, the investigator usually has to run one of the multiple regression programs discussed in Chapter 7. In this section we introduce some of these statistics in the context of simple linear regression.

Standardized regression coefficient

The **standardized regression coefficient** is the slope in the regression equation if X and Y are standardized. Standardization of X and Y is done by subtracting the respective means from each set of observations and dividing the differences by the respective standard deviations. The resulting set of standardized sample values will have a mean of zero and a standard deviation of one for both X and Y. After standardization the intercept in the regression equation will be zero, and for simple linear regression (one X variable) the standardized slope will be equal to the correlation coefficient r. In multiple regression, where several X

variables are used, the standardized regression coefficients quantify the relative contribution of each X variable (Chapter 7).

Analysis of variance table

The test for $H_0 : \beta = 0$ was discussed in Section 6.6 using the t statistic. This test allows one-sided or two-sided alternatives. When the two-sided alternative is chosen, it is possible to represent the test in the form of an **analysis of variance (ANOVA) table**. A typical ANOVA table is presented in Table 6.1. If X were useless in predicting Y, our best guess of the Y value would be $\bar{Y}$ regardless of the value of X. To measure how different our fitted line $\hat{Y}$ is from $\bar{Y}$, we calculate the sums of squares for regression as $\sum(\hat{Y} - \bar{Y})^2$, summed over each data point. (Note that $\bar{Y}$ is the average of all the $\hat{Y}$ values.) The residual mean square is a measure of how poorly or how well the regression line fits the actual data points. A large residual mean square indicates a poor fit. The F ratio is, in fact, the squared value of the t statistic described in Section 6.6 for testing $H_0 : \beta = 0$.

Table 6.1: *ANOVA table for simple linear regression*

Source of variation	Sums of squares	df	Mean square	F
Regression	$\sum(\hat{Y} - \bar{Y})^2$	1	$SS_{reg}/1$	MS_{reg}/MS_{res}
Residual	$\sum(Y - \hat{Y})^2$	$N - 2$	$SS_{res}/(N - 2)$	
Total	$\sum(Y - \bar{Y})^2$	$N - 1$		

Table 6.2: *ANOVA example from Figure 6.1*

Source of variation	Sums of squares	df	Mean square	F
Regression	16.0532	1	16.0532	50.50
Residual	47.0451	148	0.3179	
Total	63.0983	149		

Table 6.2 shows the ANOVA table for the fathers from the lung function data set. Note that the F ratio of 50.50 is the square of the t value of 7.099 obtained when testing that the population correlation was zero. It is not precisely so in this example because we rounded the results before computing the value

of t. We can compute S by taking the square root of the residual mean square, 0.3179, to get 0.5638.

Data screening in simple linear regression

For simple linear regression, a scatter diagram such as that shown in Figure 6.1 is one of the best tools for determining whether or not the data fit the basic model. Most researchers find it simplest to examine the plot of Y against X. Alternatively, the residuals $e = Y - \hat{Y}$ can be plotted against X. Table 6.3 shows the data for the hypothetical example presented in Figure 6.4. Also shown are the predicted values $\hat{Y}$ and the residuals e. Note that, as expected, the mean of the $\hat{Y}$ values is equal to $\bar{Y}$. Also, it will always be the case that the mean of the residuals is zero. The variance of the residuals from the same regression line will be discussed later in this section.

Table 6.3: *Hypothetical data example from Figure 6.4*

i	X	Y	$\hat{Y} = 6.5 + 1.8X$	$e = Y - \hat{Y} = $ Residual
1	5	14	15.5	−1.5
2	5	17	15.5	1.5
3	10	27	24.5	2.5
4	10	22	24.5	−2.5
Mean	7.5	20	20	0

Examples of three possible scatter diagrams and residual plots are illustrated in Figure 6.7. In Figure 6.7a, the idealized bivariate (X, Y) normal distribution model is illustrated, using contours similar to those in Figure 6.5. In the accompanying residual plot, the residuals plotted against X would also approximate an ellipse. An investigator could make several conclusions from Figure 6.7a. One important conclusion is that no evidence exists contradicting the **linearity of the regression** of Y or X. Also, there is no evidence for the existence of outliers (discussed later in this section). In addition, the normality assumption used when confidence intervals are calculated or statistical tests are performed is not obviously violated.

In Figure 6.7b the ellipse is replaced by a fan-shaped figure. This shape suggests that as X increases, the standard deviation of Y also increases. Note that the assumption of linearity is not obviously violated but that the assumption of homoscedasticity does not hold. In this case, the use of weighted least squares is recommended (Section 6.10).

In Figure 6.7c the ellipse is replaced by a crescent-shaped form, indicating that the regression of Y is not linear in X. One possibility for solving this problem is to fit a quadratic equation as discussed in Chapter 7. Another possibility is to transform X into $\log X$ or some other function of X, and then fit a

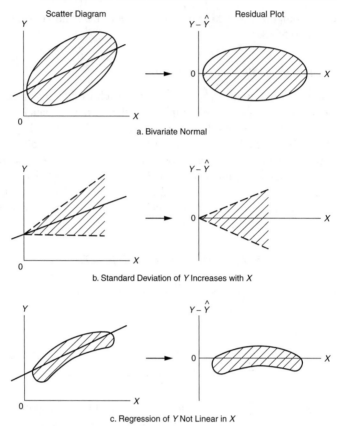

Figure 6.7: *Hypothetical Scatter Plots and Corresponding Residual Plots*

straight line to the transformed X values and Y. This concept will be discussed in Section 6.9.

Formal tests exist for testing the linearity of the simple regression equation when multiple values of Y are available for at least some of the X values. However, most investigators assess linearity by plotting either Y against X or $Y - \hat{Y}$ against X. The scatter diagram with Y plotted against X is often simpler to interpret. Lack of fit to the model may be easier to assess from the residual plots, as shown on the right-hand side of Figure 6.7. Since the residuals always have a zero mean, it is useful to draw a horizontal line through the zero point on the vertical axis, as has been done in Figure 6.7. This will aid in checking whether the residuals are symmetric around their mean (which is expected if the residuals are normally distributed). Unusual clusters of points can alert the investigator to possible anomalies in the data.

The distribution of the residuals about zero is made up of two components. One is called a random component and reflects incomplete prediction of Y from X and/or imprecision in measuring Y. But if the linearity assumption is not correct, then a second component will be mixed with the first, reflecting lack of fit to the model. Most formal analyses of residuals only assume that the first component is present. In the following discussion it will be assumed that there is no appreciable effect of nonlinearity in the analysis of residuals. It is important that the investigator assess the linearity assumption if further detailed analysis of residuals is performed.

Most multiple regression programs provide lists and plots of the residuals. The investigator can either use these programs, even though a simple linear regression is being performed, or obtain the residuals by using the transformation $Y - (A + BX) = Y - \hat{Y} = e$ and then proceed with the plots.

The raw sample regression residuals e have unequal variances and are slightly correlated. Their magnitude depends on the variation of Y about the regression line. Most multiple regression programs provide numerous adjusted residuals that have been found useful in regression analysis. For simple linear regression, the various forms of residuals have been most useful in drawing attention to important **outliers**. The detection of outliers was discussed in Chapter 3 and the simplest of the techniques discussed there, plotting of histograms, can be applied to residuals. That is, histograms of the residual values can be plotted and examined for extremes.

In recent years, procedures for the detection of outliers in regression programs have focused on three types of outliers (Chatterjee and Hadi (2006) and Fox (1991)):

1. outliers in Y from the regression line;

2. outliers in X;

3. outliers that have a large influence on the estimate of the slope coefficient.

The programs include numerous types of residuals and other statistics so that the user can detect these three types of outliers. The more commonly used ones will be discussed here. For a more detailed discussion see Belsley *et al.* (1980), Cook and Weisberg (1982), Fox (1991), and Ryan (2009). A listing of the options in packaged programs is deferred to Section 7.10 since they are mostly available in the multiple regression programs. The discussion is presented in this chapter since plots and formulas are easier to understand for the simple linear regression model.

Outliers in Y

Since the sample residuals do not all have the same variance and their magnitude depends on how closely the points lie to the straight line, they are often simpler to interpret if they are standardized. If we analyze only a single variable, Y, then it can be standardized by computing $(Y - \bar{Y})/S_Y$, which will have a mean zero and a standard deviation of one. As discussed in Chapter 3, formal tests for detection of outliers are available but often researchers simply investigate all standardized residuals that are larger than a given magnitude, say three. This general rule is based on the fact that, if the data are normally distributed, the chances of getting a value greater in magnitude than three is very small. This simple rule does not take sample size into consideration but is still widely used. A comparable standardization for a residual in regression analysis would be $(Y - \hat{Y})/S$ or e/S, but this does not take the unequal variance into account. The adjusted residual that accomplishes this is usually called a **studentized residual**:

$$\text{studentized residual} = \frac{e}{S(1-h)^{1/2}}$$

where h is a quantity called **leverage** (to be defined later when outliers in X are discussed). It is sometimes called an internally studentized residual. Standard nomenclature in this area is yet to be finalized and it is safest to read the description in the computer program used to be sure of the definition. In addition, since a single outlier can greatly affect the regression line (particularly in small samples), what are often called **deleted** studentized residuals can be obtained. These residuals are computed in the same manner as studentized residuals, with the exception that the ith deleted studentized residual is computed from a regression line fitted to all but the ith observation. Deleted residuals have two advantages. First, they remove the effect of an extreme outlier in assessing the effect of that outlier. Second, in simple linear regression, if the errors are normally distributed, the deleted studentized residuals follow a Student t distribution with $N-3$ degrees of freedom. The deleted studentized residuals also are given different names by different authors and programs, e.g., externally studentized residuals. They are called studentized residuals in Stata, RSTUDENT in SAS, and DSTRRESID in SPSS.

These and other types of residuals can either be obtained in the form of plots of the desired residual against $\hat{Y}$ or X, or in lists. The plots are useful for large samples or quick scanning. The lists of the residuals (sometimes accompanied by a simple plot of their distance from zero for each observation) are useful for identifying the actual observation that has a large residual in Y.

Once the observations that have large residuals are identified, the researcher can examine those cases more carefully to decide whether to leave them in, correct them, or declare them missing values.

Outliers in X

Possible outliers in X are measured by statistics called **leverage** statistics. One measure of leverage for simple linear regression is called h, where

$$h = \frac{1}{N} + \frac{(X - \bar{X})^2}{\Sigma(X - \bar{X})^2}$$

When X is far from $\bar{X}$, then leverage is large and *vice versa*. The size of h is limited to the range

$$\frac{1}{N} < h < 1$$

The leverage h for the ith observation tells how much Y for that observation contributes to the corresponding $\hat{Y}$. If we change Y by a quantity ΔY for the ith observation, then $h\Delta Y$ is the resulting change in the corresponding $\hat{Y}$. Observations with large leverages possess the potential for having a large effect on the slope of the line. It has been suggested for simple linear regression that a leverage is considered large if the value of h for the ith observation is greater than $4/N$ (Fox and Long, 1990). Figure 6.8 includes some observations that illustrate the difference between points that are outliers in X and in Y. Point 1 is an outlier in Y (large residual), but has low leverage since X is close to $\bar{X}$. It will affect the estimate of the intercept but not the slope. It will tend to increase the estimate of S and hence the standard error of the slope coefficient B. Point 2 has high leverage, but will not affect the estimate of the slope much because it is not an outlier in Y. Point 3 is both a high leverage point and an outlier in Y, so it will tend to reduce the value of the slope coefficient B and to tip the regression line downward. It will also affect the estimate of A and S, and thus have a large effect on the statements made concerning the regression line. Note that this is true even though the residual e is less for point 3 than it is for point 1. Thus, looking solely at residuals in Y may not tell the whole story and leverage statistics are important to examine if outliers are a concern.

Influential observations

A direct approach to the detection of outliers is to determine the **influence** of each observation on the slope coefficient B. Cook (1977) derived a function called Cook's distance, which provides a scaled distance between the value of B when all observations are used and $B(-i)$, the slope when the ith observation is omitted. This distance is computed for each of the N observations. Observations resulting in large values of Cook's distance should be examined as possible influential points or outliers. Cook (1977) suggests comparing the Cook's distance with the percentiles of the F distribution with $P + 1$ and $N - P - 1$ degrees of freedom. Here, P is the number of independent variables. Cook's distances exceeding the 50th percentile are recommended for careful examination and possible removal from the data set.

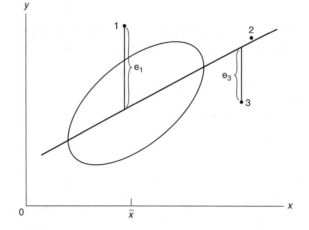

Figure 6.8: *Illustration of the Effect of Outliers*

Other distance measures have been proposed (Chatterjee and Hadi, 2006). The Welsch–Kuh distance measure (also called DFFITS) is the scaled distance between $\hat{Y}$ and $\hat{Y}(-i)$, i.e., $\hat{Y}$ derived with the ith observation deleted. Large values of DFFITS also indicate an influential observation. DFFITS tends to measure the influence on B and S^2 simultaneously.

A general lesson from the research work on outliers in regression analysis is that, when one examines either the scatter diagram of Y versus X or the plot of the residuals versus X, more attention should be given to the points that are outliers in both X and Y than to those that are only outliers in Y.

Lack of independence among residuals

When the observations can be ordered in time or place, plots of the residuals similar to those given in Figure 4.3 can be made and the discussion in Section 4.4 applies here directly. If the observations are independent, then successive residuals should not be appreciably correlated. The **serial** correlation, which is simply the correlation between successive residuals, can be used to assess lack of independence. For a sufficiently large N, the significance levels for the usual test $\rho = 0$ apply approximately to the serial correlation. Another test statistic available in some packaged programs is the Durbin–Watson statistic. The Durbin–Watson statistic is approximately equal to $2(1 - \text{serial correlation})$. Thus when the serial correlation is zero, the Durbin–Watson statistic is close to two. The Durbin–Watson statistic is used to test whether the serial correlation is zero when it is assumed that the correlation between successive residuals is restricted to a correlation between immediately adjacent residuals. Methods for dealing with correlated observations will be discussed in Chapter 18.

Normality of residuals

Some regression programs provide normal probability or quantile plots of the residuals to enable the user to decide whether the data approximate a normal distribution. If the residuals are not normally distributed, then the distribution of Y at each value of X is not normal.

For simple linear regression with the variable-X model, many researchers assess bivariate normality by examining the scatter diagram of Y versus X to see if the points approximately fall within an ellipse.

6.9 Robustness and transformations for regression

In this section we define the concept of robustness in statistical analysis, and we discuss the role of transformations in regression and correlation.

Robustness and assumptions

Regression and correlation analysis make certain assumptions about the population from which the data were obtained. A **robust analysis** is one that is useful even though all the assumptions are not met. For the purpose of fitting a straight line, we assume that the Y values are normally distributed, the population regression equation is linear in the range of concern, and the variance of Y is the same for all values of X. Linearity can be checked graphically, and transformations can help straighten out a nonlinear regression line.

The **assumption of homogeneity of variance** is not crucial for the resulting least squares line. In fact, the least squares estimates of α and β are unbiased whether or not the assumption is valid. However, if glaring irregularities of variance exist, weighted least squares can improve the fit. In this case the weights are chosen to be proportional to the inverse of the variance. For example, if the variance is a linear function of X, then the weight is $1/X$.

The **assumption of normality** of the Y values of each value of X is made only when tests of hypotheses are performed or confidence intervals are calculated. It is generally agreed in the statistical literature that slight departures from this assumption do not appreciably alter our inferences if the sample size is sufficiently large.

The **lack of randomness** in the sample can seriously invalidate our inferences. Confidence intervals are often optimistically narrow because the sample is not truly a random one from the whole population to which we wish to generalize.

In all of the preceding analyses **linearity** of the relationship between X

and Y was assumed. Thus careful examination of the scatter diagram should be the first step in any regression analysis. It is advisable to explore various transformations of Y and/or X if nonlinearity of the original measurements is apparent.

Transformations

The subject of **transformations** has been discussed in detail in the literature (e.g., Draper and Smith, 1998). In this subsection we present some typical graphs of the relationship between Y and X and some practical transformations.

In Chapter 4 we discussed the effects of transformations on the frequency distribution. There it was shown that taking the logarithm or square root of a number condensed the magnitude of larger numbers and stretched the magnitude of values less than one. Conversely, raising a number to a power greater than one stretches the large values and condenses the values less than one. These properties are useful in selecting the appropriate transformation to straighten out a nonlinear graph of one variable as a function of another.

Typical regression curves that are not linear in X can be viewed as one of the quadrants of Figure 6.9a (see the classic reference by Mosteller and Tukey, 1977). A very common case is illustrated in Figure 6.9b, which is represented by the fourth quadrant of the circle in Figure 6.9a. For example, the curve in Figure 6.9b might be made linear by transforming X to $\log X$, to $-1/X$, or to $X^{1/2}$. Another possibility would be to transform Y to Y^2. The other three cases are also indicated in Figure 6.9a. The remaining quadrants are interpreted in a similar fashion.

Other transformations could also be attempted, such as powers other than those indicated. It may also be useful to first add or subtract a constant from all values of X or Y and then take a power or logarithms. For example, sometimes taking $\log X$ does not straighten out the curve sufficiently. Subtracting a constant C (which must be smaller than the smallest X value) and then taking the logarithm has a greater effect.

The availability of packaged programs greatly facilitates the choice of an appropriate transformation. New variables can be created that are functions of the original variables, and scatter diagrams can be obtained of the new transformed variables. Visual inspection will often indicate the best transformation. Also, the magnitude of the correlation coefficient r will indicate the best linear fit since it is a measure of linear association. Attention should be paid to transformations that are commonly used in the field of application and that have a particular scientific basis or physical rationale.

Once the transformation is selected, all subsequent estimates and tests are performed in terms of the transformed values. Since the variable to be predicted is usually the dependent variable Y, transforming Y can complicate the inter-

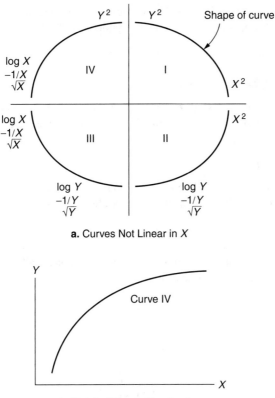

a. Curves Not Linear in X

b. Detail of Fourth Quadrant

Figure 6.9: *Choice of Transformation: Typical Curves and Appropriate Transformation*

pretation of the resulting regression equation more than if X is transformed. For example, if $\log X$ is used instead of X, the resulting equation is

$$Y = A + B \log_{10} X$$

This equation presents no problems in interpreting the predicted values of Y, and most investigators accept the transformation of $\log_{10} X$ as reasonable in certain situations.

However, if $\log Y$ is used instead of Y, the resulting equation is

$$\log_{10} Y = A + BX$$

Then the predicted value of Y, say Y^*, must be detransformed, that is,

$$Y^* = 10^{A - BX}$$

Thus slight biases in fitting log Y could be detransformed into large biases in predicting Y. For this reason most investigators look for transformations of X first.

6.10   Other types of regression

In this section we discuss three options available from computer programs: regression through the origin, weighted regression, and loess curves.

Regression through the origin

Sometimes an investigator is convinced that the **regression line** should pass **through the origin**. In this case the appropriate model is simply the mean of

$$Y = \beta X$$

That is, the intercept is forced to be zero. The programs usually give the option of using this model and estimate β as

$$B = \frac{\sum XY}{\sum X^2}$$

To test $H_0 : \beta = \beta_0$, the test statistic is

$$t = \frac{B - \beta_0}{S/(\sum X^2)^{1/2}}$$

where

$$S = \left[\frac{\sum (Y - BX)^2}{N - 1} \right]^{1/2}$$

and t has $N - 1$ degrees of freedom.

Weighted least squares regression

The investigator may also request a **weighted least squares regression line**. In weighted least squares each observation is given an individual weight reflecting its importance or degree of variability. There are three common situations in which a weighted regression line is appropriate.

1. The variance of the distribution at a given X is a function of the X value. An example of this situation was shown in Figure 6.7b.

2. Each Y observation is, in fact, the mean of several determinations, and that number varies from one value of X to another.

3. The investigator wishes to assign different levels of importance to different points. For example, data from different countries could be weighted either by the size of the population or by the perceived accuracy of the data.

In case 1 the weights are the inverse of the variances of the point. In case 2 the weights are the number of determinations at each X point. In case 3 the investigator must make up numerical weights to reflect the perception of importance or accuracy of the data points.

In weighted least squares regression the estimates of α and β and other statistics are adjusted to reflect these special characteristics of the observations. In most situations the weights will not affect the results appreciably unless they are quite different from each other. Since it is considerably more work to compute a weighted least squares regression equation, it is recommended that one of the computer programs listed in Section 6.12 be used, rather than hand calculations.

Loess curves

One use of weighted regression is to fit what is called **loess** curves. Loess is an abbreviation for local regression. Alternatively, the word **lowess** is also used. Lowess stands for locally weighted regression scatter plot smoothing. Loess curves are especially useful in illustrating the relationship between X and Y when it obviously is not a straight line, and no transformation can be found that results in a straight line or a polynomial (see Section 7.8).

Computation of loess curves requires several steps. First, the X values are ordered from smallest to largest so that X_1 is the smallest and X_N is the largest. Then, using a subset of the X (and the corresponding Y) surrounding each X_i, the programs compute a predicted $\hat{Y}_i$. We have to decide how many points surrounding each X_i to use to predict each $\hat{Y}_i$. In Stata, in the lowess command, the percent of the total number of points to be used is called the bwidth (bandwidth). Centered subsets of $N \times$ bwidth observations are used; thus bwidth $\times 100$ represents the percent of the total observations to be used. In S-PLUS, in the loess smoother command, the fraction of the corresponding parameter is called the span value. In SAS PROC LOESS the appropriate options are bucket and smooth. Both, SPSS and STATISTICA have a feature named lowess for this purpose. Note that Statistica does not provide an option or parameter to choose the bandwidth of values to be used.

Often it takes several attempts at an appropriate number of points to get the desired curve. If the curve is too jagged, we can smooth it out by taking more points. Usually a span value greater than one-fourth is recommended (see Cleveland, 1993).

For example, if we choose a bandwidth that requires nine points, the statistical program will compute a weighted regression line using the nearest nine points for each X_i in the sample. For example, for X_5 the nearest nine points are X_1 to X_9. From these nine points a single $\hat{Y}_5$ is obtained. This process is repeated for all N values of X. The result is N different values of $\hat{Y}$.

The weights are chosen such that the maximum weight in our example is assigned to X_5 and the weights decrease as the X_i are further away from X_5 in either direction (see Cleveland, 1993 for further description and the formula for the weights). Values of X_i outside the range of X_1 to X_9 receive zero weight. The final curve is obtained by connecting the successive smoothed $X_i, \hat{Y}_i$ values.

As an example, consider data describing the relationship between crude birth rate (CBR) and gross domestic product (GDP) for various countries (US Bureau of Census, 1998). The crude birth rate is the number of births during one year per 1000 persons. The gross domestic product is computed per capita and has been converted to 1995 dollars. In Figure 6.10 we use data from 20 countries and apply loess smoothing using bandwidths 0.20 (no smoothing), 0.30, 0.50, and 0.90. Figure 6.10 shows that more smoothing is obtained with higher bandwidths.

Typically, with a bandwidth of 0.20 we would expect the resulting graph to represent some smoothing. In our example, we generated the graph using Stata. For the smoothing, Stata uses k values that are smaller than X_i, X_i itself, and k values that are larger than X_i, where k is the largest integer that is smaller or equal to $(N \times \text{bwidth} - 0.5)/2$. In our example the resulting k is 1. Hence, Stata uses X_{i-1}, X_i and X_{i+1} to calculate $\hat{Y}_i$. Nevertheless, in Cleveland's formula (1993) the smallest (here X_{i-1}) and the largest observation (here X_{i+1}) to be used in the smoothing receive such a small weight that from a practical perspective the weight is zero, and hence no smoothing results with a 0.20 bwidth.

Loess curves are used when there are complex relationships and a graphic rather than an equation is desired. For example, they are useful in time series where there is a seasonal effect and results for numerous years are depicted. Some researchers use them when trying to decide if they are willing to assume a straight line. Additional information and examples can be seen in Chambers (1983), and Cleveland (1993).

6.11 Special applications of regression

In this section we discuss some important applications of regression analysis that require some caution when the investigator is using regression techniques.

Calibration

A common situation in laboratories or industry is one in which an instrument that is designed to measure a certain characteristic needs calibration. For example, the concentration of a certain chemical could be measured by an instrument. For calibration of the instrument several compounds with known con-

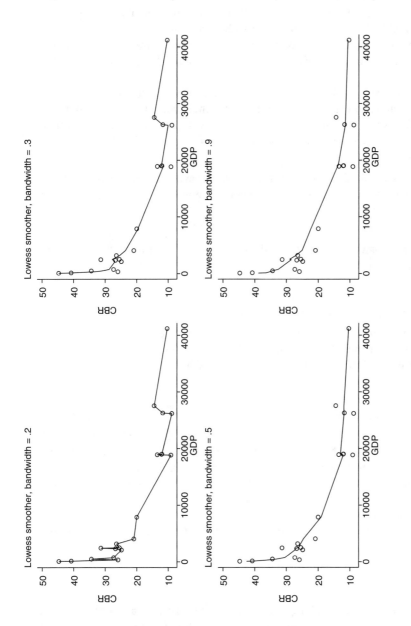

Figure 6.10: *Example of Loess Smoothing Using Bandwidths 0.20 (no smoothing), 0.30, 0.50, and 0.90*

centrations, denoted by X, could be used, and the measurements, denoted by Y, could be determined by the instrument for each compound. As a second example, a costly or destructive method of measurement that is very accurate, denoted by X, could be compared with another method of measurement, denoted by Y, that is a less costly or nondestructive method. In either situation more than one determination of the level of Y could be made for each value of X.

The object of the analysis is to derive, from the calibration data, a formula or a graph that can be used to predict the unknown value of X from the measured Y value in future determinations. We will present two methods (indirect and direct) that have been developed and a technique for choosing between them.

1. **Indirect method**. Since X is assumed known with little error and Y is the random variable, this classical method begins with finding the usual regression of Y on X,

$$\hat{Y} = A + BX$$

Then for a determination Y^* the investigator obtains the indirect estimate of X, $\hat{X}_{in}$, as

$$\hat{X}_{in} = \frac{Y^* - A}{B}$$

This method is illustrated in Figure 6.11a.

2. **Direct method**. Here we pretend that X is the dependent variable and Y is the independent variable and fit the regression of X on Y, as illustrated in Figure 6.11b. We denote the estimate of X by $\hat{X}_{dr}$; thus

$$\hat{X}_{dr} = C + DY^*$$

To compare the two methods, we compute the quantities $\hat{X}_{in}$ and $\hat{X}_{dr}$ for all of the data in the sample. Then we compute the correlation between $\hat{X}_{in}$ and X, denoted by $r(in)$, and the correlation between $\hat{X}_{dr}$ and X, denoted by $r(dr)$. We then use the method that results in the higher correlation.

It is advisable, before using either of these methods, that the investigator test whether the slope B is significantly different from zero. This test can be done by the usual t test. If the hypothesis $H_0 : \beta = 0$ is not rejected, this result is an indication that the instrument should not be used, and the investigator should not use either equation.

Forecasting

Forecasting is the technique of analyzing historical data in order to provide estimates of the future values of certain variables of interest. If data taken over time (X) approximate a straight line, forecasters might assume that this relationship would continue in the future and would use simple linear regression

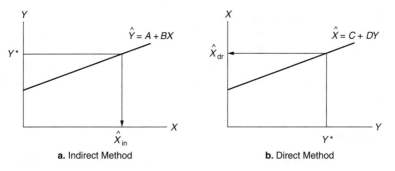

Figure 6.11: *Illustration of Indirect and Direct Calibration Methods*

to fit this relationship. This line is extended to some point in the future. The difficulty in using this method lies in the fact that we are never sure that the linear trend will continue in the future. At any rate, it is advisable not to make long-range forecasts by using this method. Methods of forecasting are often found in texts on time series. Further discussion and additional methods are found in Alwan (2000), Chatfield (2004), Diggle (1996), Yaffee and McGee (2000), and Kedem and Fokianos (2002).

Paired data

Another situation where regression analysis may be appropriate is the **paired-sample case**. A paired sample consists of observations taken at two time points (pre- and post-) on the same individual or consists of matched pairs, where one member of the pair is subject to one treatment and the other is a control or is subject to another treatment.

Many investigators apply only the paired *t* test to such data and as a result may miss an important relationship that could exist in the data. The investigator should, at least, plot a scatter diagram of the data and look for possible linear and nonlinear relationships. If a **linear relationship** seems appropriate, then the techniques of regression and correlation described in this chapter are useful in describing the relationship. If a **nonlinear relationship** exists, transformations should be considered.

Focus on residuals

Sometimes, performing the regression analysis may be only a preliminary step to obtaining the residuals, which are the quantities of interest to be saved for future analysis. For example, in the bone density study described in Chapter 1, the variable age has a major effect on predicting bone density of elderly women. One method of taking age into account in future analyses would be to fit a simple regression equation predicting bone density, with age as the only independent variable. Then future analyses of bone density could be performed using the residuals from this equation instead of the original bone density measurements. These residuals would have a mean of zero. Examination of subgroup means is then easy to interpret since their difference from zero can be easily seen.

Adjusted estimates of Y

An alternative method to using residuals is to use the slope coefficient B from the regression equation to obtain an age-adjusted bone density for each individual, which is an estimate of their bone density at the average age for the entire sample. This adjusted bone density is computed for each individual from the following equation,

$$\text{adjusted } Y = Y + B(\bar{X} - X)$$

where X and Y are the values for each individual. The adjusted Y has a mean $\bar{Y}$ and thus allows the researcher to work with numbers of the usual magnitude.

6.12 Discussion of computer programs

All the packages provide a variety of output for simple linear regression since this is a widely used analysis. This will not be the case when we discuss more complex multivariate procedures in subsequent chapters. In those chapters we will provide a table that shows whether or not each statistical package has a particular feature.

Some of the statistics mentioned in this chapter can only be obtained from some of the packages by using the multiple linear regression programs described in Section 7.10. These programs can be used to perform simple linear regression as well. This was the case with STATISTICA, the package we used to obtain the results for the fathers given in Figure 6.1 and elsewhere in this chapter. The multiple regression program was run and results that applied only to multiple regression were ignored. We could have used any of the other four statistical software packages to produce similar output.

Since STATISTICA is a Windows-based package the analysis was performed by clicking on the appropriate choices and then examining the output. First, descriptive statistics were obtained by choosing "Basic Statistics/Tables" from the STATISTICA module switcher (which appears first when STATISTICA is accessed). This was followed by clicking "Descriptive Statistics." To obtain the scatter plot, "Graph" was chosen from the Windows menu at the top of the screen and then "2D" from the pull-down list and finally "Scatter Plots." To obtain further output, "Analysis" was chosen from the Windows menu and "Other Statistics" chosen. From this the multiple linear regression option was

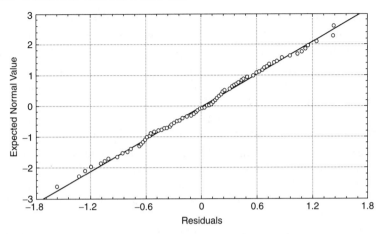

Figure 6.12: *Normal Probability Plot of the Residuals of the Regression of FEV1 on Height for Fathers*

picked. As with other Windows-based packages, the print option was chosen from the "File" menu at the top of the screen.

STATISTICA also provides numerous graphs of residuals and other regression output. For example, to check if the residuals in Y followed a normal distribution, we ran a normal probability plot of the residuals for the regression equation predicting FEV1 from height. As can be seen in Figure 6.12, the residuals appear to be normally distributed.

In the rest of the book we feature different statistical packages in different chapters.

6.13 What to watch out for

In Sections 6.8 and 6.9, methods for checking for outliers, normality, homogeneity of variance, and independence were presented along with a brief discussion of the importance of including checks in the analysis. In this section, other cautions will be emphasized.

1. **Sampling process.** In the development of the theory for linear regression, the sample is assumed to be obtained randomly in such a way that it represents the whole population. Often, convenience samples, which are samples of easily available cases, are taken for economic or other reasons. In surveys of the general population, more complex sampling methods are often employed. The net effect of this lack of simple random sampling from a defined population is likely to be an underestimate of the variance and pos-

sibly bias in the regression line. Researchers should describe their sampling procedures carefully so their results can be assessed with this in mind.

2. **Errors in variables.** Random errors in measuring Y do not bias the slope and intercept estimates but do affect the estimate of the variance. Errors in measuring X produce estimates of β that are biased towards zero. Methods for estimating regression equations that are resistant to outliers are given in Hoaglin, Mosteller and Tukey (1985) and Fuller (1987). Such methods should be considered if measurement error is likely to be large.

3. The use of nominal or ordinal, rather than interval or ratio, data in regression analysis has several possible consequences. First, often the number of distinct values can be limited. For example, in the depression data set, the self-reported health scale has only four possible outcomes: excellent, good, fair, or poor. It has been shown that when interval data are grouped into seven or eight categories, later analyses are little affected; but coarser grouping may lead to biased results. Secondly, the intervals between the numbers assigned to the scale values (1, 2, 3, and 4 for the health scale, for example) may not be properly chosen. Perhaps poor health should be given a value of 5, as it is an extreme answer and is more than one unit from fair health (Abelson and Tukey, 1963).

4. **Choice of model.** Particularly for the variable-X model, it is essential not only that the model be in the correct analytic form, i.e., linear if a linear model is assumed, but also that the appropriate X variables be chosen. This will be discussed in more detail in Chapter 8. Here, since only one X variable is used, it is possible that an important confounding variable (one correlated with both X and Y) may be omitted. This could result in a high correlation coefficient between X and Y and a well-fitting line that is caused by the omission of the confounding variable. For example, FEV1 decreases with age for adults, as does bone density. Hence, we would expect to see spurious correlation and regression between bone density and FEV1. In Section 6.11, the use of residuals to assist in solving this problem was discussed and further results will be given in the following chapters on multiple regression analysis.

5. In using the computed regression equation to predict Y from X, it is advisable to restrict the prediction to X values within the range observed in the sample, unless the investigator is certain that the same linear relationship between Y and X is valid outside of this range.

6. An observed correlation between Y and X should not be interpreted to mean a causal relationship between X and Y regardless of the magnitude of the correlation. The correlation may be due to a causal effect of X on Y, of Y on X, or of other variables on both X and Y. Causation should be inferred only after a careful examination of a variety of theoretical and experimental evidence, not merely from statistical studies based on correlation.

7. In most multivariate studies, Y is regressed on more than one X variable. This subject will be treated in the next three chapters.

8. It is important to apply the regression diagnostic procedures discussed in Section 6.8 as post-regression analysis. Honesty in reporting requires that results with and without influential observations should be discussed in the description of the analysis if influential observations are present.

6.14 Summary

In this chapter we gave a conventional presentation of simple linear regression, similar to the presentations found in many statistical textbooks. Thus, the reader familiar with the subject should be able to make the transition to the mode we follow in the remainder of the book. In addition, loess curves were introduced. These curves provide a useful graphical display when a straight line does not fit the points and no obvious transformation is known.

We made a clear distinction between random- and fixed-X variable regression. Whereas most of the statistics computed apply to both cases, certain statistics apply only to the random-X case. Computer programs are written to provide more output than is sometimes needed or appropriate. You should be aware of which model is being assumed so that you can make the proper interpretation of the results.

If packaged programs are used, it may be sufficient to run one of the simple plotting programs in order to obtain both the plot and the desired statistics. It is good practice to plot many variables against each other in the preliminary stages of data analysis. This practice allows you to examine the data in a straightforward fashion. However, it is also advisable to use the more sophisticated programs as warm-ups for the more complex data analyses discussed in later chapters.

The concepts of simple linear regression and correlation presented in this chapter can be extended in several directions. Chapter 7 treats the fitting of linear regression equations to more than one independent variable. Chapter 8 gives methods for selecting independent variables. Additional topics in regression analysis are covered in Chapter 9. These topics include missing values, dummy variables, segmented (spline) regression, and ridge regression.

6.15 Problems

6.1 In Table 8.1, financial performance data of 30 chemical companies are presented. Use growth in earnings per share, labelled EPS5, as the dependent variable and growth in sales, labelled SALESGR5, as the independent variable. (A description of these variables is given in Section 8.3.) Plot the data, compute a regression line, and test that $\beta = 0$ and $\alpha = 0$. Are earnings affected by sales growth for these chemical companies? Which company's earnings were highest, considering its growth in sales?

6.2 From the family lung function data set in Appendix A, perform a regression analysis of weight on height for fathers. Repeat for mothers. Determine the correlation coefficient and the regression equation for fathers and mothers. Test that the coefficients are significantly different from zero for both sexes. Also, find the standardized regression equation and report it. Would you suggest removing the woman who weighs 267 pounds from the data set? Discuss why the correlation for fathers appears higher than that for mothers.

6.3 In Problem 4.8, the New York Stock Exchange Composite Index and daily volume for August 9 through September 17, 1982, were presented. Describe how volume appears to be affected by the price index, using regression analysis. Describe whether or not the residuals from your regression analysis are serially correlated. Plot the index versus time and volume versus time, and describe the relationships you see in these plots.

6.4 For the data in Problem 6.3, pretend that the index increases linearly in time and use linear regression to obtain an equation to forecast the index value as a function of time. Using "volume" as a weight variable, obtain a weighted least squares forecasting equation. Does weighted least squares help the fit? Obtain a recent value of the index (from a newspaper). Does either forecasting equation predict the true value correctly (or at least somewhere near it)? Explain.

6.5 Repeat Problem 6.2 using log(weight) and log(height) in place of the original variables. Using graphical and numerical devices, decide if the transformations help.

6.6 Examine the plot you produced in Problem 6.1 and choose some transformation for X and/or Y and repeat the analysis described there. Compare the correlation coefficients for the original and transformed variables, and decide whether the transformation helped. If so, which transformation was helpful?

6.7 Using the depression data set (see Table 3.4), perform a regression analysis of depression, as measured by CESD, on income. Plot the residuals. Does the normality assumption appear to be met? Repeat using the logarithm of CESD instead of CESD. Is the fit improved?

6.8 (Continuation of Problem 6.7.) Calculate the variance of CESD for observations in each of the groups defined by income as follows: INCOME <30, INCOME between 30 and 39 inclusive, INCOME between 40 and 49 inclusive, INCOME between 50 and 59 inclusive, INCOME >59. For each observation, define a variable WEIGHT equal to 1 divided by the variance of a CESD within the income group to which it belongs. Obtain a weighted least squares regression of the untransformed variable CESD on income, using the values of WEIGHT as weights. Compare the results with the unweighted regression analysis. Is the fit improved by weighting?

6.9 From the depression data set described in Table 3.4 create a data set containing only the variables AGE and INCOME.

(a) Find the regression of income on age.

(b) Successively add and then delete each of the following points:

AGE	INCOME
42	120
80	150
180	15

and repeat the regression each time with the single extra point. How does the regression equation change? Which of the new points are outliers? Which are influential?

Use the family lung function data described in Appendix A for the next four problems.

6.10 For the oldest child, perform the following regression analyses: FEV1 on weight, FEV1 on height, FVC on weight, and FVC on height. Note the values of the slope and correlation coefficient for each regression and test whether they are equal to zero. Discuss whether height or weight is more strongly associated with lung function in the oldest child.

6.11 What is the correlation between height and weight in the oldest child? How would your answer to the last part of problem 6.10 change if $\rho = 1$? $\rho = -1$? $\rho = 0$?

6.12 Examine the residual plot from the regression of FEV1 on height for the oldest child. Choose an appropriate transformation, perform the regression with the transformed variable, and compare the results (statistics, plots) with the original regression analysis.

6.13 For the mother, perform a regression of FEV1 on weight. Test whether the coefficients are zero. Plot the regression line on a scatter diagram of MFEV1 versus MWE1. On this plot, identify the following groups of points:
group 1: ID = 12, 33, 45, 42, 94, 144;
group 2: ID = 7, 94, 105, 107, 115, 141, 141.
Remove the observations in group 1 and repeat the regression analysis. How does the line change? Repeat for group 2.

6.14 For the Parental HIV data produce a scatterplot of the age at which adolescents first started smoking versus the age at which they first started drinking alcohol. Based on the graph, do adolescents tend to start smoking before drinking alcohol, or *vice versa*? Calculate the correlation coefficient. Exclude all adolescents who had not started smoking or had not started drinking alcohol by the time they were interviewed.

6.15 For the Parental HIV data generate a variable that represents the sum of the variables describing the neighborhood where the adolescent lives (NGHB1–NGHB11). Does the age at which adolescents start smoking depend on the score describing the neighborhood?

6.16 Using the summary variable describing the neighborhood in problem 6.15, generate a loess graph to examine the relationship between this variable and the age at which adolescents started using marijuana. Interpret the graph.

Chapter 7

Multiple regression and correlation

7.1 Chapter outline

Multiple regression is performed when an investigator wishes to examine the relationship between a single dependent (outcome) variable Y and a set of independent (predictor or explanatory) variables X_1 to X_P. The dependent variable Y is of the continuous type. The X variables are also usually continuous, although they can be discrete.

In Section 7.2 the two basic models used for multiple regression are introduced and a data example is given in the next section. The background assumptions, model, and necessary formulas for the fixed-X model are given in Section 7.4, while Section 7.5 provides the assumptions and additional formulas for statistics that can be used for the variable-X model. Tests to assist in interpreting the fixed-X model are presented in Section 7.6 and similar information is given for the variable-X model in Section 7.7. Section 7.8 discusses the use of residuals to evaluate whether the model is appropriate and to find outliers. Three methods of changing the model to make it more appropriate for the data are given in that section: transformations, polynomial regression, and interaction terms. Multicollinearity is defined and methods for recognizing it are also explained in Section 7.8. Section 7.9 presents several other options available when performing regression analysis, namely, regression through the origin, weighted regression, and testing whether two subgroups' regressions are equal. Section 7.10 discusses how the numerical results in this chapter were obtained using Stata and SAS, and a table is included that summarizes the options available in the software programs used in this book. Section 7.11 explains what to watch out for when performing a multiple regression analysis.

There are two additional chapters in this book on multiple linear regression analysis. Chapter 8 presents methodology used to choose independent or predictor variables when the investigator is uncertain which variables to include in the model. Chapter 9 discusses missing values and dummy variables (used when some of the independent variables are discrete), and gives methods for handling multicollinearity.

7.2   When are regression and correlation used?

The methods described in this chapter are appropriate for studying the relationship between several X variables and one Y variable. By convention the X variables are often called **independent variables**, although they do not have to be statistically independent and are permitted to be intercorrelated. They are also called predictor or explanatory variables. The Y variable is called the **dependent variable** or outcome variable.

As in Chapter 6, the data for multiple regression analysis can come from one of two situations.

1. **Fixed-X case.** The levels of the various X's are selected by the researcher or dictated by the nature of the situation. For example, in a chemical process an investigator can set the temperature, the pressure, and the length of time that a vat of material is agitated, and then measure the concentration of a certain chemical. A regression analysis can be performed to **describe** or **explain** the relationship between the independent variables and the dependent variable (the concentration of a certain chemical) or to **predict** the dependent variable.

2. **Variable-X case.** A random sample of individuals is taken from a population, and the X variables and the Y variable are measured on each individual. For example, a sample of adults might be taken from Los Angeles and information obtained on their age, income, and education in order to see whether these variables predict their attitudes toward air pollution. Regression and correlation analyses can be performed in the variable-X case. Both **descriptive** and **predictive information** can be obtained from the results.

Multiple regression is one of the most commonly used statistical methods, and an understanding of its use will help you comprehend other multivariate techniques.

7.3   Data example

In Chapter 6 the data for fathers from the lung function data set were analyzed. These data fit the variable-X case. Height was used as the X variable in order to predict FEV1, and the following equation was obtained:

$$FEV1 = -4.087 + 0.118 \text{ (height in inches)}$$

However, FEV1 tends to decrease with age for adults, so we should be able to predict it better if we use both height and age as independent variables in a multiple regression equation. We expect the slope coefficient for age to be negative and the slope coefficient for height to be positive.

A geometric representation of the simple regression of FEV1 on age and height, respectively, is shown in Figures 7.1a and 7.1b. The multiple regression equation is represented by a plane, as shown in Figure 7.1c. Note that the plane slopes upward as a function of height and downward as a function of age. A hypothetical individual whose FEV1 is large relative to his age and height appears above both simple regression lines as well as above the multiple regression plane.

For illustration purposes Figure 7.2 shows how a plane can be constructed. Constructing such a regression plane involves the following steps.

1. Draw lines on the X_1, Y wall, setting $X_2 = 0$.

2. Draw lines on the X_2, Y wall, setting $X_1 = 0$.

3. Drive nails in the walls at the lines drawn in steps 1 and 2.

4. Connect the pairs of nails by strings and tighten the strings.

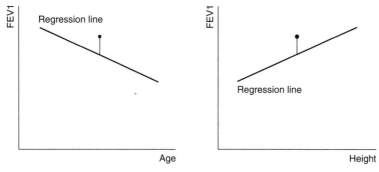

a. Simple Regression of FEV1 on Age   b. Simple Regression of FEV1 on Height

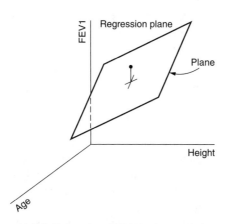

c. Multiple Regression of FEV1 on Age and Height

Figure 7.1: *Hypothetical Representation of Simple and Multiple Regression Equations of FEV1 on Age and Height*

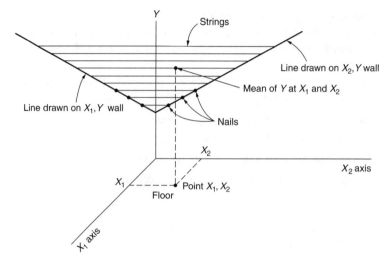

Figure 7.2: *Visualizing the Construction of a Plane*

The resulting strings in step 4 form a plane. This plane is the regression plane of Y on X_1 and X_2. The mean of Y at a given X_1 and X_2 is the point on the plane vertically above the point X_1, X_2.

Data for the fathers in the lung function data set were analyzed by Stata, and the following descriptive statistics were printed:

Variable	Obs	Mean	Std.Dev	Minimum	Maximum
Age	150	40.13	6.89	26.00	59.00
Height	150	69.26	2.78	61.00	76.00
FEV1	150	4.09	0.65	2.50	5.85

The "regress" command from Stata produced the following regression equation:

$$\widehat{FEV1} = -2.761 - 0.027(\text{age}) + 0.114(\text{height})$$

As expected, there is a positive slope associated with height and a negative slope with age. For predictive purposes we would expect a 30-year-old male whose height was 70 inches to have an FEV1 value of 4.41 liters.

Note that a value of $-4.087 + 0.118(70) = 4.17$ liters would be obtained for a father in the same sample with a height of 70 inches using the first equation (not taking age into account). In the single predictor equation the coefficient for height was 0.118. This value is the rate of change of FEV1 for fathers as a function of height when no other variables are taken into account. With two predictors the coefficient for height is 0.114, which is interpreted as the rate of change of FEV1 as a function of height **after adjusting for age**. The latter slope is also called the **partial regression coefficient** of FEV1 on height after adjusting for age. Even when both equations are derived from the same sample, the simple and partial regression coefficients may not be equal.

The output of this program includes other items to be discussed later in this chapter.

7.4 Regression methods: fixed-X case

In this section we present the background assumptions, model, and formulas necessary for an understanding of multiple linear regression for the fixed-X case. Computations can become tedious when there are several independent variables, so we assume that you will obtain output from a packaged computer program. Therefore we present a minimum of formulas and place the main emphasis on the techniques and interpretation of results. This section is slow reading and requires concentration.

Since there is more than one X variable, we use the notation $X_1, X_2, \ldots, X_P$ to represent P possible variables. In packaged programs these variables may appear in the output as $\mathbf{X}(1)$, $\mathbf{X}1$, $\mathbf{VAR}$ 1, etc. For the fixed-X case, values of the X variables are assumed to be fixed in advance in the sample. At each combination of levels of the X variables, we conceptualize a distribution of the values of Y. This distribution of Y values is assumed to have a mean value equal to $\alpha + \beta_1 X_1 + \beta_2 X_2 + \cdots + \beta_P X_P$ and a variance equal to σ^2 at given levels of $X_1, X_2, \ldots, X_P$.

When $P = 2$, the surface is a plane, as depicted in Figure 7.1c or 7.2. The parameter β_1 is the rate of change of the mean of Y as a function of X_1, where the value of X_2 is held fixed. Similarly, β_2 is the rate of change of the mean of Y as a function of X_2 when X_1 is fixed. Thus, β_1 and β_2 are the slopes of the regression plane with regard to X_1 and X_2.

When $P > 2$, the regression plane generalizes to a so-called **hyperplane**, which cannot be represented geometrically on two-dimensional paper. Some people conceive the vertical axis as always representing the Y variable, and they think of all of the X variables as being represented by the horizontal plane. In this situation the hyperplane can still be imagined, as in Figure 7.1c. The parameters $\beta_1, \beta_2, \ldots, \beta_P$ represent the slope of the regression hyperplane with respect to $X_1, X_2, \ldots, X_P$, respectively. The betas are called **partial regression coefficients**. For example, β_1 is the rate of change of the mean of Y as a function of X_1 when the levels of $X_2, \ldots, X_P$ are held fixed. In this sense it represents the change of the mean of Y as a function of X_1 after adjusting for $X_2, \ldots, X_P$.

Again we assume that the variance of Y is homogeneous over the range of concern of the X variables. Usually, for the fixed-X case, P is rarely larger than three or four.

Least squares method

As for simple linear regression, the method of least squares is used to obtain estimates of the parameters. These estimates, denoted by $A, B_1, B_2, \ldots, B_P$, are printed in the output of any multiple regression program. The estimate A is usually labelled "intercept," and $B_1, B_2, \ldots, B_P$ are usually given in tabular form under the label "coefficient" or "regression coefficient" by variable name. The formulas for these estimates are mathematically derived by minimizing the sums of the squared vertical deviations.

The **predicted value** $\hat{Y}$ for a given set of values $X_1^*, X_2^*, \ldots, X_P^*$ is then calculated as

$$\hat{Y} = A + B_1 X_1^* + B_2 X_2^* + \cdots + B_P X_P^*$$

The estimate of σ^2 is the **residual mean square**, obtained as

$$S^2 = \text{RES.MS.} = \frac{\sum (Y - \hat{Y})^2}{N - P - 1}$$

The square root of the residual mean square is called the **standard error of the estimate**. It represents the variation unexplained by the regression plane.

Packaged programs usually print the value of S^2 and the standard errors for the regression coefficients (and sometimes for the intercept). They also print the P values for the test that α and each β are zero. The standard errors and the residual mean square are useful in computing confidence intervals and prediction intervals around the computed $\hat{Y}$. These intervals can be computed from the output in the following manner.

Prediction and confidence intervals

Some regression programs can compute confidence and prediction intervals for Y at specified values of the X variables. If this is not an option in the program, then we would need access to the estimated correlations among the regression coefficients. These can be found in the regression programs of R, S-PLUS, SAS, SPSS, and STATISTICA or in the correlate program in Stata. Given this information, we can compute the estimated variance of $\hat{Y}$ as

$$
\begin{aligned}
\text{Var } \hat{Y} = \frac{S^2}{N} \quad &+ \quad [(X_1^* - \bar{X}_1)^2 \text{Var} B_1 + (X_2^* - \bar{X}_2)^2 \text{Var} B_2 + \cdots \\
&+ \quad (X_P^* - \bar{X}_P)^2 \text{Var } B_P] \\
&+ \quad [2(X_1^* - \bar{X}_1)(X_2^* - \bar{X}_2)\text{Cov}(B_1, B_2) \\
&+ \quad 2(X_1^* - \bar{X}_1)(X_3^* - \bar{X}_3)\text{Cov}(B_1, B_3) + \cdots]
\end{aligned}
$$

The variances (Var) of the various B_i are computed as the squares of the standard errors of the B_i, i going from 1 to P. The covariances (Cov) are computed

from the standard errors and the correlations (Corr) among the regression coefficients. For example,

$$\mathrm{Cov}(B_1, B_2) = (\text{standard error } B_1)(\text{standard error } B_2)[\mathrm{Corr}(B_1, B_2)]$$

If $\hat{Y}$ is interpreted as an estimate of the mean of Y at $X_1^*, X_2^*, \ldots, X_P^*$, then the
confidence interval for this mean is computed as

$$\mathrm{CI}(\text{mean } Y \text{ at } X_1^*, X_2^*, \ldots, X_P^*) = \hat{Y} \pm t(\mathrm{Var}\,\hat{Y})^{1/2}$$

where t is the $100(1 - \alpha/2)$ percentile of the t distribution with $N - P - 1$
degrees of freedom.

When $\hat{Y}$ is interpreted as an estimate of the Y value for an individual, then
the variance of $\hat{Y}$ is increased by S^2, similar to what is done in simple linear
regression. Then the prediction interval (PI) is

$$\mathrm{PI}(\text{individual } Y \text{ at } X_1^*, X_2^*, \ldots, X_P^*) = \hat{Y} \pm t(S^2 + \mathrm{Var}\,\hat{Y})^{1/2}$$

where t is the same as it is for the confidence interval above. Note that these
intervals require the additional assumption that for any set of levels of X the
values of Y are normally distributed.

In summary, for the fixed-X case we presented the model for multiple linear
regression analysis and discussed estimates of the parameters. Next, we present
the variable-X model.

7.5 Regression and correlation: variable-X case

For this model the X's and the Y variable are random variables measured on
cases that are randomly selected from a population. An example is the lung
function data set where FEV1, age, height, and other variables were measured
on individuals in households. As in Chapter 6, the previous discussion and
formulas given for the fixed-X case apply to the variable-X case. (This result is
justifiable theoretically by conditioning on the X variable values that happen to
be oriented in the sample.) Furthermore, since the X variables are also random
variables, the joint distribution of $Y, X_1, X_2, \ldots, X_P$ is of interest.

When there is only one X variable, it was shown in Chapter 6 that the bivariate distribution of X and Y can be characterized by its region of concentration.
These regions are ellipses if the data come from a bivariate normal distribution. Two such ellipses and their corresponding regression lines are illustrated
in Figures 7.3a and 7.3b.

When two X variables exist, the regions of concentration become three-
dimensional forms. These forms take the shape of ellipsoids if the joint distribution is multivariate normal. In Figure 7.3c such an ellipsoid with the corresponding regression plane of Y on X_1 and X_2 is illustrated. The regression
plane passes through the point representing the population means of the three

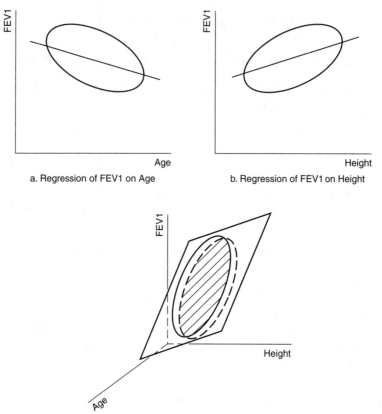

a. Regression of FEV1 on Age   b. Regression of FEV1 on Height

c. Regression of FEV1 on Age and Height

Figure 7.3: *Hypothetical Regions of Concentration and Corresponding Regression Lines and Planes for the Population Variable-X Model*

variables and intersects the vertical axis at a distance α from the origin. Its position is determined by the slope coefficients β_1 and β_2. When $P > 2$, we can imagine the horizontal plane representing all the X variables and the vertical plane representing Y. The ellipsoid of concentration becomes the so-called **hyperellipsoid**.

Estimation of parameters

For the variable-X case several additional parameters are needed to characterize the joint distribution. These include the population means, $\mu_1, \mu_2, \ldots, \mu_P$ and μ_Y, the population standard deviations $\sigma_1, \sigma_2, \ldots, \sigma_P$ and σ_Y, and the covariances of the X and Y variables.

The variances and covariances are usually arranged in the form of a square array called the **covariance matrix**. For example, if $P = 3$, the covariance matrix is a four-by-four array of the form

	X_1	X_2	X_3	Y
X_1	σ_1^2	σ_{12}	σ_{13}	σ_{1Y}
X_2	σ_{12}	σ_2^2	σ_{23}	σ_{2Y}
X_3	σ_{13}	σ_{23}	σ_3^2	σ_{3Y}
Y	σ_{1Y}	σ_{2Y}	σ_{3Y}	σ_Y^2

A single horizontal line is included to separate the dependent variable Y from the independent variables. The estimates of the means, variances, and covariances are available in the output of most regression programs. The estimated variances and covariances are denoted by S instead of σ.

In addition to the estimated covariance matrix, an estimated **correlation matrix** is also available, which is given in the same format:

	X_1	X_2	X_3	Y
X_1	1	r_{12}	r_{13}	r_{1Y}
X_2	r_{12}	1	r_{23}	r_{2Y}
X_3	r_{13}	r_{23}	1	r_{3Y}
Y	r_{1Y}	r_{2Y}	r_{3Y}	1

Since the correlation of a variable with itself must be equal to 1, the diagonal elements of the correlation matrix are equal to 1. The off-diagonal elements are the simple correlation coefficients described in Section 6.5. As before, their numerical values always lie between $+1$ and -1, inclusive.

As an example, the following covariance matrix is obtained from the lung function data set for the fathers. For the sake of illustration, we include a third X variable, weight, given in pounds.

	Age	Height	Weight	FEV1
Age	47.47	−1.08	−3.65	−1.39
Height	−1.08	7.72	34.70	0.91
Weight	−3.65	34.70	573.80	2.07
FEV1	−1.39	0.91	2.07	0.42

Note that this matrix is symmetric; that is, if a mirror were placed along the diagonal, then the elements in the lower triangle would be mirror images of those in the upper triangle. For example, the covariance of age and height is −1.08, the same as the covariance between height and age.

Symmetry holds also for the correlation matrix that follows:

	Age	Height	Weight	FEV1
Age	1.00	−0.06	−0.02	−0.31
Height	−0.06	1.00	0.52	0.50
Weight	−0.02	0.52	1.00	0.13
FEV1	−0.31	0.50	0.13	1.00

Note that age is not highly correlated with height or weight but that height and weight have a substantial positive correlation, as might be expected. The largest correlation between an independent variable and FEV1 is between height and FEV1. Usually the correlations are easier to interpret than the covariances since they always lie between $+1$ and -1, and zero signifies no correlation.

All the correlations are computed from the appropriate elements of the covariance matrix. For example, the correlation between age and height is

$$r_{12} = \frac{S_{12}}{(S_1 S_2)^{1/2}}$$

or

$$-0.06 = \frac{-1.08}{(47.47)^{1/2}(7.72)^{1/2}}$$

Note that the correlations are computed between Y and each X as well as among the X variables. We will return to interpreting these simple correlations in Section 7.7.

Multiple correlation

So far, we have discussed the concept of correlation between two variables. It is also possible to describe the strength of the linear relationship between Y and a set of X variables by using the **multiple correlation coefficient**. In the population we will denote this multiple correlation by $\mathcal{R}$. It represents the simple correlation between Y and the corresponding point on the regression plane for all possible combinations of the X variables. Each individual in the population has a Y value and a corresponding point on the plane computed as

$$Y' = \alpha + \beta_1 X_1 + \beta_2 X_2 + \cdots + \beta_P X_P$$

Correlation $\mathcal{R}$ is the population simple correlation between all such Y and Y' values. The numerical value of $\mathcal{R}$ cannot be negative. The maximum possible value for $\mathcal{R}$ is 1.0, indicating a perfect fit of the plane to the points in the population.

Another interpretation of the $\mathcal{R}$ coefficient for multivariate normal distributions involves the concept of the **conditional distribution**. This distribution describes all of the Y values of individuals whose X values are specified at certain levels. The variance of the conditional distribution is the variance of the Y values about the regression plane in a vertical direction. For multivariate normal distributions this variance is the same at all combinations of levels of the X variables and is denoted by σ^2. The following fundamental expression relates $\mathcal{R}$ to σ^2 and σ_Y^2:

$$\sigma^2 = \sigma_Y^2(1 - \mathcal{R}^2)$$

Rearrangement of this expression shows that

$$\mathcal{R}^2 = \frac{\sigma_Y^2 - \sigma^2}{\sigma_Y^2} = 1 - \frac{\sigma^2}{\sigma_Y^2}$$

When the variance about the plane, or σ^2, is small relative to σ_Y^2, then the squared multiple correlation $\mathcal{R}^2$ is close to 1. When the variance σ^2 about the plane is almost as large as the variance σ_Y^2 of Y, then $\mathcal{R}^2$ is close to zero. In this case the regression plane does not fit the Y values much better than μ_Y. Thus the multiple correlation squared suggests the proportion of the variation accounted for by the regression plane. As in the case of simple linear regression, another interpretation of $\mathcal{R}$ is that $(1 - \mathcal{R}^2)^{1/2} \times 100$ is the percentage of σ_Y **not** explained by X_1 to X_P.

Note that σ_Y^2 and σ^2 can be estimated from a computer output as follows:

$$S_Y^2 = \frac{\Sigma(Y - \bar{Y})^2}{N - 1} = \frac{\text{total sum of squares}}{N - 1}$$

and

$$S^2 = \frac{\Sigma(Y - \hat{Y})^2}{N - P - 1} = \frac{\text{residual sum of squares}}{N - P - 1}$$

Partial correlation

Another correlation coefficient is useful in measuring the degree of dependence between two variables after adjusting for the linear effect of one or more of the other X variables. For example, suppose an educator is interested in the correlation between the total scores of two tests, T_1 and T_2, given to twelfth graders. Since both scores are probably related to the student's IQ, it would be reasonable to first remove the linear effect of IQ from both T_1 and T_2 and then find the correlation between the adjusted scores. The resulting correlation coefficient is called a **partial correlation coefficient** between T_1 and T_2 given IQ.

For this example we first derive the regression equations of T_1 on IQ and T_2 on IQ. These equations are displayed in Figures 7.4a and 7.4b. Consider an individual whose IQ is IQ* and whose actual scores are T_1^* and T_2^*. The test scores defined by the population regression lines are $\tilde{T}_1$ and $\tilde{T}_2$, respectively. The adjusted test scores are the residuals $T_1^* - \tilde{T}_1$ and $T_2^* - \tilde{T}_2$. These adjusted scores are calculated for each individual in the population, and the simple correlation between them is computed. The resulting value is defined to be the population partial correlation coefficient between T_1 and T_2 with the linear effects of IQ removed.

Formulas exist for computing partial correlations directly from the population simple correlations or other parameters without obtaining the residuals. In

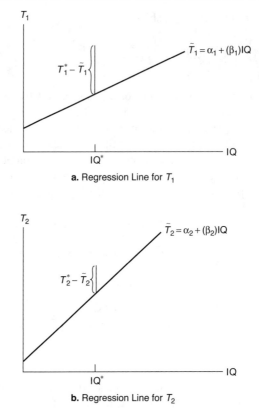

Figure 7.4: *Hypothetical Population Regressions of T_1 and T_2 Scores, Illustrating the Computation of a Partial Correlation Coefficient*

the above case suppose that the simple correlation for T_1 and T_2 is denoted by ρ_{12}, for T_1 and IQ by ρ_{1q}, and T_2 and IQ by ρ_{2q}. Then the partial correlation of T_1 and T_2 given IQ is derived as

$$\rho_{12.q} = \frac{\rho_{12} - \rho_{1q}\rho_{2q}}{\left[\left(1 - \rho_{1q}^2\right)\left(1 - \rho_{2q}^2\right)\right]^{1/2}}$$

In general, for any three variables denoted by i, j, and k,

$$\rho_{ij.k} = \frac{\rho_{ij} - \rho_{ik}\rho_{jk}}{\left[\left(1 - \rho_{ik}^2\right)\left(1 - \rho_{jk}^2\right)\right]^{1/2}}$$

Sample correlation coefficients are used in place of the population correlation coefficients given in the previous formulas.

It is also possible to compute partial correlations between any two variables after removing the linear effect of two or more other variables. In this case the residuals are deviations from the regression planes on the variables whose effects are removed. The partial correlation is the simple correlation between the residuals.

Computer software programs discussed in Section 7.10 calculate estimates of multiple and partial correlations. We note that the formula given above can be used with sample simple correlations to obtain sample partial correlations.

This section has covered the basic concepts in multiple regression and correlation. Additional items found in output will be presented in Section 7.8 and in Chapters 8 and 9.

7.6 Interpretation: fixed-X case

In this section we discuss interpretation of the estimated parameters and present some tests of hypotheses for the fixed-X case.

As mentioned previously, the regression analysis might be performed for prediction: deriving an equation to predict Y from the X variables. This situation was discussed in Section 7.4. The analysis can also be performed for description: an understanding of the relationship between the Y and X variables.

In the fixed-X case the number of X variables is generally small. The values of the slopes describe how Y changes with changes in the levels of the X variables. In most situations a plane does not describe the response surface for the whole possible range of X values. Therefore it is important to define the region of concern for which the plane applies. The magnitudes of the β's and their estimates depend on this choice. In the interpretation of the relative magnitudes of the estimated B's, it is useful to compute $B_1\bar{X}_1 + B_2\bar{X}_2, \ldots, B_P\bar{X}_P$ and compare the resulting values. A large (relative) value of the magnitude of $B_i\bar{X}_i$ indicates a relatively important contribution of the variable X_i. Here the $\bar{X}$'s represent typical values within the region of concern.

If we restrict the range of concern, the **additive model**, $\alpha + \beta_1 X_1 + \beta_2 X_2 + \cdots + \beta_P X_P$, is often an appropriate description of the underlying relationship. It is also sometimes useful to incorporate interactions or nonlinearities. This step can partly be achieved by transformations and will be discussed in Section 7.8. Examination of residuals can help you assess the adequacy of the model, and this analysis is also discussed in Section 7.8.

Analysis of variance

For a test of whether the regression plane is at all helpful in predicting the values of Y, the ANOVA table presented in most regression outputs can be

Table 7.1: *ANOVA table for multiple regression*

Source of variation	Sums of squares	df	Mean square	F
Regression	$\sum(\hat{Y} - \bar{Y})^2$	P	SS_{reg}/P	MS_{reg}/MS_{res}
Residual	$\sum(Y - \hat{Y})^2$	$N - P - 1$	$SS_{res}/(N - P - 1)$	
Total	$\sum(Y - \bar{Y})^2$	$N - 1$		

Table 7.2: *ANOVA example from the lung function data (fathers)*

Source of variation	Sums of squares	df	Mean square	F
Regression	21.0570	2	10.5285	36.81
Residual	42.0413	147	0.2860	
Total	63.0983			

used. The null hypothesis being tested is

$$H_0 : \beta_1 = \beta_2 = \cdots = \beta_P = 0$$

That is, the mean of Y is as accurate in predicting Y as the regression plane. The ANOVA table for multiple regression has the form given in Table 7.1, which is similar to that of Table 6.1.

When the fitted plane differs from a horizontal plane in a significant fashion, then the term $\sum(\hat{Y} - \bar{Y})^2$ will be large relative to the residuals from the plane $\sum(Y - \hat{Y})^2$. This result is the case for the fathers where $F = 36.81$ with $P = 2$ degrees of freedom in the numerator and $N - P - 1 = 147$ degrees of freedom in the denominator (Tables 7.1 and 7.2). $F = 36.81$ can be compared with the printed F value in standard tables with v_1 (numerator) degrees of freedom across the top of the table and v_2 (denominator) degrees of freedom in the first column of the table for an appropriate α. For example, for $\alpha = 0.05, v_1 = 2$, and $v_2 = 120$ (147 degrees of freedom are not tabled, so we take the next lower level), the tabled $F(1 - \alpha) = F(0.95) = 3.07$, from the body of the table. Since the computed F of 36.81 is much larger than the tabled F of 3.07, the null hypothesis is rejected.

The P value can be determined more precisely. In fact, it is printed in the output as $P = 0.0000$: this could be reported as $P < 0.0001$. Thus the variables age and height together help significantly in predicting an individual's FEV1.

If this blanket hypothesis is rejected, then the degree to which the regression equation fits the data can be assessed by examining a quantity called the **coefficient of determination**, defined as

$$\text{coefficient of determination} = \frac{\text{sum of squares regression}}{\text{sum of squares total}}$$

If the regression sum of squares is not in the program output, it can be obtained by subtraction, as follows:

$$\text{regression sum of squares} = \text{total sum of squares} - \text{residual sum of squares}$$

For the fathers,

$$\text{coefficient of determination} = \frac{21.0570}{63.0983} = 0.334$$

This value is an indication of the reduction in the variance of Y achieved by using X_1 and X_2 as predictors. In this case the variance around the regression plane is 33.4% less than the variance of the original Y values. Numerically, the coefficient of determination is equal to the square of the multiple correlation coefficient, and therefore it is called RSQUARE in the output of some computer programs. Although the multiple correlation coefficient is not meaningful in the fixed-X case, the interpretation of its square is valid.

Other tests of hypotheses

Tests of hypotheses can be used to assess whether variables are contributing significantly to the regression equation. It is possible to test these variables either singularly or in groups. For any specific variable X_i we can test the null hypothesis

$$H_0 : \beta_i = 0$$

by computing

$$t = \frac{B_i - 0}{SE(B_i)}$$

and performing a one- or two-sided t test with $N - P - 1$ degrees of freedom. These t statistics are often printed in the output of packaged programs. Other programs print the corresponding F statistics ($t^2 = F$), which can test the same null hypothesis against a two-sided alternative.

When this test is performed for each X variable, the joint significance level cannot be determined. A method designed to overcome this uncertainty makes use of the so-called **Bonferroni inequality**. In this method, to compute the appropriate joint P value for any test statistic, we multiply the single P value obtained from the printout by the number of X variables we wish to test. The joint P value is then compared with the desired overall level.

For example, to test that age alone has an effect, we use

$$H_0 : \beta_1 = 0$$

and from the computer output

$$t = \frac{-0.0266 - 0}{0.00637} = -4.18$$

with 147 degrees of freedom or $P = 0.00005$. This result is also highly significant, indicating that the equation using age and height is significantly better than an equation using height alone.

Most computer outputs will display the P level of the t statistic for each regression coefficient. To get joint P values according to the Bonferroni inequality, we multiply each individual P value by the number of X variables before we compare it with the overall significance level α. For example, if $P = 0.015$ and there are two X's, then $2(0.015) = 0.03$, which could be compared with an α level of 0.05.

7.7   Interpretation: variable-X case

In the variable-X case all of the Y and X variables are considered to be random variables. The means and standard deviations printed by packaged programs are used to estimate the corresponding population parameters. Frequently, the covariance and/or the correlation matrices are printed. In the remainder of this section we discuss the interpretation of the various correlation and regression coefficients found in the output.

Correlation matrix

Most people find the correlation matrix more useful than the covariance matrix since all of the correlations are limited to the range -1 and $+1$. Note that because of the symmetry of these matrices, some programs will print only the terms on one side of the diagonal. The diagonal terms of a correlation matrix will always be 1 and thus sometimes are not printed.

Initial screening of the correlation matrix is helpful in obtaining a preliminary impression of the interrelationships among the X variables and between Y and each of the X variables. One way of accomplishing this screening is to first determine a cutoff point on the magnitude of the correlation, for example, 0.3 or 0.4. Then each correlation greater in magnitude than this number is underlined or highlighted. The resulting pattern can give a visual impression of the underlying interrelationships; i.e., highlighted correlations are indications of possible relationships. Correlations near $+1$ and -1 among the X variables indicate that the two variables are nearly perfect functions of each other, and the investigator should consider dropping one of them since they convey nearly the same information. Also, if the physical situation leads the investigator to expect larger correlations than those found in the printed matrix, then the presence of outliers in the data or of nonlinearities among the variables may be causing the discrepancies.

If tests of significance are desired, the following test can be performed. To test the hypothesis

$$H_0 : \rho = 0$$

use the statistic given by

$$t = \frac{r(N-2)^{1/2}}{(1-r^2)^{1/2}}$$

where ρ is a particular population simple correlation and r is the estimated simple correlation. This statistic can be compared with the t table value with df $= N - 2$ to obtain one-sided or two-sided P values. Again, $t^2 = F$ with 1, and $N - 2$ degrees of freedom can be used for two-sided alternatives.

Some programs, such as the SPSS procedure CORRELATIONS, will print the P value for each correlation in the matrix. Since several tests are being performed in this case, it may be advisable to use the Bonferroni correction to the P value, i.e., each P is multiplied by the number of correlations (say M). This number of correlations can be determined as follows:

$$M = \frac{(\text{no. of rows})(\text{no. of rows} - 1)}{2}$$

After this adjustment to the P values is made, they can be compared with the nominal significance level α.

Standardized coefficients

The interpretations of the regression plane, the associated regression coefficients, and the standard error around the plane are the same as those in the fixed-X case. All the tests presented in Section 7.6 apply here as well.

Another way to interpret the regression coefficients is to examine **standardized coefficients**. These are printed in the output of many regression programs and can be computed easily as

$$\text{standardized } B_i = B_i \left(\frac{\text{standard deviation of } X_i}{\text{standard deviation of } Y} \right)$$

These coefficients are the ones that would be obtained if the Y and X variables were standardized prior to performing the regression analysis. When the X variables are uncorrelated or have very weak correlation with each other, the standardized coefficients of the various X variables can be directly compared in order to determine the relative contribution of each to the regression line. The larger the magnitude of the standardized B_i, the more X_i contributes to the prediction of Y. Comparing the unstandardized B directly does not achieve this result because of the different units and degrees of variability of the X variables. The regression equation itself should be reported for future use in terms of the unstandardized coefficients so that prediction can be made directly from the raw X variables. When standardized coefficients are used, the standardized slope coefficient is the amount of change in the mean of the standardized Y values when the value of X is increased by one standard deviation, keeping the other X variables constant.

If the X variables are intercorrelated, then the usual standardized coefficients are difficult to interpret (Bring, 1994). The standard deviation of X_i used in computing the standardized B_i should be replaced by a partial standard deviation of X_i which is adjusted for the multiple correlation of X_i with the other X variables included in the regression equation. The partial standard deviation is equal to the usual standard deviation multiplied by the square root of one minus the squared multiple correlation of X_i with the other X's and then multiplied by the square root of $(N-1)/(N-P)$. The quantity one minus the squared multiple correlation of X_i with the other X's is called tolerance (its inverse is called the **variance inflation factor**) and is available from some programs (see Section 7.9 where multicollinearity is discussed). For further discussion of adjusted standardized coefficients see Bring (1994).

Multiple and partial correlations

As mentioned earlier in Section 7.5, the multiple correlation coefficient is a measure of the strength of the linear relationship between Y and the set of variables $X_1, X_2, \ldots, X_P$. The multiple correlation has another useful property: it is the highest possible simple correlation between Y and any linear combination of X_1 to X_P. This property explains why R (the computed correlation) is never negative. In this sense the least squares regression plane maximizes the correlation between the set of X variables and the dependent variable Y. It therefore presents a numerical measure of how well the regression plane fits the Y values. When the multiple correlation R is close to zero, the plane barely predicts Y better than simply using $\bar{Y}$ to predict Y. A value of R close to 1 indicates a very good fit.

As in the case of simple linear regression, discussed in Chapter 6, R^2 is an estimate of the proportional reduction in the variance of Y achieved by fitting the plane. Again, the proportion of the standard deviation of Y around the plane is estimated by $(1-R^2)^{1/2}$. If a test of significance is desired, the hypothesis

$$H_0 : \mathcal{R} = 0$$

can be tested by the statistic

$$F = \frac{R^2/P}{(1-R^2)/(N-P-1)}$$

which is compared with a tabled F with P and $N - P - 1$ degrees of freedom. Since this hypothesis is equivalent to

$$H_0 : \beta_1 = \beta_2 = \cdots = \beta_P = 0$$

the F statistic is equivalent to the one calculated in Table 7.1. (See Section 8.4 for an explanation of adjusted multiple correlations.)

As mentioned earlier, the partial correlation coefficient is a measure of the linear relationship between two variables after adjusting for the linear effect of a group of other variables. For example, suppose a regression of Y on X_1 and X_2 is fitted. The square of the partial correlation of Y on X_3 after adjusting for X_1 and X_2 is the proportion of the variance of Y reduced by using X_3 as an additional X variable. The hypothesis

$$H_0 : \text{partial } \rho = 0$$

can be tested by using a test statistic similar to the one for simple r, namely,

$$t = \frac{(\text{partial } r)(N-Q-2)^{1/2}}{[1-(\text{partial } r^2)]^{1/2}}$$

where Q is the number of variables adjusted for. The value of t is compared with the tabled t value with $N - Q - 2$ degrees of freedom. The square of the t statistic is an F statistic with df $= 1$ and $N - Q - 2$; the F statistic may be used to test the same hypothesis against a two-sided alternative.

In the special case where Y is regressed on X_1 to X_P, a test that the partial correlation between Y and any of the X variables, say X_i, is zero after adjusting for the remaining X variables is equivalent to testing that the regression coefficient β_i is zero.

In Chapter 8, where variable selection is discussed, partial correlations will play an important role.

7.8 Regression diagnostics and transformations

In this section we discuss the use of residual analysis for finding outliers, transformations, polynomial regression, and the incorporation of interaction terms into the equation. For further discussion of residuals and transformations see diagnostic plots for the effect of an additional variable in Section 8.9. Initial screening can be done by examining scatter plots of each X_i variable one at a time against the outcome or dependent variable.

Residuals

The use of residuals in the case of simple linear regression was discussed in Section 6.8. Residual analysis is even more important in multiple regression analysis because the scatter diagrams of Y and all the X variables simultaneously cannot be portrayed on two-dimensional paper or computer screens. If there are only two X variables and a single Y variable, programs do exist for showing three-dimensional plots. But unless these plots can be easily rotated, finding an outlier is often difficult and lack of independence and the need for transformations can be obscured.

Fortunately, the use of residuals changes little conceptually between simple and multiple linear regression. The information given on types of residuals (raw, standardized, studentized, and deleted) and the various statistics given in Section 6.8 all apply here. The references cited in that section are also useful in the case of multiple regression. The formulas for some of the statistics, for example h (a leverage statistic), become more complicated, but basically the same interpretations apply. The values of h are obtained from the diagonal of a matrix called the "hat" matrix (hence the use of the symbol h) but a large value of h still indicates an outlier in at least one X variable.

The three types of outliers discussed in Section 6.8 (outliers in Y, outliers in X or h, and influential points) can be detected by residual analysis in multiple regression analysis. Scatter plots of the various types of residuals against $\hat{Y}$ are the simplest to do and we recommend that, as a minimum, one of each of the three types be plotted against $\hat{Y}$ whenever any new regression analysis is performed. Plots of the residuals against each X are also helpful in showing whether a large residual is associated with a particular value of one of the X variables. Since an outlier in Y is more critical when it is also an outlier in X, many investigators plot deleted studentized residuals on the vertical axis versus leverage on the horizontal axis and look for points that are in the upper right-hand corner of the scatter plot.

If an outlier is found then it is important to determine which case(s) has the suspected outlier. Examination of a listing of the three types of outliers in a spreadsheet format is a common way to determine the case while at the same time allowing the investigator to see the values of Y and the various X variables. Sometimes this process provides a clue that an outlier has occurred in a certain subset of the cases. In some of the newer interactive programs, it is also possible to identify outliers using a mouse by clicking on the suspected outlier in a graph and have the information transferred to the data spreadsheet. Removal of outliers can be chosen as an option, thus enabling the user to quickly see the effect of discarding an outlier on the regression plane. In any case, we recommend performing the multiple regression both with and without suspected outlier(s) to see the actual difference in the multiple regression equation.

The methods used for finding an outlier were originally derived to find a single outlier. These same methods can be used to find multiple outliers, although in some cases the presence of an outlier can be partially hidden by the presence of other outliers. Note also that the presence of several outliers in the same area will have a larger effect than one outlier alone, so it is important to examine multiple outliers for possible removal. For this purpose, we suggest use of the methods for single outliers, deleting the worst one first and then repeating the process in subsequent iterations (see Chatterjee and Hadi, 1988, for additional methods).

As discussed in Section 6.8, Cook's distance or DFFITS should be examined in order to identify influential cases. Chatterjee and Hadi (1988) suggest that a case may be considered influential if its DFFITS value is greater than $2[(P+1)/(N-P-1)]^{\frac{1}{2}}$ in absolute value, or if its Cook's distance is greater than the 50th percentile of the F distribution with $P+1$ and $N-P-1$ degrees of freedom. As was mentioned in Section 6.8, an observation is considered an outlier in X if its leverage is high and an outlier in Y if its residual or studentized residual is large. Reasonable cutoff values are $2(P+1)/N$ for the ith diagonal leverage value h_{ii} and 2 for the studentized residual. An observation is considered influential if it is an outlier in both X and Y. We recommend examination of the regression equations with and without the influential observations.

The Durbin–Watson statistic or the serial correlation of the residuals can be used to assess independence of the residuals (Section 6.8). Normal probability or quantile plots of the residuals can be used to assess the normality of the error variable.

Transformations

In examining the adequacy of the multiple linear regression model, the investigator may wonder whether the transformations of some or all of the X variables might improve the fit. (See Section 6.9 for review.) For this purpose the residual plots against the X variables may suggest certain transformations. For example, if the residuals against X_i take the humped form shown in Figure 6.7c, it may be useful to transform X_i. Reference to Figure 6.9a suggests that $\log(X_i)$ may be the appropriate transformation. In this situation $\log(X_i)$ can be used in place of or together with X_i in the same equation. When both X_i and $\log(X_i)$ are used as predictors, this method of analysis is an instance of attempting to improve the fit by including an additional variable. Other commonly used additional variables are the square X_i^2 and square root $X_i^{1/2}$ of appropriate X_i variables; other candidates may be new variables altogether. Plots against candidate variables should be obtained in order to check whether the residuals are related to them. We note that these plots are most informative when the X variable considered is not highly correlated with other X variables.

Caution should be taken at this stage not to be deluged by a multitude of additional variables. For example, it may be preferable in certain cases to use

$\log(X_i)$ instead of X_i and X_i^2; the interpretation of the equation becomes more difficult as the number of predictors increases. The subject of variable selection will be discussed in Chapter 8.

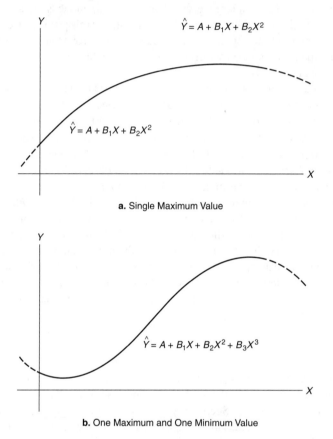

a. Single Maximum Value

b. One Maximum and One Minimum Value

Figure 7.5: *Hypothetical Polynomial Regression Curves*

Polynomial regression

A subject related to variable transformations is the so-called **polynomial regression**. For example, suppose that the investigator starts with a single variable X and that the scatter diagram of Y on X indicates a curvilinear function, as shown in Figure 7.5a. An appropriate regression equation in this case has the form

$$\hat{Y} = A + B_1 X + B_2 X^2$$

This equation is, in effect, a multiple regression equation with $X_1 = X$ and $X_2 = X^2$. Thus the multiple regression programs can be used to obtain the estimated curve. Similarly, for Figure 7.5b the regression curve has the form

$$\hat{Y} = A + B_1 X + B_2 X^2 + B_3 X^3$$

which is a multiple regression equation with $X_1 = X, X_2 = X^2$, and $X_3 = X^3$. Both of these equations are examples of polynomial regression equations with degrees two and three, respectively.

Interactions

Another issue in model fitting is to determine whether the X variables interact. If the effects of two variables X_i and X_j are **not** interactive, then they appear as $B_i X_i + B_j X_j$ in the regression equation. In this case the effects of the two variables are said to be **additive**. Assume for now that X_i and X_j are the only variables in the model. When interpreting the coefficients B_i and B_j we would infer that an increase in X_i of one unit is estimated to change Y by B_i given the comparison is made at a fixed level of X_j. Similarly, an increase in X_j of one unit is estimated to change Y by B_j given the comparison is made at a fixed level of X_i. As an example, if X_i represents height measured in inches and X_j represents age measured in years, B_i and B_j could be interpreted as follows. An increase in height of one inch is estimated to change Y by B_i when comparing individuals of the same age. For this comparison it would not matter what age the individuals are as long as they are of the same age and of an age seen in the data. Similarly, B_j would be interpreted as the estimated change in Y for an increase in age of one year when comparing individuals of the same height. If the effects of the two variables are additive we can infer that an increase in X_i of one unit combined with an increase in X_j of one unit is estimated to change Y by $B_i + B_j$. Or, in our example, if we were to compare an individual with a specific height and a specific age to an individual who is one inch shorter and one year younger then we would estimate the difference in Y to be $B_i + B_j$. Another way of expressing this concept is to say that there is **no interaction** between X_i and X_j; thus when estimating the combined effects we can obtain the estimate by adding the individual effects.

If the additive terms for these variables do not completely specify their effects on the dependent variable Y, then **interaction** of X_i and X_j is said to be present. This phenomenon can be observed in many situations. For example, in chemical processes the additive effects of two agents are often not an accurate reflection of their combined effect since synergies or catalytic effects are often present. Similarly, in studies of economic growth the interactions between the two major factors labor and capital are important in predicting outcome. Thus when estimating the combined effect of X_i and X_j in the presence of an interaction we cannot simply add the individual effects. It is therefore often advisable to include additional variables in the regression equation to represent interactions. A commonly used practice is to add the product $X_i X_j$ to the set of X variables to represent the interaction between X_i and X_j. This and further methods of incorporating interactions will be discussed in more detail in Section 9.3.

Table 7.3: *Percent positive residuals: No interaction*

| X_i | $X_j(\%)$ | | |
	Low	Medium	High
Low	50	52	48
Medium	52	48	50
High	49	50	51

Table 7.4: *Percent positive residuals: Interactions present*

| X_i | $X_j(\%)$ | | |
	Low	Medium	High
Low	20	40	45
Medium	40	50	60
High	55	60	80

As a check for the presence of interactions, tables can be constructed from a list of residuals, as follows. The ranges of X_i and X_j are divided into intervals, as shown in the examples in Tables 7.3 and 7.4. The residuals for each cell are found from the output and simply classified as positive or negative. The percentage of **positive residuals** in each cell is then recorded. If these percentages are around 50%, as shown in Table 7.3, then no interaction term is required. If the percentages vary greatly from 50% in the cells, as in Table 7.4, then an interaction term could most likely improve the fit. Note that when both X_i and X_j are low, the percentage of positive individuals is only 20%, indicating that most of the patients lie below the regression plane. The opposite is true for the case when X_i and X_j are high. This search for interaction effects should be made after consideration of transformations and selection of additional variables.

7.9   Other options in computer programs

Because regression analysis is a very commonly used technique, packaged programs offer a bewildering number of options to the user. But do not be tempted to use options that you do not understand well. As a guide to selecting options, we briefly discuss in this section some of the more popular ones. Other options are often, but not always, described in user guides.

The options for the use of **weights for the observations** and for **regression through the origin** were discussed in Section 6.10 for simple linear regression. These options are also available for multiple regression and the discussion in Section 6.10 applies here as well. Be aware that regression through the origin means that the mean of Y is assumed to be zero when **all** X_i equal zero. This assumption is often unrealistic. For a detailed discussion of how the various regression programs handle the no-intercept regression model, see Okunade, Chang and Evans (1993).

In Section 7.5 we described the matrix of correlations among the estimated slopes, and we indicated how this matrix can be used to find the covariances among the slopes. Covariances can be obtained directly as optional output in most programs.

Multicollinearity

In practice, a problem called **multicollinearity** occurs when some of the X variables are highly intercorrelated. Here we discuss multicollinearity and the options available in computer programs to determine if it exists. In Section 9.5, methods are presented for lessening its effects. A considerable amount of effort is devoted to this problem (e.g., Greene (2008); Chatterjee and Hadi (2006); Belsey *et al.* (1980); and Fox (1991)).

When multicollinearity is present, the computed estimates of the regression coefficients are unstable and have large standard errors. Some computations may also be inaccurate. For multiple regression the squared standard error of the ith slope coefficient can be written as

$$[SE(B_i)]^2 = \frac{S^2}{(N-1)(S_i)^2} \times \frac{1}{1-(R_i)^2}$$

where S^2 is the residual mean square, S_i is the standard deviation of the ith X variable, and R_i is the multiple correlation between the ith X variable and all the other X variables. Note that as R_i approaches one, the denominator of the last part of the equation gets smaller and can become very close to zero. This results in dividing one by a number which is close to zero and therefore this fraction can be quite large. This leads to a large value for the standard error of the slope coefficient.

Some statistical packages print out this fraction, called the **variance inflation factor** (VIF). Note that the square root of VIF directly affects the size of the standard error of the slope coefficient. The inverse of the variance inflation factor is called **tolerance** (one minus the squared multiple correlation between X_i and the remaining X's) and some programs print this out. When the tolerance is small, say less than 0.01, or the variance inflation factor is large, say greater than 100, then it is recommended that the investigator use one of the methods given in Section 9.5. Note that multicollinearity is uncommon in medical research, using data on patients, or in social research, since large correlations among the variables usually do not occur.

Comparing regression planes

Sometimes it is desirable to examine the regression equations for **subgroups** of the population. For example, different regression equations for subgroups subjected to different treatments can be derived. In the program it is often convenient to designate the subgroups as various levels of a **grouping variable**.

In the lung function data set, for instance, we presented the regression equation for 150 males:

$$\text{FEV1} = -2.761 - 0.027\,(\text{age}) + 0.114\,(\text{height})$$

An equation can also be obtained for 150 females, as

$$\text{FEV1} = -2.211 - 0.020\,(\text{age}) + 0.093\,(\text{height})$$

In Table 7.5 the statistical results for males and females combined, as well the separate results, are presented.

As expected, the males tend to be older than the females (these are couples with at least one child age seven or older), taller, and have larger average FEV1. The regression coefficients for males and females are quite similar, having the same sign and similar magnitude.

In this type of analysis it is useful for an investigator to present the means and standard deviations along with the regression coefficients. This information would enable a reader of the report to try typical values of the independent variables in the regression equation in order to assess the numerical effects of the independent variables. Presentation of these results also makes it possible to assess the characteristics of the sample under study. For example, it would **not** be sensible to try to make inferences from this sample regression plane to males who were 90 years old. Since the mean age is 40 years and the standard deviation is 6.89, there is unlikely to be any male in the sample age 90 years (Section 6.13, item 5).

We can make predictions for males and females who are, say, 30 and 50 years old and 66 inches tall. From the results presented in Table 7.5 we would predict FEV1 to be as follows:

Table 7.5: *Lung function output for males and females*

	Mean	S_X	Regression coefficient	Std. error	Standardized regression coefficient
Overall					
Intercept			−6.737		
Age	38.80	6.91	−0.019	0.00444	−0.160
Height	66.68	3.69	0.165	0.00833	0.757
FEV1	3.53	0.81			
Males					
Intercept			−2.761		
Age	40.13	6.89	−0.027	0.00637	−0.282
Height	69.26	2.78	0.114	0.01579	0.489
FEV1	4.09	0.65			
Females					
Intercept			−2.211		
Age	37.56	6.71	−0.020	0.00504	−0.276
Height	64.09	2.47	0.093	0.01370	0.469
FEV1	2.97	0.49			

	Age (years)	
	30	50
Subgroup		
Males	3.96	3.43
Females	3.30	2.90

Thus, even though the equations for males and females look quite similar, the predicted FEV1 for females of the same height and age as a male is expected to be less.

The standardized regression coefficient is also useful in assessing the relative effects of the two variables. For the overall group, the ratio of the unstandardized coefficients for height and age is $0.165/0.019 = 8.68$ but the ratio of the standardized coefficient is $0.757/0.160 = 4.73$. Both show an overwhelming effect of height over age, although it is not quite as striking for the standardized coefficients.

When there are only two X variables, as in this example, it is not necessary to compute the partial standard deviation (Section 7.7) since the correlation of age with height is the same as the correlation of height with age. This simplifies the computation of the standardized coefficients.

For males alone, the ratio for the standardized coefficients is 1.73 versus 4.22 for the unstandardized coefficients. Similarly, for females, the ratio for the standardized coefficients is 1.70 versus 4.65 for the unstandardized. Thus, the relative contribution of age is somewhat closer to that of height than it appears to be from the unstandardized coefficients for males and females separately. But height is always the most important variable. Often what appears to be a major effect when you look at the unstandardized coefficients becomes quite minor when you examine the standardized coefficients, and *vice versa*.

Another interesting result is that the multiple correlation of FEV1 for the overall group ($R = 0.76$) is greater than the multiple correlations for males ($R = 0.58$) or females ($R = 0.54$). Why did this occur? Let us look at the effect of the height variable first. This can be explained partially by the greater effect of height on FEV1 for the overall group than the effect of height for males and females separately (standardized coefficients 0.757 versus 0.489 or 0.469). Secondly, the average height for males is over 5 inches more than for females (male heights are about two standard deviations greater than for females).

In Figure 6.5, it was shown that larger correlations occur when the data points fall in long thin ellipses (large ratio of major to minor axes). If we plot FEV1 against height for the overall group we would get a long thin ellipse. This is because the ellipse for men lies to the right and higher than the ellipse for women, resulting in an overall ellipse (when the two groups are combined) that is long and narrow. Males have higher FEV1 and are taller than women.

The effect of age on FEV1 for the combined group was much less than height (standardized coefficient of only -0.160). If we plot FEV1 versus age for males and for females, the ellipse for males is above that for females but only slightly to the right (men have higher FEV1 but are only 2.5 years older than women or about 0.4 of a standard deviation greater). This will produce a thicker or more rounded ellipse, resulting in a small correlation between age and FEV1 for the combined group.

When the effects of height and age are combined for the overall group, the effect of height predominates and results in higher multiple correlation for the overall group than for males or females separately.

If the regression coefficients are divided by their standard errors, then highly significant t values are obtained. The computer output from Stata gave values of t along with significance levels and 95% confidence intervals for the slope coefficients and the intercept. The levels of significance are often included in reports or asterisks are placed beside the regression coefficients to indicate their level of significance. The asterisks are then usually explained in footnotes to the table.

The standard errors could also be used to test the equality of the individual coefficients for the two groups. For example, to compare the regression coefficients for height for men and women and to test the null hypothesis of equal coefficients, we compute

$$Z = \frac{B_m - B_f}{[SE^2(B_m) + SE^2(B_f)]^{1/2}}$$

or

$$Z = \frac{0.114 - 0.093}{(0.01579^2 + 0.01370^2)^{1/2}} = 1.005$$

The computed value of Z can be compared with the percentiles of the normal distribution from standard tables to obtain an approximate P value for large samples. In this case, the coefficients are not significantly different at the usual $P = 0.05$ level.

The standard deviation around the regression plane (i.e., the square root of the residual mean square) is useful to readers in making comparisons with their own results and in obtaining confidence intervals at the point where all the X's take on their mean value (Section 7.4).

As mentioned earlier, the grouping option is useful in comparing relevant subgroups such as males and females or smokers and nonsmokers. An F test is available as optional output from the SAS GLM procedure to check whether the regression equations derived from the different subgroups are significantly different from each other. It is also possible to do the test using information provided by most of the programs, as we will demonstrate. This test is an example of the general likelihood ratio F test that is often used in statistics and is given in its usual form in Section 8.5.

For our lung function data example, the null hypothesis H_0 is that a single population plane for both males and females combined is the same as the true plane for each group separately. The alternative hypothesis H_1 is that different planes should be fitted to males and females separately. The residual sums of squares obtained from $\sum(Y - \hat{Y})^2$ where the two separate planes are fitted will always be less than or equal to the residual sum of squares from the single plane for the overall sample. These residual sums of squares are printed in the analysis of variance table that accompanies regression programs. Table 7.2 is an example of such a program for males.

The general F test compares the residual sums of squares when a single plane is fitted to what is obtained when two planes are fitted. If these two residuals are not very different, then the null hypothesis cannot be rejected. If fitting two separate planes results in a much smaller residual sum of squares, then the null hypothesis is rejected. The smaller residual sum of squares for the two planes indicates that the regression coefficients differ beyond chance between the two groups. This difference could be a difference in either the intercepts or the slopes or both. In this test, normality of errors is assumed.

The F statistic for the test is

$$F = \frac{[SS_{res}(H_0) - SS_{res}(H_1)]/[df(H_0) - df(H_1)]}{SS_{res}(H_1)/df(H_1)}$$

In the lung function data set, we have

$$F = \frac{[82.6451 - (42.0413 + 25.0797)]/[297 - (147 + 147)]}{(42.0413 + 25.0797)/(147 + 147)}$$

where 82.6451 is the residual sum of squares for the overall plane with 297 degrees of freedom, 42.0413 is the residual sum of squares for males with 147 degrees of freedom, and 25.0797 is the residual sum of squares for females with 147 degrees of freedom also. Thus,

$$F = \frac{(82.6451 - 67.1210)/3}{67.1210/294} = \frac{5.1747}{0.2283} = 22.67$$

with 3 and 294 degrees of freedom. This F value corresponds to $P < 0.0001$ and indicates that a better prediction can be obtained by fitting separate planes for males and females.

Other topics in regression analysis and corresponding computer options will be discussed in the next two chapters. In particular, selecting a subset of predictor variables is given in Chapter 8.

7.10   Discussion of computer programs

Almost all general purpose statistical packages and some spreadsheet software have multiple regression programs. Although the number of features varies, most of the programs have a wide choice of options (see Table 7.6). In this section we show how we obtained the computer output in this chapter, and summarize the multiple regression output options for the statistical packages used in this book. The lung function data set is a rectangular data set where the rows are the families and the columns are the variables, given first for males, then repeated for females and for up to three children. There are 150 families; the unit of analysis is the family. We can use the data set as it is to analyze the information for the males or for the females separately by choosing the variables for males (FSEX, FAGE,...) when analyzing fathers and those for females (MSEX, MAGE,...) when analyzing mothers (see Appendix A). But when we compare the two regression planes as was done in Section 7.9, we need to make each parent a unit of analysis so the sample size becomes 300 for mothers and fathers combined. To do this the data set must be "unpacked" so that the females are listed below the males.

As examples, we will give the commands used to do this in SAS and Stata. These examples makes little sense unless you either read the manuals or are familiar with the package. Note that any of the six packages will perform this operation. It is particularly simple in the true Windows packages as the copy-and-paste options of Windows can be used.

To unpack a data set in SAS, we can use multiple output statements in the data step. In the following SAS program, each iteration of the data step reads in a single record, and the two output statements cause two records, one for the mother and one for the father, to be written to the output data set. The second data step writes the new file as ASCII.

```
**Lung-function data from Detels ***;
** Unpack data for mother & fathers ***;
data c.monsdads;
infile ''c:\work\aavc\lung1.dat'';
input id area fsex fage fheight fweight ffvc ffev1
msex mage mheight mweight mfvc mfev1
ocsex ocage ocheight ocweight ocfvc ocfev1
mcsex mcage mcheight mcweight mcfvc mcfev1
ycsex ycage ycheight ycweight ycfvc ycfev1;
*output father **;
sex = fsex; age = fage; height = fheight; weight = fweight;
fvc = ffvc; fev1 = ffev1;
output;
*output mother **;
sex = msex; age = mage; height = mheight; weight = mweight;
fvc = mfvc; fev1 = mfev1;
output;
drop fsex--ycfev1;
run;
data temp;
* output as ascii **;
set c.momsdads
file ''c:\work\aavc\momsdads.dat'';
put id area sex age height weight fvc fev1;
run;
proc print;
title ''Lung function data from Detels'';
run;
```

In Stata, the packed and unpacked forms of the data set are called the wide and long forms. We can convert between the wide and long forms by using the reshape command. The following commands unpack five records for each of the records in the lung function data set. Note that there is information not only for males and females, but also for three children, thus giving five records for each family. The following instructions will result in a data set with (five records per family) × (150 families) = 750 cases. If the family has fewer than three children, then only the data for the actual children will be included; the rest will be missing.

```
. dictionary using lung1.dat{
* Lung function data from Detels.
id area sex1 age1 height1 weight1 fvc1 fev11
sex2 age2 height2 weight2 fvc2 fev12
sex3 age3 height3 weight3 fvc3 fev13
sex4 age4 height4 weight4 fvc4 fev14
sex5 age5 height5 weight5 fvc5 fev15
}
(151 observations read)
. reshape groups 1--5
. reshape vars sex age height weight fvc fev1
. reshape cons id area
. reshape long
. reshape query
```

Four options must be specified when using the reshape command. In the above example, the group options tells Stata that the variables of the five family members can be distinguished by the suffix 1–5 in the original data set and that these suffixes are to be saved as a variable called "member" in the new data set. The vars option tells which of these variables are repeated for each family member and thus need to be unpacked. The cons (constant) option specifies which of the variables are constant within the family and are to be output to the new data set. To perform the conversion, either the long (for unpacking) or the wide (for packing) option must be stated. Finally, the query option can be used to confirm the results.

To obtain the means and standard deviations in Stata, we used the statement

```
. by member: summarize
```

This gave the results for each of the five family members separately. Similarly,

```
. by member: correlate age height weight fev1
```

produced the correlation matrix for each family member separately. The statement

```
. regress fev1 age height if member < 3
```

produced the regression output for the fathers (members = 1) and mothers (members = 2) combined ($N = 300$). To obtain regressions for the fathers and mothers separately, we used the statement:

```
. by member: regress fev1 age height if member < 3
```

The regression options for the six statistical packages are summarized in Table 7.6. The lm or linear model option is chosen in S-PLUS. Reg is the main regression program for SAS. We will emphasize the regression output from that program. Concise instructions are typed to tell the program what regression analysis you wish to perform. Similarly, REGRESSION is the general regression program for SPSS. For Stata, the regress program is used but the user also has to use other programs to get the desired output. The index of Stata is helpful in finding the various options you want. The instructions that need to be typed in Stata are very brief. In STATISTICA, the user chooses Multiple Regression.

The most common output that is available in all the regression programs has been omitted from Table 7.6. This includes the slope and intercept coefficients, the standard error of A and B, the test that α and β are zero, the ANOVA table, the multiple correlation, and the test that it is zero. These features are in all the programs and are in most other multiple regression programs.

It should be noted that all the programs have a wide range of options for multiple linear regression. These same programs include other output that will be discussed in Chapters 8 and 9.

Note also that in regression analysis you are fitting a model to the data without being 100% certain what the "true" model is. We recommend that you carefully review your objectives before performing the regression analysis to determine just what is needed. A second run can always be done to obtain additional output. Once you find the output that answers your objectives, we recommend writing out the result in English or highlighting the printed output while this objective is still fresh in your mind.

Most investigators aim for the simplest model or regression equation that will provide a good prediction of the dependent variable. In Chapter 8, we will discuss the choice of the X variables for the case when you are not sure which X variables to include in your model.

7.11 What to watch out for

The discussion and cautionary remarks given in Section 6.13 also apply to multiple regression analysis. In addition, because it is not possible to display all the results in a two-dimensional scatter diagram in the case of multiple regression, checks for some of the assumptions (such as independent normally distributed error terms) are somewhat more difficult to perform. But since regression programs are so widely used, they contain a large number of options to assist in checking the data. At a minimum, outliers and normality should routinely be checked. At least one measure of the three types (outliers in Y, outliers in X,

Table 7.6: Software commands and output for multiple linear regression

	S-PLUS/R	SAS	SPSS	Stata	STATISTICA
Matrix output					
Covariance matrix of Xs	var	CORR	REGRESSION	correlate	Multiple Regression
Correlation matrix of Xs	cor	REG	REGRESSION	correlate	Multiple Regression
Correlation matrix of Bs	lm	REG	REGRESSION	estat vce	Multiple Regression
Regression equation output and					
Standardized coefficients		REG	REGRESSION	regress	Multiple Regression
Partial correlations		REG	REGRESSION	pcorr	Multiple Regression
Options					
Weighted regression	lm	REG	REGRESSION	regress	Multiple Regression
Regression through zero	lm	REG	REGRESSION	regress	Multiple Regression
Regression for subgroups	lm	REG	REGRESSION	if, by	Multiple Regression
Outliers in Y					
Raw residuals	lm	REG	REGRESSION	predict	Multiple Regression
Deleted studentized residuals		REG	REGRESSION	predict	Multiple Regression
Other types of residuals	ls.diag	REG	REGRESSION	predict	Multiple Regression
Outliers in X					
Leverage or h	lm.influence	REG	REGRESSION	predict	General Linear Models
Influence measures					
Cooks distance D	ls.diag	REG	REGRESSION	predict	Multiple Regression
DFFITS	ls.diag	REG	REGRESSION	predict	General Linear Models
Other influence measures		REG	REGRESSION	predict	
Checks for other assumptions					
Tolerance	lm.fit	REG	REGRESSION	estat vif	Multiple Regression
Variance inflation factor		REG	REGRESSION	estat vif	Multiple Regression
Serial correlation of residuals		REG		estat dwatson	Multiple Regression
Durban–Watson statistic	durbinWatson (S-PLUS only)	REG	REGRESSION	estat dwatson	Multiple Regression
Test homogeneity of variance	bartlett.test (R only)	REG	ONEWAY	oneway	ANOVA/MANOVA
Loess	loess	LOESS	LOWESS	lowess	lowess

and influential points) should be obtained along with a normal probability plot of the residuals against $\hat{Y}$ and against each X.

Often two investigators collecting similar data at two locations obtain slope coefficients that are not similar in magnitude in their multiple regression equations. One explanation of this inconsistency is that if the X variables are not independent, the magnitudes of the coefficients depend on which other variables are included in the regression equation. If the two investigators do not have the same variables in their equations then they should not expect the slope coefficients for a specific variable to be the same. This is one reason why they are called **partial regression coefficients**. For example, suppose Y can be predicted very well from X_1 and X_2 but investigator A ignores X_2. The expected value of the slope coefficient B_1 for investigator A will be different from β_1, the expected value for the second investigator. Hence, they will tend to get different results. The difference will be small if the correlation between the two X variables is small, but if they are highly correlated even the sign of the slope coefficient can change. In order to get results that can be easily interpreted, it is important to include the proper variables in the regression model. In addition, the variables should be included in the model in the proper form such that the linearity assumption holds.

Some investigators discard all the variables that are not significant at a certain P value, say $P < 0.05$, but this can sometimes lead to problems in interpreting other slope coefficients, as they can be altered by the omission of that variable. More on variable selection will be given in Chapter 8.

Another problem that can arise in multiple regression is multicollinearity, mentioned in Section 7.9. This is more apt to happen in economic data than in data on individuals, due to the high correlation among the variables. The programs will warn the user if this occurs. The problem is sometimes solved by discarding a variable or, if this does not make sense because of the problems discussed in the above paragraph, special techniques exist to obtain estimates of the slope coefficients when there is multicollinearity (Chapter 9).

Another practical problem in performing multiple regression analyses is shrinking sample size. If many X variables are used and each has some missing values, then many cases may be excluded because of missing values. The programs themselves can contribute to this problem. Suppose you specify a series of regression programs with slightly different X variables. To save computer time the SAS REG procedure only computes the covariance or correlation matrix once using all the variables. Hence, any variable that has a missing value for a case causes that case to be excluded from **all** of the regression analyses, not just from the analyses that include that variable.

When a large portion of the sample is missing, then the problem of making inferences regarding the parent population can be difficult. In any case, the sample size should be "large" relative to the number of variables in order to obtain stable results. Some statisticians believe that if the sample size is not at least five to ten times the number of variables, then interpretation of the results is risky.

7.12 Summary

In this chapter we presented a subject in which the use of the computer is almost a must. The concepts underlying multiple regression were presented along with a fairly detailed discussion of packaged regression programs. Although much of the material in this chapter is found in standard textbooks, we have emphasized an understanding and presentation of the output of packaged programs. This philosophy is one we will follow in the rest of the book. In fact, in future chapters the emphasis on interpretation of output will be even more pronounced.

Certain topics included in this chapter but not covered in usual courses included how to handle interactions among the X variables and how to compare regression equations from different subgroups. We also included discussion of the examination of residuals and influential observations.

Several texts contain excellent presentations of the subject of regression analysis. We have referenced a wide selection of texts here and in Chapter 6.

7.13 Problems

7.1 Using the chemical companies' data in Table 8.1, predict the price earnings (P/E) ratio from the debt to equity (D/E) ratio, the annual dividends divided by the 12-months' earnings per share (PAYOUTR1), and the percentage net profit margin (NPM1). Obtain the correlation matrix, and check the correlations between the variables. Summarize the results, including appropriate tests of hypotheses.

7.2 Fit the regression plane for the fathers using FFVC as the dependent variable and age and height as the independent variables.

7.3 Write the results for Problem 7.2 so they would be suitable for inclusion in a report. Include table(s) that present the results the reader should see.

7.4 Fit the regression plane for mothers with MFVC as the dependent variable and age and height as the independent variables. Summarize the results in a tabular form. Test whether the regression results for mothers and fathers are significantly different.

7.5 From the depression data set described in Table 3.4, predict the reported level of depression as given by CESD, using INCOME, SEX, and AGE as independent variables. Analyze the residuals and decide whether or not it is reasonable to assume that they follow a normal distribution.

7.6 Search for a suitable transformation for CESD if the normality assumption in Problem 7.5 cannot be made. State why you are not able to find an ideal transformation if that is the case.

7.7 Using a statistical package of your choice, create a hypothetical data set which you will use for exercises in this chapter and some of the following chapters. Begin by generating 100 independent cases for each of ten variables using the standard normal distribution (means $= 0$ and variances $= 1$). Call the new variables $X1, X2, \ldots, X9, Y$. Use the following seeds in generating these 100 cases.

```
X1 seed 36541
X2 seed 43893
X3 seed 45671
X4 seed 65431
X5 seed 98753
X6 seed 78965
X7 seed 67893
X8 seed 34521
X9 seed 98431
Y   seed 67895
```

You now have ten independent, random, normal numbers for each of 100 cases. The population mean is 0 and the population standard deviation is 1. Further transformations are done to make the variables intercorrelated. The transformations are accomplished by making some of the variables functions of other variables, as follows:

```
X1 = 5*X1
X2 = 3*X2
X3 = X1 + X2 + 4*X3
X4 = X4
X5 = 4*X5
X6 = X5 - X4 + 6*X6
X7 = 2*X7
X8 = X7 + 2*X8
X9 = 4*X9
Y  = 5 + X1 + 2*X2 + X3 + 10*Y
```

We now have created a random sample of 100 cases on 10 variables: $X1, X2, X3, X4, X5, X6, X7, X8, X9, Y$. The population distribution is multivariate normal. It can be shown that the population means and variances are as follows:

	$X1$	$X2$	$X3$	$X4$	$X5$	$X6$	$X7$	$X8$	$X9$	Y
Mean	0	0	0	0	0	0	0	0	0	5
Var.	25	9	50	1	16	53	4	8	16	297

The population correlation matrix is as follows:

	$X1$	$X2$	$X3$	$X4$	$X5$	$X6$	$X7$	$X8$	$X9$	Y
$X1$	1	0	0.71	0	0	0	0	0	0	0.58
$X2$	0	1	0.42	0	0	0	0	0	0	0.52
$X3$	0.71	0.42	1	0	0	0	0	0	0	0.76
$X4$	0	0	0	1	0	-0.14	0	0	0	0
$X5$	0	0	0	0	1	0.55	0	0	0	0
$X6$	0	0	0	-0.14	0.55	1	0	0	0	0
$X7$	0	0	0	0	0	0	1	0.71	0	0
$X8$	0	0	0	0	0	0	0.71	1	0	0
$X9$	0	0	0	0	0	0	0	0	1	0
Y	0.58	0.52	0.76	0	0	0	0	0	0	1

The population squared multiple correlation coefficient between Y and $X1$ and $X9$ is 0.34, between Y and $X1, X2, X3$ is 0.66, and between Y and $X4$ and $X9$ is zero. Also, the population regression line of Y on X1 to X9 has $\alpha = 5, \beta_1 = 1, \beta_2 = 2, \beta_3 = 1, \beta_4 = \beta_5 = \cdots = \beta_9 = 0$.

Now, using the data you have generated, obtain the sample statistics from a computer packaged program, and compare them with the parameters given above. Using the Bonferroni inequality, test the simple correlations, and determine which are significantly different from zero. Comment.

7.8 Repeat Problem 7.7 using another statistical package and see if you get the same sample.

7.9 (Continuation of Problem 7.7.) Calculate the population partial correlation coefficient between $X2$ and $X3$ after removing the linear effect of $X1$. Is it larger or smaller than ρ_{23}? Explain. Also, obtain the corresponding sample partial correlation. Test whether it is equal to zero.

7.10 (Continuation of Problem 7.7.) Using a multiple regression program, perform an analysis, with the dependent variable $= Y$ and the independent variables $= X1$ to $X9$, on the 100 generated cases. Summarize the results and state whether they came out the way you expected them to, considering how the data were generated. Perform appropriate tests of hypotheses. Comment.

7.11 (Continuation of Problem 7.5.) Fit a regression plane for CESD on IN-COME and AGE for males and females combined. Test whether the regression plane is helpful in predicting the values of CESD. Find a 95% prediction interval for a female with INCOME $= 17$ and AGE $= 29$ using this regression. Do the same using the regression calculated in Problem 7.5, and compare.

7.12 (Continuation of Problem 7.11.) For the regression of CESD on INCOME and AGE, choose 15 observations that appear to be influential or outlying.

State your criteria, delete these points, and repeat the regression. Summarize the differences in the results of the regression analyses.

7.13 For the lung function data described in Appendix A, find the regression of FEV1 on weight and height for the fathers. Divide each of the two explanatory variables into two intervals: greater than, and less than or equal to the respective median. Is there an interaction between the two explanatory variables?

7.14 (Continuation of Problem 7.13.) Find the partial correlation of FEV1 and age given height for the oldest child, and compare it to the simple correlation between FEV1 and age of the oldest child. Is either one significantly different from zero? Based on these results and without doing any further calculations, comment on the relationship between OCHEIGHT and OCAGE in these data.

7.15 (Continuation of Problem 7.13.) (a) For the oldest child, find the regression of FEV1 on (i) weight and age; (ii) height and age; (iii) height, weight, and age. Compare the three regression equations. In each regression, which coefficients are significantly different from zero? (b) Find the correlation matrix for the four variables. Which pair of variables is most highly correlated? least correlated? Heuristically, how might this explain the results of part (a)?

7.16 Repeat Problem 7.15(a) for fathers' measurements instead of those of the oldest children. Are the regression coefficients more stable? Why?

7.17 For the Parental HIV data generate a variable that represents the sum of the variables describing the neighborhood where the adolescent lives (NGHB1–NGHB11). Is the age at which adolescents start smoking different for girls compared to boys, after adjusting for the score describing the neighborhood?

Chapter 8

Variable selection in regression

8.1 Chapter outline

The variable selection techniques provided in statistical programs can assist in making the choice among numerous independent or predictor variables. Section 8.2 discusses when these techniques are used and Section 8.3 introduces an example chosen because it illustrates clearly the differences among the techniques. Section 8.4 presents an explanation of the criteria used in deciding how many and which variables to include in the regression model. Section 8.5 describes the general F test, a widely used test in regression analysis that provides a criterion for variable selection. Most variable selection techniques consist of successive steps. Several methods of selecting the variable that is "best" at each step of the process are given in Section 8.6. Section 8.7 describes the options available to find the "best" variables using subset regression. Section 8.8 presents a discussion of obtaining the output given in this chapter by SAS and summarizes what output the six statistical packages provide. Data screening to ensure that the data follow the regression model is discussed in Section 8.9. Partial regression plots are also presented there. Since investigators often overestimate what data selection methods will do for them, Section 8.10 describes what to watch out for in performing these variable selection techniques. Section 8.11 contains the summary.

8.2 When are variable selection methods used?

In Chapter 7 it was assumed that the investigators knew which variables they wished to include in the model. This is usually the case in the fixed-X model, where often only two to five variables are being considered. It also is the case frequently when the investigator is working from a theoretical model and is choosing the variables that fit this model. But sometimes the investigator has only partial knowledge or is interested in finding out what variables can be used to best predict a given dependent variable.

Variable selection methods are used mainly in exploratory situations where many independent variables have been measured and a final model explaining the dependent variable has not been reached. Variable selection methods are useful, for example, in a survey situation in which numerous characteristics and responses of each individual are recorded in order to determine some characteristic of the respondent, such as level of depression. The investigator may have prior justification for using certain variables but may be open to suggestions for the remaining variables. For example, age and gender have been shown to relate to levels of the CESD (the depression variable given in the depression study described in Chapter 3). An investigator might wish to enter age and gender as independent variables and try additional variables in an exploratory fashion. The set of independent variables can be broken down into logical subsets. First, the usual demographic variables that will be entered first, namely, age and gender. Second, a set of variables that other investigators have shown to affect CESD, such as perceived health status. Finally, the investigator may wish to explore the relative value of including another set of variables after the effects of age, gender, and perceived health status have been taken into account. In this case, it is partly a model-driven regression analysis and partly an exploratory regression analysis. The programs described in this chapter allow analysis that is either partially or completely exploratory. The researcher may have one of two goals in mind:

1. To use the resulting regression equation to identify variables that best explain the level of the dependent variable — a **descriptive** or **explanatory purpose**;

2. To obtain an equation that predicts the level of the dependent variable with as little error as possible — a **predictive purpose**.

The variable selection methods discussed in this chapter can sometimes serve one purpose better than the other, as will be discussed after the methods are presented.

8.3   Data example

Table 8.1 presents various characteristics reported by the 30 largest chemical companies; the data are taken from a January 1981 issue of *Forbes*. This data set will henceforth be called the chemical companies data. It is included in the book's web site along with a codebook.

The variables listed in Table 8.1 are defined as follows (Brigham and Houston, 2009):

- **P/E:** price-to-earnings ratio, which is the price of one share of common stock divided by the earnings per share for the past year. This ratio shows the dollar amount investors are willing to pay for the stock per dollar of current earnings of the company.

- **ROR5:** percent rate of return on total capital (invested plus debt) averaged over the past five years.

Table 8.1: Chemical companies' financial performance: Source: Forbes, vol. 127, no. 1 (January 5, 1981)

Company	P/E	ROR5(%)	D/E	SALESGR5(%)	EPS5(%)	NPM1(%)	PAYOUTR1
Diamond Shamrock	9	13.0	0.7	20.2	15.5	7.2	0.43
Dow Chemical	8	13.0	0.7	17.2	12.7	7.3	0.38
Stauffer Chemical	8	13.0	0.4	14.5	15.1	7.9	0.41
E.I. du Pont	9	12.2	0.2	12.9	11.1	5.4	0.57
Union Carbide	5	10.0	0.4	13.6	0.8	6.7	0.32
Pennwalt	6	9.8	0.4	12.1	14.5	3.8	0.51
W.R. Grace	10	9.9	0.5	10.2	7.0	4.8	0.38
Hercules	9	10.3	0.3	11.4	8.7	4.5	0.48
Monsanto	11	9.5	0.4	13.5	5.9	3.5	0.57
American Cyanamid	9	9.9	0.4	12.1	4.2	4.6	0.49
Celanese	7	7.9	0.4	10.8	16.0	3.4	0.49
Allied Chemical	7	7.3	0.6	15.4	4.9	5.1	0.27
Rohm & Haas	7	7.8	0.4	11.0	3.0	5.6	0.32
Reichhold Chemicals	10	6.5	0.4	18.7	−3.1	1.3	0.38
Lubrizol	13	24.9	0.0	16.2	16.9	12.5	0.32
Nalco Chemical	14	24.6	0.0	16.1	16.9	11.2	0.47
Sun Chemical	5	14.9	1.1	13.7	48.9	5.8	0.10
Cabot	6	13.8	0.6	20.9	36.0	10.9	0.16
Interm. Minerals & Chemical	10	13.5	0.5	14.3	16.0	8.4	0.40
Dexter	12	14.9	0.3	29.1	22.8	4.9	0.36
Freeport Minerals	14	15.4	0.3	15.2	15.1	21.9	0.23
Air Products & Chemicals	13	11.6	0.4	18.7	22.1	8.1	0.20
Mallinckrodt	12	14.2	0.2	16.7	18.7	8.2	0.37
Thiokol	12	13.8	0.1	12.6	18.0	5.6	0.34
Witco Chemical	7	12.0	0.5	15.0	14.9	3.6	0.36
Ethyl	7	11.0	0.3	12.8	10.8	5.0	0.34
Ferro	6	13.8	0.2	14.9	9.6	4.4	0.31
Liquid Air of North America	12	11.5	0.4	15.4	11.7	7.2	0.51
Williams Companies	9	6.4	0.7	16.1	−2.8	6.8	0.22
Akzona	14	3.8	0.6	6.8	−11.1	0.9	1.00
Mean	9.37	12.01	0.42	14.91	12.93	6.55	0.39
Standard deviation	2.80	4.50	0.23	4.05	11.15	3.92	0.16

- **D/E:** debt-to-equity (invested capital) ratio for the past year. This ratio indicates the extent to which management is using borrowed funds to operate the company.

- **SALESGR5:** percent annual compound growth rate of sales, computed from the most recent five years compared with the previous five years.

- **EPS5:** percent annual compound growth in earnings per share, computed from the most recent five years compared with the preceding five years.

- **NPM1:** percent net profit margin, which is the net profits divided by the net sales for the past year, expressed as a percentage.

- **PAYOUTR1:** annual dividend divided by the latest 12-month earnings per share. This value represents the proportion of earnings paid out to shareholders rather than retained to operate and expand the company.

The P/E ratio is usually high for growth stocks and low for mature or troubled firms. Company managers generally want high P/E ratios, because high ratios make it possible to raise substantial amounts of capital for a small number of shares and make acquisitions of other companies easier. Also, investors consider the P/E ratios of companies, both over time and in relation to similar companies, as a factor in valuation of stocks for possible purchase and/or sale. Therefore it is of interest to investigate which of the other variables reported in Table 8.1 influence the level of the P/E ratio. In this chapter we use the regression of the P/E ratio on the remaining variables to illustrate various variable selection procedures. Note that all the firms of the data set are from the chemical industry; such a regression equation could vary appreciably from one industry to another.

To get a preliminary impression of the data, examine the means and standard deviations shown at the bottom of Table 8.1. Also examine Table 8.2, which shows the simple correlation matrix. Note that P/E, the dependent variable, is most highly correlated with D/E. The association, however, is negative. Thus investors tend to pay less per dollar earned for companies with relatively heavy indebtedness. The sample correlations of P/E with ROR5, NPM1, and PAYOUTR1 range from 0.32 to 0.35 and thus are quite similar. Investors tend to pay more per dollar earned for stocks of companies with higher rates of return on total capital, higher net profit margins, and higher proportion of earnings paid out to them as dividends. Similar interpretations can be made for the other correlations.

If a single independent variable were to be selected for "explaining" P/E, the variable of choice would be D/E since it is the most highly correlated. But clearly there are other variables that represent differences in management style, dynamics of growth, and efficiency of operation that should also be considered.

It is possible to derive a regression equation using all the independent variables, as discussed in Chapter 7. However, some of the independent variables are strongly interrelated. For example, from Table 8.2 we see that growth

Table 8.2: *Correlation matrix of chemical companies data*

	P/E	ROR5	D/E	SALES-GR5	EPS5	NPM1	PAY-OUTR1
P/E	1.00						
ROR5	0.32	1.00					
D/E	−0.47	−0.46	1.00				
SALESGR5	0.13	0.36	−0.02	1.00			
EPS5	−0.20	0.56	0.19	0.39	1.00		
NPM1	0.35	0.62	−0.22	0.25	0.15	1.00	
PAYOUTR1	0.33	−0.30	−0.16	−0.45	−0.57	−0.43	1.00

of earnings (EPS5) and growth of sales (SALESGR5) are positively correlated, suggesting that both variables measure related aspects of dynamically
growing companies. Further, growth of earnings (EPS5) and growth of sales
(SALESGR5) are positively correlated with return on total capital (ROR5),
suggesting that dynamically growing companies show higher returns than mature, stable companies. In contrast, growth of earnings (EPS5) and growth of
sales (SALESGR5) are both negatively correlated with the proportion of earnings paid out to stockholders (PAYOUTR1), suggesting that earnings must be
plowed back into operations to achieve growth. The profit margin (NPM1)
shows the highest positive correlation with rate of return (ROR5), suggesting
that efficient conversion of sales into earnings is consistent with high return on
total capital employed.

Since the independent variables are interrelated, it may be better to use
a subset of the independent variables to derive the regression equation. Most
investigators prefer an equation with a small number of variables since such
an equation will be easier to interpret. For future predictive proposes it is often
possible to do at least as well with a subset as with the total set of independent
variables. Methods and criteria for subset selection are given in the subsequent
sections. It should be noted that these methods are fairly sensitive to gross
outliers. The data sets should therefore be carefully edited prior to analysis.

8.4 Criteria for variable selection

In many situations where regression analysis is useful, the investigator has
strong justification for including certain variables in the equation. The justification may be to produce results comparable to previous studies or to conform
to accepted theory. But often the investigator has no preconceived assessment
of the importance of some or all of the independent variables. It is in the latter
situation that variable selection procedures can be useful.

Any variable selection procedure requires a criterion for deciding how many and which variables to select. As discussed in Chapter 7, the least squares method of estimation minimizes the **residual sum of squares** (RSS) about the regression plane [RSS = $\sum(Y - \hat{Y})^2)$]. Therefore an implicit criterion is the value of RSS. In deciding between alternative subsets of variables, the investigator would select the one producing the smaller RSS if this criterion were used in a mechanical fashion. Note, however, that

$$\text{RSS} = \sum(Y - \bar{Y})^2(1 - R^2)$$

where R is the multiple correlation coefficient. Therefore minimizing RSS is equivalent to maximizing the multiple correlation coefficient. If the criterion of maximizing R were used, the investigator would always select all of the independent variables, because the value of R will never decrease by including additional variables.

Since the multiple correlation coefficient, on the average, overestimates the population correlation coefficient, the investigator may be misled into including too many variables. For example, if the population multiple correlation coefficient is, in fact, equal to zero, the average of all possible values of R^2 from samples of size N from a multivariate normal population is $P/(N-1)$, where P is the number of independent variables (Wishart, 1931). An estimated multiple correlation coefficient that reduces the bias is the **adjusted multiple correlation coefficient**, denoted by $\bar{R}$. It is related to R by the following equation:

$$\bar{R}^2 = R^2 - \frac{P(1 - R^2)}{N - P - 1}$$

where P is the number of independent variables **in the equation**. Note that in this chapter the notation P will sometimes signify less than the total number of available independent variables. Note also that in many texts and computer manuals, the formulas are written with a value, say P', that signifies the number of parameters being estimated. In the usual case where an intercept is included in the regression model, there are $P + 1$ parameters being estimated, one for each of the P variables plus the intercept; thus $P' = P + 1$. This is the reason the formulas in this and other texts may not be precisely the same.

The investigator may proceed now to select the independent variables that maximize $\bar{R}^2$. Note that excluding some independent variables may, in fact, result in a higher value of $\bar{R}^2$. As will be seen, this result occurs for the chemical companies' data. Note also that if N is very large relative to P, then $\bar{R}^2$ will be approximately equal to R^2. Conversely, maximizing $\bar{R}^2$ can give different results from those obtained in maximizing R^2 when N is small.

Another method suggested by statisticians is to minimize the **residual mean square**, which is defined as

$$\text{RMS or RES. MS.} = \frac{\text{RSS}}{N - P - 1}$$

This quantity is related to the adjusted $\bar{R}^2$ as follows:

$$\bar{R}^2 = 1 - \frac{(\text{RMS})}{S_Y^2}$$

where

$$S_Y^2 = \frac{\sum(Y - \bar{Y})^2}{N - 1}$$

Since S_Y^2 does not involve any independent variables, minimizing RMS is equivalent to maximizing $\bar{R}^2$.

Another quantity used in variable selection and found in standard computer packages is the so-called C_p **criterion**. The theoretical derivation of this quantity is beyond the scope of this book (Mallows, 1973). However, C_p can be expressed as follows:

$$C_p = (N - P - 1)\left(\frac{\text{RMS}}{\hat{\sigma}^2} - 1\right) + (P + 1)$$

where RMS is the residual mean square based on the P selected variables and $\hat{\sigma}^2$ is the RMS derived from the total set of independent variables. The quantity C_p is the sum of two components, $P + 1$, and the remainder of the expression. While $P + 1$ increases as we choose more independent variables, the other part of C_p will tend to decrease. When all variables are chosen, $P + 1$ is at its maximum but the other part of C_p is zero since RMS$= \hat{\sigma}^2$. Many investigators recommend selecting those independent variables that minimize the value of C_p.

Another recommended criterion for variable selection is **Akaike's information criterion** (AIC). This criterion was developed to find an optimal and parsimonious model (not entering unnecessary variables). It has been shown to be equivalent to C_p for large sample sizes (asymptotically equivalent under some conditions); see Nishii (1984). This means that for very large samples, both criteria will result in selecting the same variables. However, they may lead to different results in small samples. For a general development of AIC see Bozdogan (1987). The formula for estimating AIC given by Nishii (1984) and used by SAS is

$$\text{AIC} = N \ln \frac{\text{RSS}}{N} + 2(P + 1)$$

where RSS is the residual sum of squares. As more variables are added the fraction part of the formula gets smaller but $P + 1$ gets larger. A minimum value of AIC is desired. Another formula used by some programs such as S-PLUS is given by Atkinson (1980):

$$\text{AIC}^* = \frac{\text{RSS}}{\hat{\sigma}^2} + 2(P + 1)$$

where $\hat{\sigma}^2$ is an estimate of σ^2.

Often the above criteria will lead to similar results, but not always. Other criteria have been proposed that tend to allow either fewer or more variables in the equation. It is not recommended that any of these criteria be used mechanically under all circumstances, particularly when the sample size is small. Practical methods for using these criteria will be discussed in Sections 8.6 and 8.7. Hocking (1976) compared these criteria (except for AIC) and others, and recommended the following uses:

1. In a given sample the value of the unadjusted R^2 can be used as a measure of data description. Many investigators exclude variables that add only a very small amount to the R^2 value obtained from the remaining variables.

2. If the object of the regression is extrapolation or estimation of the regression parameters, the investigator is advised to select those variables that maximize the adjusted R^2 (or, equivalently, minimize RMS). This criterion may result in too many variables if the purpose is prediction (Stuart and Ord, 1994.)

3. For prediction, finding the variables that make the value of C_p approximately equal to $P+1$ or minimizing AIC is a reasonable strategy. (For extensions to AIC that have been shown to work well in simulation studies, see Bozdogan, 1987.)

The above discussion was concerned with criteria for judging among alternative subsets of independent variables. The numerical values of some of these criteria are used to decide which variables enter the model in the statistical packages. In other cases, the value of the criterion is simply printed out but is not used to choose which variables enter directly by the program. Note that investigators can use the printed values to form their own choice of independent variables. How to select the candidate subsets is our next concern. Before describing methods to perform the selection, though, we first discuss the use of the F test for determining the effectiveness of a subset of independent variables relative to the total set.

8.5   A general F test

Suppose we are convinced that the variables $X_1, X_2, \ldots, X_p$ should be used in the regression equation. Suppose also that measurements on Q additional variables $X_{P+1}, X_{P+2}, \ldots, X_{P+Q}$ are available. Before deciding whether any of the additional variables should be included, we can test the hypothesis that, as a group, the Q variables do not improve the regression equation.

If the regression equation in the population has the form

$$Y = \alpha + \beta_1 X_1 + \beta_2 X_2 + \cdots + \beta_P X_P + \beta_{P+1} X_{P+1}$$
$$+ \cdots + \beta_{P+Q} X_{P+Q} + e$$

we test the hypothesis H_0: $\beta_{P+1} = \beta_{P+2} = \cdots = \beta_{P+Q} = 0$. To perform the test, we first obtain an equation that includes all the $P + Q$ variables, and we obtain the residual sum of squares (RSS_{P+Q}). Similarly, we obtain an equation that includes only the first P variables and the corresponding residual sum of squares (RSS_p). Then the test statistic is computed as

$$F = \frac{(\mathrm{RSS}_P - \mathrm{RSS}_{P+Q})/Q}{\mathrm{RSS}_{P+Q}/(N - P - Q - 1)}$$

The numerator measures the improvement in the equation from using the additional Q variables. This quantity is never negative. The hypothesis is rejected if the computed F exceeds the tabled $F(1 - \alpha)$ with Q and $N - P - Q - 1$ degrees of freedom.

This very general test is sometimes referred to as the **generalized linear hypothesis test**. Essentially, this same test was used in Section 7.9 to test whether or not it is necessary to report the regression analyses by subgroups. The quantities P and Q can take on any integer values greater than or equal to one. For example, suppose that six variables are available. If we take P equal to 5 and Q equal to 1, then we are testing H_0: $\beta_6 = 0$ in the equation $Y = \alpha + \beta_1 X_1 + \beta_2 X_2 + \beta_3 X_3 + \beta_4 X_4 + \beta_5 X_5 + \beta_6 X_6 + e$. This test is the same as the test that was discussed in Section 7.6 for the significance of a single regression coefficient.

As another example, in the chemical companies' data it was already observed that D/E is the best single predictor of the P/E ratio. A relevant hypothesis is whether the remaining five variables improve the prediction obtained by D/E alone. Two regressions were run (one with all six variables and one with just D/E), and the results were as follows:

$$\mathrm{RSS}_6 = 103.06$$
$$\mathrm{RSS}_1 = 176.08$$
$$P = 1, Q = 5$$

Therefore the test statistic is

$$F = \frac{(176.08 - 103.06)/5}{103.06/(30 - 1 - 5 - 1)} = \frac{14.60}{4.48} = 3.26$$

with 5 and 23 degrees of freedom. Comparing this value with the value found in standard statistical tables, we find the P value to be less than 0.025. Thus at the 5% significance level we conclude that one or more of the other five variables significantly improves the prediction of the P/E ratio. Note, however, that the D/E ratio is the most highly correlated variable with the P/E ratio. This selection process affects the inference so that the true P value of the test is unknown, but it is perhaps greater than 0.025. Strictly speaking, the general linear hypothesis test is valid only when the hypothesis is determined prior to examining the data.

The generalized linear hypothesis test is the basis for several selection procedures, as will be discussed in the next two sections.

8.6   Stepwise regression

The variable selection problem can be described as considering certain subsets of independent variables and selecting that subset that either maximizes or minimizes an appropriate criterion. Two obvious subsets are the **best single variable** and the **complete set of independent variables**, as considered in Section 8.5. The problem lies in selecting an **intermediate subset** that may be better than both these extremes. We will want to use a method that both tells us how many predictor or independent variables to use and which ones to use. We do not want to use too many independent variables since they may add little to our ability to predict the outcome or dependent variable. Also, when a large number of independent variables are used with a different sample, they may actually yield a worse prediction of the dependent variable than a regression equation with fewer independent variables. We want to choose a set of independent variables that both will yield a good prediction and also be as few in number as possible. In this section we discuss some methods for making this choice.

Forward selection method

Selecting the best single variable is a simple matter: we choose the variable with the highest absolute value of the simple correlation with Y. In the chemical companies' data example D/E is the best single predictor of P/E because the correlation between D/E and P/E is -0.47, which has a larger magnitude than any other correlation with P/E (Table 8.2). Note that D/E also has the largest standardized coefficient and the largest t value if we test that each regression coefficient is equal to zero.

To choose a second variable to combine with D/E, we could naively select NPM1 since it has the second highest absolute correlation with P/E (0.35). This choice may not be wise, however, since another variable together with D/E may give a higher multiple correlation than D/E combined with NPM1. One strategy therefore is to search for that variable that maximizes multiple R^2 when combined with D/E. Note that this procedure is equivalent to choosing the second variable to minimize the residual sum of squares given that D/E is kept in the equation. This procedure is also equivalent to choosing the second variable to maximize the magnitude of the partial correlation with P/E after removing the linear effect of D/E. The variable thus selected will also maximize the F statistic for testing that the above-mentioned partial correlation is zero (Section 7.7). This F test is a special application of the general linear hypothesis test described in Section 8.5. In other words, we will choose a second variable that takes into account the effects of the first variable.

In the chemical companies' data example the partial correlations between P/E and each of the other variables after removing the linear effect of D/E are as follows:

	Partial correlations
ROR5	0.126
SALESGR5	0.138
EPS5	−0.114
NPM1	0.285
PAYOUTR1	0.286

Therefore, the method described above, called the **forward selection method**, would choose D/E as the first variable and PAYOUTR1 as the second variable. Note that it is almost a toss-up between PAYOUTR1 and NPM1 as the second variable. The computer programs implementing this method will ignore such reasoning and select PAYOUTR1 as the second variable because 0.286 is greater than 0.285. But to the investigator the choice may be more difficult since NPM1 and PAYOUTR1 measure quite different aspects of the companies.

Similarly, the third variable chosen by the forward selection procedure is the variable with the highest absolute partial correlation with P/E after removing the linear effects of D/E and PAYOUTR1. Again, this procedure is equivalent to maximizing multiple R and F and minimizing the residual mean square, given that D/E and PAYOUTR1 are kept in the equation.

The forward selection method proceeds in this manner, each time adding one variable to the variables previously selected, until a specified **stopping rule** is satisfied. The most commonly used stopping rule in packaged programs is based on the F test of the hypothesis that the partial correlation of the variable entered is equal to zero. One version of the stopping rule terminates entering variables when the computed value of F is less than a specified value. This cutoff value is often called the **minimum F-to-enter**.

Equivalently, the P value corresponding to the computed F statistic could be calculated and the forward selection stopped when this P value is greater than a specified level. Note that here also the P value is affected by the fact that the variables are selected from the data and therefore should not be used in the hypothesis-testing context.

Bendel and Afifi (1977) compared various levels of the minimum F-to-enter used in forward selection. A recommended value is the F percentile corresponding to a P value equal to 0.15. For example, if the sample size is large, the recommended minimum F-to-enter is the 85th percentile of the F distribution with 1 and ∞ degrees of freedom, or 2.07.

Any of the criteria discussed in Section 8.4 could also be used as the basis for a stopping rule. For instance, the multiple R^2 is always presented in the output of the commonly used programs whenever an additional variable is selected. The user can examine this series of values of R^2 and stop the process when the increase in R^2 is a very small amount. Alternatively, the series of adjusted $\bar{R}^2$ values can be examined. The process stops when the adjusted $\bar{R}^2$ is maximized.

As an example, data from the chemical companies analysis are given in Table 8.3. Using the P value of 0.15 (or the minimum F of 2.07) as a cutoff, we would enter only the first four variables since the computed F-to-enter for the fifth variable is only 0.38. In examining the multiple R^2, we note that going from five to six variables increased R^2 only by 0.001. It is therefore obvious that the sixth variable should not be entered. However, other readers may disagree about the practical value of a 0.007 increase in R^2 obtained by including the fifth variable. We would be inclined not to include it. Finally, in examination of the adjusted multiple $\bar{R}^2$ we note that it is also maximized by including only the first four variables.

This selection of variables by the forward selection method for the chemical companies' data agrees well with our understanding gained from study of the correlations between variables. As we noted in Section 8.3, return on total capital (ROR5) is correlated with NPM1, SALESGR5, and EPS5 and so adds little to the regression when they are already included. Similarly, the correlation between SALESGR5 and EPS5, both measuring aspects of company growth, corroborates the result shown in Table 8.3 that EPS5 adds little to the regression after SALESGR5 is already included.

It is interesting to note that the computed F-to-enter follows no particular pattern as we include additional variables. Note also that even though R^2 is always increasing, the amount by which it increases varies as new variables are added. Finally, this example verifies that the adjusted multiple $\bar{R}^2$ increases to a maximum and then decreases.

The other criteria discussed in Section 8.4 are C_P and AIC. These criteria are also recommended for stopping rules. Specifically, the combination of variables minimizing their values can be chosen. A numerical example for this procedure will be given in Section 8.7.

Computer programs may also terminate the process of forward selection when the tolerance level (Section 7.9) is smaller than a specified minimum tolerance.

Table 8.3: *Forward selection of variables for chemical companies*

Variables added	Computed F-to-enter	Multiple R^2	Multiple $\bar{R}^2$
1. D/E	8.09	0.224	0.197
2. PAYOUTR1	2.49	0.290	0.237
3. NPM1	9.17	0.475	0.414
4. SALESGR5	3.39	0.538	0.464
5. EPS5	0.38	0.545	0.450
6. ROR5	0.06	0.546	0.427

Backward elimination method

An alternative strategy for variable selection is the **backward elimination method**. This technique begins with all of the variables in the equation and proceeds by eliminating the least useful variables one at a time.

As an example, Table 8.4 lists the regression coefficient (both standardized and unstandardized), the P value, and the corresponding computed F for testing that each coefficient is zero for the chemical companies' data. The F statistic here is called the **computed F-to-remove**.

Since ROR5 has the smallest computed F-to-remove, it is a candidate for removal. The user must specify the **maximum** F-to-remove; if the computed F-to-remove is less than that maximum, ROR5 is removed. No recommended value can be given for the maximum F-to-remove. We suggest, however, that a reasonable choice is the 70th percentile of the F distribution (or, equivalently, $P = 0.30$). For a large value of the residual degrees of freedom, this procedure results in a maximum F-to-remove of 1.07. In Table 8.4, since the computed F-to-remove for ROR5 is 0.06, this variable is removed first. Note that ROR5 also has the smallest standardized regression coefficient and the largest P value.

The backward elimination procedure proceeds by computing a new equation with the remaining five variables and examining the computed F-to-remove for another likely candidate. The process continues until no variable can be removed according to the stopping rule.

In comparing the forward selection and the backward elimination methods, we note that one advantage of the former is that it involves a smaller amount of computation than the latter. However, it may happen that two or more variables can together be a good predictive set while each variable taken alone is not very effective. In this case backward elimination would produce a better equation than forward selection. Neither method is expected to produce the best possible equation for a given number of variables to be included (other than one or the total set).

Table 8.4: *Backward elimination: Coefficients, F and P values for chemical companies*

	Coefficients			
Variable	Unstandardized	Standardized	F-to-remove	P
Intercept	1.24	–	–	–
NPM1	0.35	0.49	3.94	0.06
PAYOUTR1	9.86	0.56	12.97	0.001
SALESGR1	0.20	0.29	3.69	0.07
D/E	−2.55	−0.21	3.03	0.09
EPS5	−0.04	−0.15	0.38	0.55
ROR5	0.04	0.07	0.06	0.81

Stepwise procedure

One very commonly used technique that combines both of the above methods is called the **stepwise procedure**. In fact, the forward selection method is often called the **forward stepwise method**. In this case a step consists of adding a variable to the predictive equation. At step 0 the only "variable" used is the mean of the Y variable. At that step the program normally prints the computed F-to-enter for each variable. At step 1 the variable with the highest computed F-to-enter is entered, and so forth.

Similarly, the backward elimination method is often called the **backward stepwise method**. At step 0 the computed F-to-remove is calculated for each variable. In successive steps the variables are removed one at a time, as described above.

The standard stepwise regression programs do forward selection with the option of removing some variables already selected. Thus at step 0 only the Y mean is included. At step 1 the variable with the highest computed F-to-enter is selected. At step 2 a second variable is entered if any variable qualifies (i.e., if at least one computed F-to-enter is greater than the minimum F-to-enter). After the second variable is entered, the F-to-remove is computed for both variables. If either of them is lower than the maximum F-to-remove, that variable is removed. If not, a third variable is included if its computed F-to-enter is large enough. In successive steps this process is repeated. For a given equation, variables with small enough computed F-to-remove values are removed, and the variables with large enough computed F-to-enter values are included. The process stops when no variables can be deleted or added.

The choice of the **minimum** F-to-enter and the **maximum** F-to-remove affects both the nature of the selection process and the number of variables selected. For instance, if the maximum F-to-remove is much smaller than the minimum F-to-enter, then the process is essentially forward selection. In any case the minimum F-to-enter must be larger than the maximum F-to-remove; otherwise, the same variable will be entered and removed continuously. In many situations it is useful for the investigator to examine the full sequence until all variables are entered. This step can be accomplished by setting the minimum F-to-enter equal to a small value, such as 0.1 (or a corresponding P value of 0.99). In that case the maximum F-to-remove must then be smaller than 0.1. After examining this sequence, the investigator can make a second run, using other F values.

Values that have been found to be useful in practice are a minimum F-to-enter equal to 2.07 and a maximum F-to-remove of 1.07. With these recommended values, for example, the results for the chemical companies' data are identical to those of the forward selection method presented in Table 8.3 up to step 4. At step 4 the variables in the equation and their computed F-to-remove are as shown in Table 8.5. Also shown in the table are the variables not in the equation and their computed F-to-enter. Since all the computed F-to-remove values are larger than the minimum F-to-remove of 1.07, none of the variables is removed. Also, since both of the computer F-to-enter values are smaller than the minimum F-to-enter of 2.07, no new variables are entered. The process terminates with four variables in the equation.

There are many situations in which the investigator may wish to include certain variables in the equation. For instance, the theory underlying the subject of the study may dictate that a certain variable or variables be used. The investigator may also wish to include certain variables in order for the results to be comparable with previously published studies. Similarly, the investigator may wish to consider two variables when both provide nearly the same computed F-to-enter. If the variable with the slightly lower F-to-enter is the preferred variable from other viewpoints, the investigator may force it to be included. This preference may arise from cost or simplicity considerations. Most stepwise computer programs offer the user the option of **forcing variables** at the beginning of or later in the stepwise sequence, as desired.

We emphasize that none of the procedures described here is guaranteed or even expected to produce the best possible regression equation. This comment applies particularly to the intermediate steps where the number of variables selected is more than one but less than $P - 1$. Programs that examine the best possible subsets of a given size are also available. These programs are discussed next.

8.7 Subset regression

The availability of the computer makes it possible to compute the multiple R^2 and the regression equation for all possible subsets of variables. For a specified subset size the "best" subset of variables is the one that has the largest multiple

Table 8.5: *Step 4 of the stepwise procedure for chemical companies' data example*

Variables in equation	Computed F-to-remove	Variables not in equation	Computed F-to-remove
D/E	3.03		
PAYOUTR1	13.34		
NPM1	9.51		
SALESGR5	3.39		
		EPS5	0.38
		ROR5	0.06

R^2. The investigator can thus compare the best subsets of different sizes and choose the preferred subset. The preference is based on a compromise between the value of the multiple R^2 and the subset size. In Section 8.4 three criteria were discussed for making this comparison, namely, adjusted $\bar{R}^2, C_p$, and AIC.

For a small number of independent variables, say three to six, the investigator may indeed be well advised to obtain all possible subsets. This technique allows detailed examination of all the regression models so that an appropriate model can be selected.

SAS can evaluate all possible subsets. By using the BEST option, SAS will print out the best n sets of independent variables for $n = 1, 2, 3, \dots$ etc. SAS uses R^2, adjusted R^2, and C_p for the selection criterion. For a given size, these three criteria are equivalent. But if size is not specified, they may select subsets of different sizes.

Table 8.6 includes part of the output produced by running SAS REG for the chemical companies' data. As before, the best single variable is D/E, with a multiple R^2 of 0.224. Included in Table 8.6 are the next two best candidates if only one variable is to be used. Note that if either NPM1 or PAYOUTR1 is chosen instead of D/E, then the value of R^2 drops by about half. The relatively low value of R^2 for any of these variables reinforces our understanding that any one independent variable alone does not do well in "explaining" the variation in the dependent variable P/E.

The best combination of two variables is NPM1 and PAYOUTR1, with a multiple R^2 of 0.408. This combination does not include D/E, as would be the case in stepwise regression (Table 8.3). In stepwise regression the two variables selected were D/E and PAYOUTR1 with a multiple R^2 of 0.290. Here stepwise regression does not come close to selecting the best combination of two variables. Here stepwise regression resulted in the fourth-best choice. The third-best combination of two variables is essentially as good as the second best. The best combination of two variables, NPM1 and PAYOUTR1, as chosen by SAS REG, is interesting in light of our earlier interpretation of the variables in Section 8.3. Variable NPM1 measures the efficiency of the opera-

Table 8.6: *Best three subsets for one, two, three, or four variables for chemical companies' data selected by stepwise procedure*

Number of variables	Names of variables	R^2	$\bar{R}^2$	C_p	AIC
1	D/E	0.224	0.197	13.30	57.1
1	NPM1	0.123	0.092	18.40	60.8
1	PAYOUTR1	0.109	0.077	19.15	61.3
2	NPM1, PAYOUTR1	0.408	0.364	6.00	51.0
2	PAYOUTR1, ROR5	0.297	0.245	11.63	56.2
2	ROR5, EPS5	0.292	0.240	11.85	56.3
3	PAYOUTR1, NPM1, SALESGR5	0.482	0.422	4.26	49.0
3	PAYOUTR1, NPM1, D/E	0.475	0.414	4.60	49.4
3	PAYOUTR1, NPM1, ROR5	0.431	0.365	6.84	51.8
4	PAYOUTR1, NPM1, SALESGR5, D/E	0.538	0.464	3.42	47.6
4	PAYOUTR1, NPM1, SALESGR5, EPS5	0.498	0.418	5.40	50.0
4	PAYOUTR1, NPM1, SALESGR5, ROR5	0.488	0.406	5.94	50.6
Best 5	PAYOUTR1, NPM1, SALESGR5, D/E, EPS5	0.545	0.450	5.06	49.1
All 6		0.546	0.427	7.00	51.0

tion in converting sales to earnings, while PAYOUTR1 measures the intention to plow earnings back into the company or distribute them to stockholders. These are quite different aspects of **current** company behavior. In contrast, the debt-to-equity ratio D/E may be, in large part, a **historical** carry-over from past operations or a reflection of management style.

For the best combination of three variables the value of the multiple R^2 is 0.482, only slightly better than the stepwise choice. If D/E is a lot simpler to obtain than SALESGR5, the stepwise selection might be preferred since the loss in the multiple R^2 is negligible. Here the investigator, when given the option of different subsets, might prefer the first (NPM1, PAYOUTR1, SALESGR5) on theoretical grounds, since it is the only option that explicitly includes a measure of growth (SALESGR5). (You should also examine the four-variable combinations in light of the above discussion.)

Summarizing the results in the form of Table 8.6 is advisable. Then plotting the best combination of one, two, three, four, five, or six variables helps the investigator decide how many variables to use. For example, Figure 8.1 shows a plot of the multiple R^2 and the adjusted $\bar{R}^2$ versus the number of variables included in the best combination in the data set. Note that R^2 is a nondecreasing function. However, it levels off after four variables. The adjusted $\bar{R}^2$ reaches its maximum with four variables (D/E, NPM1, PAYOUTR1, and SALESGR5) and decreases with five and six variables.

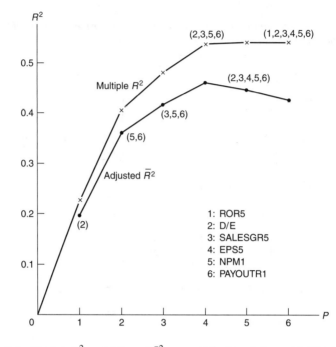

Figure 8.1: *Multiple R^2 and Adjusted $\bar{R}^2$ versus P for Best Subset with P Variables for Chemical Companies' Data*

Figure 8.2 shows C_p versus P (the number of variables) for the best combinations for the chemical companies' data. The same combination of four variables selected by the $\bar{R}^2$ criterion minimizes C_p. A similar graph in Figure 8.3 shows that AIC is also minimized by the same choice of four variables.

In this particular example all three criteria agree. However, in other situations the criteria may select different numbers of variables. In such cases the investigator's judgment must be used in making the final choice. Even in this case an investigator may prefer to use only three variables, such as NPM1, PAYOUTR1, and SALESGR5. The value of having these tables and figures is that they inform the investigator of how much is being given up, as estimated by the sample on hand.

You should be aware that variable selection procedures are highly dependent on the particular sample being used. Also, any tests of hypotheses should be viewed with extreme caution since the significance levels are not valid when variable selection is performed. For further information on subset regression see Miller (2002).

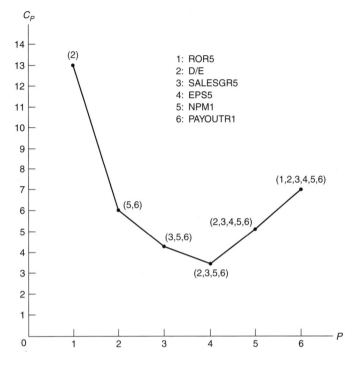

Figure 8.2: C_p versus P for Best Subset with P Variables for Chemical Companies' Data

8.8 Discussion of computer programs

Table 8.7 gives a summary of the options available in the six statistical packages. Note that only R, SAS, S-PLUS, and STATISTICA offer all or best subset selection. A user written command for best subset selection is available for Stata. By repeatedly selecting which variables are in the regression model manually, you can perform subset regression for yourself with any program. This would be quite a bit of work if you have numerous independent variables. Also, by relaxing the entry criterion for forward stepwise you can often reduce the candidate variables to a smaller set and then try backward stepwise on the reduced set.

SAS was used to obtain the forward stepwise and best subset results operating on a data file called "chemco." The model was written out with the proper dependent and independent variables under consideration. The basic proc statements for stepwise regression using SAS REG are as follows:

```
proc reg data = chemco outest = chem1;
model pe = ror5 de salesgr5 eps5 npm1 payoutr1
```

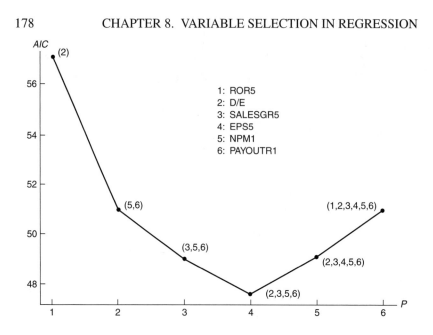

Figure 8.3: *AIC versus P for Best Subset with P Variables for Chemical Companies'*
Data

```
/selection = stepwise;
run;
```
The data file is called chemco and the stepwise option is used for selection.
For other data sets if too few variables enter the model, the entry levels can
be changed from the default of 0.15 for stepwise entry to 0.25 by using SLE
= 0.25 in the model statement.

The subset selection method was used to obtain subset regression and to get
the partial regression residual plots (Section 8.9) by adding the word "partial."
"Details" was added to get additional output as follows:
```
proc reg data = chemco outest = chem2;
model pe = ror5 de salesgr5 eps5 npm1 payoutr1
/selection = rsquare adjrsq cp aic partial details;
run;
```
The data file called chemco had been created prior to the run and the vari-
ables were named so they could be referred to by name. Four selection criteria
were listed to obtain the results for Table 8.6. The manual furnishes numerous
examples that are helpful in understanding instructions.

All the statistical packages have tried to make their stepwise programs attractive to users as these are widely used options.

Table 8.7: *Software commands and output for various variable selection methods*

	S-PLUS/R	SAS	SPSS	Stata	STATISTICA
Variable selection statistic used					
F-to-enter or remove	stepwise (S-PLUS only)	REG	REGRESSION		General Regression Models
Alpha for F			REGRESSION	sw: regress	General Regression Models
R^2	leaps	REG			General Regression Models
Adjusted R^2	leaps	REG		vselect[a]	General Regression Models
C_P	leaps	REG		vselect[a]	General Regression Models
Printed variable selection criteria					
R^2	leaps	REG	REGRESSION	sw: regress	General Regression Models
Adjusted R^2	leaps	REG	REGRESSION	sw: regress	General Regression Models
C_P	leaps	REG	REGRESSION	vselect[a]	General Regression Models
AIC	step	REG	REGRESSION	vselect[a]	Generalized Linear/Nonlinear Models
Selection method					
Forward	step	REG	REGRESSION	sw: regress	General Regression Models
Backward	step	REG	REGRESSION	sw: regress	General Regression Models
Stepwise	step	REG	REGRESSION	sw: regress	General Regression Models
Forcing variable entry	leaps	REG	REGRESSION	sw: regress	General Regression Models
Best subsets	leaps	REG		vselect[a]	General Regression Models
All subsets	leaps	REG		vselect[a]	General Regression Models
Effects of variable plots					
Partial residual plots			REGRESSION	cprplot	Multiple Regression
Partial regression plots		REG	REGRESSION	avplot	
Augmented partial residual plot				acprplot	

[a] User written command

8.9   Discussion of strategies

The process leading to a final regression equation involves several steps and tends to be iterative. The investigator should begin with data screening, eliminating blunders and possibly some outliers and deleting variables with an excessive proportion of missing values. Initial regression equations may be forced to include certain variables as dictated by the underlying situation. Other variables are then added on the basis of the selection procedures outlined in the chapter. The choice of procedure depends on which computers and programs are accessible to the investigator.

The results of this phase of the computation will include various alternative equations. For example, if the forward selection procedure is used, a regression equation at each step can be obtained. Thus in the chemical companies' data the equations resulting from the first four steps are as given in Table 8.8. Note that the coefficients for given variables could vary appreciably from one step to another. In general, coefficients for variables that are independent of the remaining predictor variables tend to have stable coefficients. The effect of a variable whose coefficient is highly dependent on the other variables in the equation is more difficult to interpret than the effect of a variable with stable coefficients.

Table 8.8: *Regression coefficients from the first four steps for the chemical companies*

Steps	Intercept	D/E	PAYOUTR1	NPM1	SALESGR5
1	11.81	−5.86			
2	9.86	−5.34	4.46		
3	5.08	−3.44	8.61	0.36	
4	1.28	−3.16	10.75	0.35	0.19

A particularly annoying situation occurs when the sign of a coefficient changes from one step to the next. A method for avoiding this difficulty is to impose restrictions on the coefficients (Chapter 9).

For some of the promising equations in the sequence given in Table 8.6, another program could be run to obtain extensive residual analysis and to determine influential cases. Examination of this output may suggest further data screening. The whole process could then be repeated.

Several procedures can be used. One strategy is to obtain plots of outliers in Y, outliers in X or leverage, and influential points after each step. This can result in a lot of graphical output; so, often investigators will just use those plots for the final equation selected. Note that if you decide not to use the regression equation as given in the final step, the diagnostics associated with the final step do not apply to your regression equation and the program should be re-run to obtain the appropriate diagnostics.

Another procedure that works well when variables are added one at a time is to examine what are called **added variable** or **partial plots**. These plots are useful in assessing the linearity of the candidate variable to be added, given the effects of the variable(s) already in the equation. They are also useful in detecting influential points. Here we will describe three types of partial plots that vary in their usefulness to detect nonlinearity or influential points (Chatterjee and Hadi (1988); Fox and Long (1990); Draper and Smith (1998); Cook and Weisberg (1982)). To clarify the plots, some formulas will be given.

Suppose we have already fitted an equation given by

$$Y = A + BX$$

where BX refers to one or more X's already in the equation, and we want to evaluate a candidate variable X_k. First, several residuals need to be defined. The residual

$$e(Y.X) = Y - (A + BX)$$

is the usual residual in Y when the equation with the X variables(s) only is used. Next, the residual in Y when both the X variables(s) and X_k are used is given by

$$e(Y.X, X_k) = Y - (A' + B'X + B_k X_k)$$

Then, using X_k as if it were a dependent variable being predicted by the other X variable(s), we have

$$e(X_k.X) = X_k - (A'' + B''X)$$

Partial regression plots (also called **added variable plots**) have $e(Y.X)$ plotted on the vertical axis and $e(X_k.X)$ plotted on the horizontal axis. Partial regression plots are used to assess the nonlinearity of the X_k variable and influential points. If the points do not appear to follow a linear relationship, then lack of linearity of Y as a function of X_k is possible (after taking the effect of X into account). In this case, transformations of X_k should be considered. Also, points that are distant from the majority of the points are candidates for being outliers. Points that are outliers on both the vertical and horizontal axes have the most influence.

Figure 8.4 presents such a plot where P/E is Y, the variables NPM1 and PAYOUTR1 are the X variables, and SALESGR5 is X_k. This plot was obtained using Stata. A straight line is fitted to the points in Figure 8.4 with a slope B_k. The residuals from such a line are $e(Y.X, X_k)$.

In examining Figure 8.4, it can be seen that the means of $e(Y.X)$ and $e(X_k.X)$ are both zero. If a straight line were fitted to the points it would have a positive slope. There is no indication of nonlinearity so there is no reason to

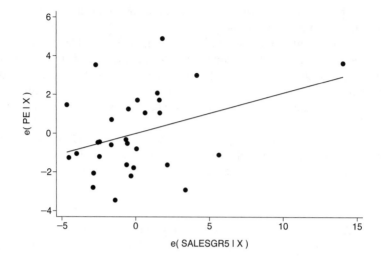

Figure 8.4: *Partial Regression Residual Plot*

consider transforming the variable SALESGR5. There is one point to the right that is somewhat separated from the other points, particularly in the horizontal direction. In other words, using NPM1 and PAYOUTR1 to predict SALESGR5 we have one company that had a larger SALESGR5 than might be expected.

In general, if X_k is statistically independent of the X variable(s) already in the equation, then its residuals would have a standard deviation that is approximately equal to the standard deviation of the original data for variable X_k. Otherwise the residuals would have a smaller standard deviation. In this example SALESGR5 has a correlation of -0.45 with PAYOUTR1 and 0.25 with NPM1 (Table 8.2) so we would expect the standard deviation of the residuals to be smaller, and it is.

The one isolated point does not appear to be an outlier in terms of the Y variable or P/E. Examining the raw data in Table 8.1 shows one company, Dexter, that appears to have a large SALESGR5 relative to its size of NPM1 and PAYOUTR1. Since it is not an outlier in Y, we did not remove it.

Partial residual plots (also called **component plus residual plots**) have $e(Y.X, X_k) + B_k X_k$ on the vertical axis and X_k on the horizontal axis. A straight line fitted to this plot will also have a slope of B_k and the residuals are $e(Y.X, X_k)$. However, the appearance of partial regression and partial residual plots can be quite different. The latter plot may also be used to assess the linearity of X_k and warn the user when a transformation is needed. When there is a strong multiple correlation between X_k and the X variable(s) already entered, this plot should not be used to assess the relationship between Y and X_k (taking into account the other X variable(s)) as it can give an incorrect impression. It may be difficult to assess influential points from this plot.

Mallows (1986) suggested an additional component to be added to partial residual plots and this plot is called an **augmented partial residual plot**. The square of X_k is added to the equation and the residuals from this augmented equation plus $B_k X_k + B_{k+1} X_k^2$ are plotted on the vertical axis and X_k is again plotted on the horizontal axis. If linearity exists, this augmented partial residual plot and the partial residual plot will look similar. Mallows includes some plots that demonstrate the ease of interpretation with augmented partial residual plots. When nonlinearity exists, the augmented plot will look different from the partial residual plot and is simpler to assess.

An alternative approach to variable selection is the so-called **stagewise regression procedure**. In this method the first stage (step) is the same as in forward stepwise regression, i.e., the ordinary regression equation of Y on the most highly correlated independent X variable is computed. Residuals from this equation are also computed. In the second step the X variable that is most highly correlated with these residuals is selected. Here the residuals are considered the "new" Y variables. The regression coefficient of these residuals on the X variable selected in step 2 is computed. The constant and the regression coefficient from the first step are combined with the regression coefficient from the second step to produce the equation with two X variables. This equation is the two-stage regression equation. The process can be continued to any number of desired stages.

The resulting **stagewise regression equation** does not fit the data as well as the least squares equation using the same variables (unless the X variables are independent in the sample). However, stagewise regression has some desirable advantages in certain applications. In particular, econometric models often use stagewise regression to adjust the data source factor, such as a trend or seasonality. Another feature is that the coefficients of the variables already entered are preserved from one stage to the next. For example, for the chemical companies' data the coefficient for D/E would be preserved over the successive stage of a stagewise regression. This result can be contrasted with the changing coefficient in the stepwise process summarized in Table 8.9. In behavioral science applications the investigator can determine whether the addition of a variable reflecting a score or an attitude scale improves prediction of an outcome over using only demographic variables.

8.10 What to watch out for

The variable selection techniques described in this chapter are used for the variable-X regression model. Therefore, the areas to watch out for discussed in Chapters 6 and 7 also apply to this chapter. Only possible problems connected with variable and model selection will be mentioned here. For a general discussion of these subjects, see McQuarrie and Tsai (1998) or Claeskens and Hjort (2008). Areas of special concern are as follows.

1. Selection of the best possible variables to be checked for inclusion in the model is extremely important. As discussed in Section 7.11, if an appropriate set of variables is not chosen, then the coefficients obtained for the "good" variables will be biased by the lack of inclusion of other needed variables (if the "good" variables and the needed variables are correlated). Lack of inclusion of needed or missing variables also results in an overestimate of the standard errors of the slope of the variables used. Also, the predicted Y is biased unless the missing variables are uncorrelated with the variables included or have zero slope coefficients (Chatterjee and Hadi, 2006). If the best candidate variables are not measured in the first place, then no method of variable selection can make up for it later.

2. Since the programs selected the "best" variable or set of variables for testing in the various variable selection methods, the level of significance of the resulting F test cannot be read from a table of the F distribution. Any levels of significance should be viewed with caution. The test should be viewed as a screening device, not as a test of significance.

3. Minor differences can affect the order of entry of the variables. For example, in forward stepwise selection the next variable chosen is the one with the largest F-to-enter. If the F-to-enter is 4.444444 for variable one and 4.444443 for variable two, then variable one enters first unless some method of forcing is used or the investigator runs the program in an interactive mode and intervenes. The rest of the analysis can be affected greatly by the result of this toss-up situation.

4. Spurious variables can enter, especially for small samples. These variables may enter due to outliers or other problems in the data set. Whatever the cause, investigators will occasionally obtain a variable from mechanical use of variable selection methods that no other investigator can duplicate and which would not be expected for any theoretical reason. Flack and Chang (1987) discuss situations in which the frequency of selecting such "noise" variables is high.

5. Methods using C_p and AIC, which can depend on a good estimate of σ^2, may not actually be better in practice than other methods if a proper estimate for σ^2 does not exist.

6. Especially in forward stepwise selection, it may be useful to perform regression diagnostics on the entering variable given the variables already entered, as described in the previous section.

7. Another reason for being cautious about the statements made concerning the results of stepwise or best subsets regression analysis is that usually we

are using a single sample to estimate the population regression equation. The results from another sample from the same population may be different enough to result in different variables entering the equation or a different order to entry. If they have a large sample, some investigators try to check this by splitting their sample in half, running separate regression analyses on both halves, and then comparing the results. If they differ markedly, then the reason for this difference should be explored.

8. Another test is to split the sample, using three-fourths of the sample to obtain the regression equation, and then applying the results of that equation to the other fourth. That is, the unused X's and Y's from the one-fourth sample are used in the equation and the simple correlation of the Y's and the predicted Y's is computed and compared to the multiple correlation from the same data. This simple correlation is usually smaller than the multiple correlation from the original regression equation computed from the three-fourths sample. It can also be useful to compare the distribution of the residuals from the three-fourths sample to that of the unused one-fourth sample. If the multiple correlation and the residuals are markedly different in the one-fourth sample, using the multiple regression for prediction in the future may be questionable.

9. Numerous other methods exist for adjusting the inferences after variable selection (see Greene (2008) for example). One problem is that the standard errors obtained may be too small, either because of variable selection or because of invalid assumptions. A method called *bootstrapping* provides reasonable estimates of standard errors and P values that do not depend on unrealistic assumptions. Although this topic is not discussed here, we urge the interested reader to refer to Efron and Tibshirani (1994) or Davison and Hinkley (1997).

10. Investigators sometimes use a large number of possible predictor variables at the start so that many variables have a chance of being included. If there are missing data in the data set, this can result in having many missing values in the final regression equation even if variables that are left out are the ones with the missing data.

11. When both approaches, variable selection and imputation for missing values (see Section 9.2) are used, we have two options: 1) impute missing data first and apply variable selection procedures next, or 2) apply variable selection procedures to only the complete cases and then apply imputation techniques to the final model. These two options can lead to very different results and conclusions. Unfortunately, there is no easy solution to this chicken and egg type of problem and you may wish to try both options and compare the results.

8.11   Summary

Variable selection techniques are helpful in situations in which many independent variables exist and no standard model is available. The methods we described in this chapter constitute one part of the process of making the final selection of one or more equations.

Although these methods are useful in exploratory situations, they should not be used as a substitute for modeling based on the underlying scientific problem. In particular, the variables selected by the methods described here depend on the particular sample on hand, especially when the sample size is small. Also, significance levels are not valid in variable selection situations.

These techniques can, however, be an important component in the modeling process. They frequently suggest new variables to be considered. They also help cut down the number of variables so that the problem becomes less cumbersome both conceptually and computationally. For these reasons variable selection procedures are among the most popular of the packaged options. The ideas underlying variable selection have also been incorporated into other multivariate programs (see e.g., Chapters 11, 12, and 13).

In the intermediate stage of analysis the investigator can consider several equations as suggested by one or more of the variable selection procedures. The equations selected should be considered tentative until the same variables are chosen repeatedly. Consensus will also occur when the same variables are selected by different subsamples within the same data set or by different investigators using separate data sets.

8.12   Problems

8.1 Use the depression data set described in Table 3.4. Using CESD as the dependent variable, and age, income, and level of education as the independent variables, run a forward stepwise regression program to determine which of the independent variables predict level of depression for women.

8.2 Repeat Problem 8.1 using subset regression, and compare the results.

8.3 *Forbes* gives, each year, the same variables listed in Table 8.1 for the chemical industry. The changes in lines of business and company mergers resulted in a somewhat different list of chemical companies in 1982. We have selected a subset of 13 companies that are listed in both years and whose main product is chemicals. Table 8.9 includes data for both years (*Forbes*, vol. 127, no. 1 (January 5, 1981) and *Forbes*, vol. 131, no. 1 (January 3, 1983)). Do a forward stepwise regression analysis, using P/E as the dependent variable and ROR5, D/E, SALESGR5, EPS5, NPM1, and PAYOUTR1 as independent variables, on both years' data and compare the results. Note that this period showed little growth for this subset of companies, and the variable(s) entered should be evaluated with that idea in mind.

Table 8.9: *Chemical companies' financial performance as of 1980 and 1982*

Company	Symbol	P/E	ROR5	D/E	SALESGR5	EPS5	NPM1	PAYOUTR1
1980								
Diamond Shamrock	dia	9	13.0	0.7	20.2	15.5	7.2	0.43
Dow Chemical	dow	8	13.0	0.7	17.2	12.7	7.3	0.38
Stauffer Chemical	stf	8	13.0	0.4	14.5	15.1	7.9	0.41
E.I. du Pont	dd	9	12.2	0.2	12.9	11.1	5.4	0.57
Union Carbide	uk	5	10.0	0.4	13.6	0.8	6.7	0.32
Pennwalt	psm	6	9.8	0.5	12.1	14.5	3.8	0.51
W.R. Grace	gra	10	9.9	0.5	10.2	7.0	4.8	0.38
Hercules	hpc	9	10.3	0.3	11.4	8.7	4.5	0.48
Monsanto	mtc	11	9.5	0.4	13.5	5.9	3.5	0.57
American Cyanamid	acy	9	9.9	0.4	12.1	4.2	4.6	0.49
Celanese	cz	7	7.9	0.4	10.8	16.0	3.4	0.49
Allied Chemical	acd	7	7.3	0.6	15.4	4.9	5.1	0.27
Rohm & Haas	roh	7	7.8	0.4	11.0	3.0	5.6	0.32
1982								
Diamond Shamrock	dia	8	9.8	0.5	19.7	-1.4	5.0	0.68
Dow Chemical	dow	13	10.3	0.7	14.9	3.5	3.6	0.88
Stauffer Chemical	stf	9	11.5	0.3	10.6	7.6	8.7	0.46
E.I. du Pont	dd	9	12.7	0.6	19.7	11.7	3.1	0.65
Union Carbide	uk	9	9.2	0.3	10.4	4.1	4.6	0.56
Pennwalt	psm	10	8.3	0.4	8.5	3.7	3.1	0.74
W.R. Grace	gra	6	11.4	0.6	10.4	9.0	5.5	0.39
Hercules	hpc	14	10.0	0.4	10.3	9.0	3.2	0.71
Monsanto	mtc	10	8.2	0.3	10.8	-0.2	5.6	0.45
American Cyanamid	acy	12	9.5	0.3	11.3	3.5	4.0	0.60
Celanese	cz	15	8.3	0.6	10.3	8.9	1.7	1.23
Applied Corp	ald	6	8.2	0.3	17.1	4.8	4.4	0.36
Rohm & Haas	roh	12	10.2	0.3	10.6	16.3	4.3	0.45

8.4 For adult males it has been demonstrated that age and height are useful in predicting FEV1. Using the data described in Appendix A, determine whether the regression plane can be improved by also including weight.

8.5 Using the data given in Table 8.1, repeat the analyses described in this chapter with $(P/E)^{1/2}$ as the dependent variable instead of P/E. Do the results change much? Does it make sense to use the square root transformation?

8.6 Use the data you generated from Problem 7.7, where $X1, X2, \ldots, X9$ are the independent variables and Y is the dependent variable. Use the generalized linear hypothesis test to test the hypothesis that $\beta_4 = \beta_5 = \cdots = \beta_9 = 0$. Comment in light of what you know about the population parameters.

8.7 For the data from Problem 7.7, perform a variable selection analysis, using the methods described in this chapter. Comment on the results in view of the population parameters.

8.8 In Problem 7.7 the population multiple $\mathcal{R}^2$ of Y on $X4, X5, \ldots, X9$ is zero. However, from the sample alone we don't know this result. Perform a variable selection analysis on $X4$ to $X9$, using your sample, and comment on the results.

8.9 (a) For the lung function data set described in Appendix A with age, height, weight, and FVC as the candidate independent variables, use subset regression to find which variables best predict FEV1 in the oldest child. State the criteria you use to decide. (b) Repeat, using forward selection and backward elimination. Compare with part (a).

8.10 Force the variables you selected in Problem 8.9(a) into the regression equation with OCFEV1 as the dependent variable, and test whether including the FEV1 of the parents (i.e., the variables MFEV1 and FFEV1 taken as a pair) in the equation significantly improves the regression.

8.11 Using the methods described in this chapter and the family lung function data described in Appendix A, and choosing from among the variables OCAGE, OCWEIGHT, MHEIGHT, MWEIGHT, FHEIGHT, and FWEIGHT, select the variables that best predict height in the oldest child. Show your analysis.

8.12 From among the candidate variables given in Problem 8.11, find the subset of three variables that best predicts height in the oldest child, separately for boys and girls. Are the two sets the same? Find the best subset of three variables for the group as a whole. Does adding OCSEX into the regression equation improve the fit?

8.13 Using the Parental HIV data find the best model that predicts the age at which adolescents started drinking alcohol. Since the data were collected retrospectively, only consider variables which might be considered representative of the time before the adolescent started drinking alcohol.

Chapter 9

Special regression topics

9.1 Chapter outline

In this chapter we present brief descriptions of several topics in regression analysis—some that might occur because of problems with the data set and some that are extensions of the basic analysis. Section 9.2 discusses methods used to alleviate the effects of missing values. Section 9.3 defines dummy variables and shows how to code them as well as a brief introduction to regression trees. It also includes a discussion of the interpretation of the regression equation when dummy variables are used and when interactions are present. In Section 9.4 the use of linear constraints on the regression coefficients is explained and an example is given on how it can be used for spline regression. In Section 9.5 methods for checking for multicollinearity are discussed. Some easy methods that are sometimes helpful are given. Finally, ridge regression, a method for doing regression analysis when multicollinearity is present, is presented in Section 9.6.

9.2 Missing values in regression analysis

We included in Section 3.4 an introductory discussion of missing values due to unit nonresponse and item nonresponse. Here we focus on **item nonresponse**. Missing values due to missing data for some variables (item nonresponse) can be an aggravating problem in regression analysis since often too many cases are excluded because of missing values in standard packaged programs. The reason for this is the way statistical software excludes cases. SAS is typical in how missing values are handled. If you include variables in the VAR statement in SAS that have any missing values for some cases, those cases will be excluded from the regression analysis even when you do **not** include the particular variables with missing values in the regression equation. The REG procedure assumes you may want to use all the variables and deletes cases with any missing values. Sometimes, by careful selection of the X variables included in the VAR statement for particular regression equations, dramatically more cases with complete values will be included. In general, all the programs use complete case analysis (listwise deletion) as a default.

The subject of missing values in regression analysis has received wide attention from statisticians. Afifi and Elashoff (1966) present a review of the literature up to 1966, and Hartley and Hocking (1971) present a simple taxonomy of incomplete data problems (see also Little, 1982). Little (1992), Little and Rubin (2002), Schafer (1997) and Molenberghs and Kenward (2007) present informative reviews of methods for handling missing X's in regression analysis. Allison (2001), McKnight *et al.* (2007) and Enders (2010) have presentations that are at approximately the same level as this book. Despite the fact that methodological research relating to missing values has been very active since the 1980s, adoption of these methods in the applied literature has been slow (Molenberghs, 2007). Initially, one reason for this was the lack of statistical software capabilities. However, most mainstream statistical packages now have features available to handle various missing data procedures, most prominently, multiple imputations and appropriately adjusted summary analyses.

Types of missing values

Missing values are often classified into three types (Little and Rubin, 1990, 2002). The first of these types is called **missing completely at random** (MCAR). For example, if a chemist knocks over a test tube or loses a part of a record, then these observations could be considered MCAR. Here, we assume that being missing is independent of both X and Y. In other words, the missing cases are a random subsample of the original sample. In such a situation there should be no major difference between those cases with missing values and those without, on either the X or the Y variable. This possibility can be checked by dividing the cases into groups based on whether or not they have missing values. The groups can then be compared in terms of the data observed in both groups using a series of t tests or by more complicated multivariate analyses (e.g., the analysis given in Chapter 12). But it should be noted that lack of a significant difference does not necessarily prove that the data are missing completely at random. Sometimes the chance of making a beta error, i.e., accepting a false null hypothesis, is quite large. For example, if the original sample is split into two samples, one with a small number of cases, then the power of the test could be low.

It is always recommended that the investigator examine the pattern of the missing values, particularly if there are numerous missing values. Clusters of missing data may be an indication of lack of MCAR. This can be done by examining the data array of any of the statistical or spreadsheet programs and noting where the missing values fall. Programs such as SOLAS print an array where the values that are missing are shaded grey and the data that are present are left blank so it is easy to recognize the patterns. Sometimes when a case has a value for one variable missing, values for other variables will also be missing. If variable A is only missing when variable B is missing, the pattern is called a monotone pattern. In order to see if patterns of missing values such as monotone patterns exist in a data set, it may be necessary to change the order of the variables and of the cases. This can either be done by the investigator using, e.g., cut and paste methods or it can be done automatically using a program such as SOLAS.

The second type of missing value is called **missing at random** (MAR). Here the probability that Y is missing depends on the value of X but not Y. In this case the observed values of Y are not a random subsample of the original sample of Y, but they are a random sample of the subsamples within subclasses defined by values of X. For example, poorly educated responders may have trouble answering some questions. Thus although they are included as cases, they may have missing values on the difficult questions. Now, suppose that education is an X variable and the answer to a difficult question is the Y variable. If the uneducated who respond were similar to the uneducated in the population, then the data would be missing at random. We note that if the missing data are MAR, and if the mechanism causing the data to be missing is unrelated to the parameters to be estimated, then we say that this is an **ignorable mechanism**.

The third type of missing data or **nonrandom missing data** occurs when being missing is related to both X and Y. Nonrandom missing data mechanisms are sometimes called **nonignorable mechanisms**. For example, if we were predicting levels of depression and both being depressed (Y) and being poor (X) influence whether or not a person responded, then the third type of missing data has occurred. Nonrandom missing data may show up as "don't knows" or missing values that are difficult or impossible to relate clearly to any variable. In practice, missing values often occur for bizarre reasons that are difficult to fit into any statistical model. Even careful data screening can cause missing values, since observations that are declared outliers might be treated as missing values.

Options for treating missing values

The strategies that are used to handle missing values in regression analysis depend on what types of missing data occur. For nonrandom missing data, the methods for treating nonresponse depend on the particular response bias that has occurred and it may be difficult to determine what is happening. The methods require that investigators have knowledge of how the nonreponse occurred. These methods will not be covered here (see Allison, 2001, Rubin, 1987, Schafer, 1999, or Molenberghs and Kenward, 2007). In this case, some investigators analyze the complete cases only and place caveats about the population to which inferences can be drawn in their report. In such a situation, the best hope is that the number of incomplete cases is small.

In contrast, when the data are missing completely at random, several options are available since the subject of randomly missing observations in regression analysis has received wide attention from statisticians. The simplest option is to use only the cases with complete data (**complete case analysis**). Using this option when the data are MCAR will result in unbiased estimates of the regression parameters. The difficulties with this approach are the loss of information from discarding cases and possible unequal sample sizes from one analysis to another. The choice among the techniques given in this section usually depends on the magnitude of the missing data. If only Y values are missing, there is usually no advantage in doing anything other than using the complete cases. The confusion in this matter arises when you have large data sets and numerous analyses are performed using the data set. Then, what is a Y variable for one analysis may become an X variable for another analysis.

When the data are missing at random (MAR), then using complete cases does not result in unbiased estimates of all the parameters. For example, suppose that simple linear regression is being done and the scatter diagram of points is missing more points than would be expected for some range of X values—perhaps more values are missing for small X values. Then the estimates of the mean and variance of Y may be quite biased but the estimate of the regression line may be satisfactory (Little and Rubin, 2002). Since most investigators use only data with no missing values in their regression analyses, this remains a recommended procedure for MAR data especially if there are not many missing values.

We will mention three other methods of handling missing values in regression analysis. These are pairwise deletion, maximum likelihood, and a number of imputation methods. We will discuss imputation in more detail and will present single imputation first and then multiple imputation.

Several statistical packages provide the option of **pairwise deletion** of cases. Pairwise deletion is an alternative method for determining which variables should be deleted in computing the variance–covariance or the correlation array. As stated earlier, most statistical packages delete a case if any of the variables listed for possible inclusion has a missing value. Pairwise deletion is done separately for each pair of variables actually used in the calculation of the covariances. The result is that any two covariances (or correlations) could be computed using somewhat different cases. A case is not removed because it is missing data in, say variable number one; only the covariances or correlations computed using that variable are affected.

We do not recommend the pairwise deletion methods unless the investigator performs runs with other methods also as a check. In simulations it has been shown to work in some instances and to lead to problems in others. Pairwise deletion tends to work better if the data are MCAR than if they are MAR. It may result in making the computations inaccurate or impossible to do because of the inconsistencies introduced by different sample sizes for different correlations. But it is available in several statistical packages so it is not difficult to perform. Note that not all statistical packages use the same sample size for subsequent data analyses when pairwise deletion is done.

Another method, which applies to regression analysis with normally distributed X variables, was developed by Beale and Little (1975). This method is based on the theoretical principle of **maximum likelihood** used in statistical estimation and it has good theoretical properties. It involves direct analysis of the incomplete data by maximum likelihood methods. Further explanation of this method is beyond the scope of this text (see Allison, 2001, Enders, 2001, or Little and Rubin, 2002).

The **imputation** method is commonly used, particularly in surveys where missing values may be quite prevalent. Various methods are used to fill in, or *impute*, the missing data and then the regression analysis is performed on the completed data set. In most cases these methods are used on the X variables, although they can also be used on the Y variable.

One method of filling in the missing values is **mean substitution**. Here the missing values for a given variable are replaced by the mean value for that variable. As mentioned in Section 3.4, we do not recommend this method unless there are very few missing values. It results in biased estimates of the variances and covariances and leads to biased estimates of the slope coefficients.

A more complicated imputation method involves first computing a regression equation where the variable with missing value(s) is considered temporarily to be a dependent variable with the independent variables being ones that are good predictors of it. Then this equation is used to "predict" the missing values. For example, suppose that for a sample of schoolchildren some heights are missing. The investigator could first derive the regressions of height on age for boys and girls and then use these regressions to predict the missing heights. If the children have a wide range of ages, this method presents an improvement over mean substitution (for simple regression, see Afifi and Elashoff, 1969a, 1969b). The variable height could then be used as an X variable in a future desired regression model or as a Y variable. It is also possible to use outcome variables as predictor variables in fitting the regression equation. This method is sometimes known as **conditional mean imputation** since we are conditioning the imputed value on another variable(s), while mean substitution is sometimes called a **deterministic** method.

A variation on the above method is to add to the predicted value of the temporary dependent variable a residual value chosen at random from the residuals from the regression equation. This method is called a **stochastic** substitution. One way of estimating a residual is to assume that the points are normally distributed about the regression plane. Then, take a sample of size one from a normal distribution that has a mean zero and a standard deviation of one. Note that random normal numbers (z_i) are available from many statistical packages. The imputed value is then computed as $\hat{Y}_i + z_i S$ where S is the standard error of the estimate $\hat{Y}_i$ given in Section 7.4 (see Kalton and Kasprzyk (1986) for additional methods of choosing the residuals). If the variable with the missing value is highly skewed, then the investigator should first make a transformation to reduce the skewness before fitting the regression equation. After the residual is added to the regression line or plane the results should be retransformed into the original scale (see Allison, 2001).

When a regression equation is used without adding a randomly chosen residual, then all the missing values are filled in with values that are right on the regression line (or plane). Estimation of a point on the regression line plus a residual is an attractive method but it can take considerable time if numerous variables have missing values since a regression equation and residuals will have to be found for each variable. This same technique of adding a residual can be used with mean substitution where the "residuals" are residuals or deviations from the mean.

Hot deck imputation, where the missing data are replaced by a result from a case that is not missing and which is similar on other variables to the nonresponder that has the missing data, can also be used (David *et al.*, 1986, Groves *et al.*, 2002). There are numerous variations on how hot deck imputation is performed. One method is to first sort the responders and nonresponders using several variables for which there are complete data and that are known to relate to the variable that has missing data. For example, if gender and age are thought to relate to the variable with missing data, then the responders and nonresponders could be sorted on both gender and age. An imputation class c is then formed of responders who are close to the nonresponder in the sorted data set (for example, close in age and the same gender). Then a value for one of the responders in the imputation class is substituted for the missing value for the nonresponder. One method of choosing the responder whose value is taken is to take a random sample of the cases in the imputation class. Another method called "nearest neighbor" is sometimes used. Here the missing value is replaced by the response from the case that is closest to the nonrespondent in the ordered data set. Other methods have been used.

One of the advantages of the hot deck method is that it can be used with discrete or continuous data. No theoretical distributions need to be assumed. By the same token, it is difficult to evaluate whether it has been successful or not.

The advantages of using single imputation is that the user now has a data set that is complete. Cases will not be dropped by a statistical package because one or more variables have missing values. But this has come at a cost. One cost is the effort involved in doing the imputation. The second cost is the lack of information on the effect the imputation might have on the results, particularly possible biases of the parameter estimates, estimated standard errors, and P

values that are smaller and confidence intervals that are narrower than they should be.

One method that has been used mainly in the survey field and in biomedical studies to improve on single imputation is **multiple imputation**. For additional reading see Allison (2001), Little (1992), Little and Rubin (2002), Rubin (1987), Schafer (1997) or Molenberghs and Kenward (2007) for a more complete discussion and numerous additional references. Multiple imputation involves replacing each missing value with several, say m, imputed values not just one. Instead of just imputing one value for each missing value, m imputed values are obtained. These are placed in the data set one at a time so that m complete data sets are obtained (one for each set of imputed values). A fundamental point is that these imputed values have added to them random errors that are based on the distribution of the observed values. This process, in turn, helps produce realistic values of the standard errors of the estimated parameters.

For example, if we wished to perform a simple linear regression analysis and had missing values in the X variable for k individuals, we first impute m values of X for each of the k individuals. This process results in m completed data sets. The second step is to obtain m values of the regression equation, one for each data set. The third step involves combining the results from the m regression equations. From the m values of the slope coefficient we could obtain a mean slope $\overline{B} = \sum B_i / m$ for a combined estimate from all the m imputations. We also compute the between imputation variance of the estimated slopes B_i

$$B = \frac{\sum_{i=1}^{m} (B_i - \overline{B})^2}{m - 1}.$$

The within imputation variance, or $\overline{W}$, is the average of the squared standard errors of the estimated B_i from the m slope coefficients. To estimate the variance of the combined m imputations, we have

$$T = \overline{W} + (1 + m^{-1}) B.$$

When the response rate is high, we would expect the within variance to be large relative to the between variance. If the response rate is low then the opposite is true. Hence the investigator can get some insight into the effect of performing the multiple imputation. If this effect is very large, a different method of imputation can be tried to see if it reduces the size of the between variation.

For example, a multiple hot deck imputation can be employed. For each class c, we may have n_{obs} respondents and n_{mis} nonrespondents out of a total of n cases in the class. Instead of simple random sampling of m values, Rubin (1987) suggests a modification that tends to generate appropriate between imputation variability. For each of the m imputations, first select n_{obs} possible values at random with replacement and from these n_{obs} values draw n_{mis} values at random with replacement.

There are several advantages of multiple imputation over single imputation (see Rubin, 1987 and Schafer, 1999). An important one is that we can get a sense of the effect of the imputation. We can see how much variation there is in what we are estimating from imputation to imputation. Secondly, we can estimate the total variability. This may not be a perfect estimate but it is better than ignoring it. Finally, using the mean of m imputations instead of a single value reduces the bias of the estimation.

One question that arises when using multiple imputation is how many imputations to perform. If there are only a few missing values, five to seven imputations might be sufficient. But for a high percentage of missing data and for certain specific missing data structures, 20 to 50 or more imputations might be recommended (Royston, 2004 and Royston, Carlin and White, 2009).

In spite of this large selection of possible techniques available for handling missing values if the data are **missing completely at random**, we recommend considering using only the cases with complete observations in some instances, for example, if the percentage of incomplete cases is small. We believe it is the most understandable analysis, it certainly is the simplest to do, and in that case it leads to unbiased results. If this is done, the investigator should report the number of cases with missing values.

If the sample size is small, or for some other reason it is desirable to adjust for missing values when the data are MCAR, we would suggest the maximum likelihood method if the data are normally distributed (Little and Rubin, 1990, say that this method does well even when the data are not normal and present methods to alleviate lack of normality). S-PLUS has a missing data library that supports the maximum likelihood EM approach and there also is a program called Amos. Multiple imputation is also recommended.

When the data are **missing at random**, we would still recommend analyzing available data (complete case analysis) if the number of missing values is small relative to the sample size. If having the same sample size for each analysis of the same data set is highly desirable, then either single or multiple imputation could be considered if the number of missing values is very small. If the percent of missing items is sizable, then either the maximum likelihood method or multiple imputation should be considered. The additional work involved in multiple imputation over single imputation is worth the effort, because it enables the user to see the effects of the imputation on the results. Since performing multiple imputation does involve more effort, it is recommended that the investigator study the use of this method in other books on the subject, for example, the book by Allison (2001), which is particularly accessible for readers with little mathematical background.

There are several statistical packages that can reduce the time it takes to perform multiple imputation (see Table 9.1). A summary of these packages is given in Horton and Lipsitz (2001). SAS 8.2 has two procedures, PROC MI that performs the imputation and PROC MIANALYZE that analyzes the results of the imputation. S-PLUS has a missing data library to support data augmentation (DA) algorithms that can be used to generate multiple imputations. S-PLUS also distributes Multiple Imputation for Chained Equations (MICE) that provides several imputation methods as well as Schafer's free software macros for S-PLUS. Stata can perform various multiple imputations methods and combine the results with a set of mi commands. STATISTICA offers the possibility of mean substitution and pairwise deletion. SPSS includes mean and regression imputation as well as pairwise deletion. SOLAS is a software package designed specifically for analysis of data sets with missing values. It includes the ability to visualize the pattern of missing values, create multiple data sets, and calculate the final results.

It should be noted that the imputation methods mentioned in this section can also be used to obtain a complete data set that can be used in the multivariate techniques given in the remainder of this book. An additional method for handling missing values in categorical variables will be given in Section 12.10 of the Logistics Regression chapter.

9.3 Dummy variables

Often X variables desired for inclusion in a regression model are not continuous. In terms of Stevens's classification, they are neither interval nor ratio. Such variables could be either nominal or ordinal. Ordinal measurements represent variables with an underlying scale. For example, the severity of a burn could be classified as mild, moderate, or severe. These burns are commonly called first-, second-, and third-degree burns, respectively. The X variable representing these categories may be coded 1, 2, or 3. The data from ordinal variables could be analyzed by using the numerical coding for them and treated as though they were continuous (or interval data). This method takes advantage of the underlying order of the data. On the other hand, this analysis assumes equal intervals between successive values of the variable. Thus in the burn example, we would assume that the difference between a first- and a second-degree burn is equivalent to the difference between a second- and third-degree burn. As was discussed in Chapter 5, an appropriate choice of scale may be helpful in assigning numerical values to the outcome that would reflect the varying lengths of intervals between successive responses.

In contrast, the investigator may ignore the underlying order for ordinal variables and treat them as though they were nominal variables. In this section we discuss methods for using one or more nominal X variables in regression analysis along with the role of interactions.

Table 9.1: *Software commands and output for missing values*

	S-PLUS/R	SAS	SOLAS	SPSS	Stata	STATISTICA
Method						
Pairwise deletion	mice	CORR	Single Imputation	CORRELATION	pwcorr	Correlation matrices
Mean substitution		STANDARD	Single Imputation	BASE		Miss Data Replace
Hot deck imputation					ice[a]	
Regression method	mice	MI	Multiple Imputation	MULT. IMP.*	mi impute	Time Series
Propensity score		MI	Multiple Imputation			
EM algorithm	missing	MI		MULT. IMP.*	mi impute	
MCMC method	missing/mice	MI		MULT. IMP.*	mi impute	
Multiple imputation	missing/mice	MI	Multiple Imputation	MULT. IMP.*	mi impute	
Output						
Missing pattern	missing/mice	MI	Missing Pattern	MULT. IMP.*	misstable	Miss Data Range Plots
Multi imputation var info	missing	MI	Combined	MULT. IMP.*	mi impute	
Fraction of missing info	missing/mice	MI	Combined	MULT. IMP.*	mi estimate	
Multiple imputed data sets		MI	Multiple Imputation	MULT. IMP.*	mi	
Analysis multi imputed data						
Two sample *t* test	missing	TTEST	t & Nonpar Tests	T-TEST	mi estimate	
Nonparametric test	missing	NPAR1WAY	t & Nonpar Tests	NPTESTS	mi estimate	
Linear regression	missing/mice	REG	Multiple Regression	REGRESSION	mi estimate	
Logistic regression	missing/mice	LOGISTIC	Combined	LOGISTIC	mi estimate	
Combined results	missing/mice	MIANALYZE		SET MIOUTPUT	mi estimate	

[a] User written command
*Abbreviation for MULTIPLE IMPUTATION

A single binary variable

We will begin with a simple example to illustrate the technique. Suppose that the dependent variable Y is yearly income in dollars and the independent variable X is the sex of the respondent (male or female). To represent sex, we create a **dummy** or **indicator variable** $D = 0$ if the respondent is male and $D = 1$ if the respondent is female. (All the major computer packages offer the option of recoding the data in this form.) The sample regression equation can then be written as $Y = A + BD$. The value of Y is

$$Y = A \qquad \text{if} \quad D = 0$$

and

$$Y = A + B \qquad \text{if} \quad D = 1$$

Since our best estimate of Y for a given group is that group's mean, A is estimated as the average income for males $(D = 1)$. The regression coefficient B is therefore $B = \overline{Y}_{\text{females}} - \overline{Y}_{\text{males}}$. In effect, males are considered the **referent group**, and females' income is measured by how much it differs from males' income.

A second method of coding dummy variables

An alternative way of coding the dummy variables is

$$D^* = -1 \quad \text{for males}$$

and

$$D^* = +1 \quad \text{for females}$$

In this case the regression equation would have the form

$$Y = A^* + B^* D^*$$

The average income for males is now

$$A^* - B^* \quad (\text{when } D^* = -1)$$

and for females it is

$$A^* + B^* \quad (\text{when } D^* = +1)$$

Thus

$$A^* = \frac{1}{2}(\overline{Y}_{\text{males}} + \overline{Y}_{\text{females}})$$

and

$$B^* = \overline{Y}_{\text{females}} - \frac{1}{2}(\overline{Y}_{\text{males}} + \overline{Y}_{\text{females}})$$

or

$$B^* = \frac{1}{2}(\overline{Y}_{\text{females}} - \overline{Y}_{\text{males}})$$

In this case neither males nor females are designated as the referent group.

A nominal variable with several categories

For another example we consider an X variable with $k > 2$ categories. Suppose income is now to be related to the religion of the respondent. Religion is classified as Catholic (C), Protestant (P), Jewish (J), and other (O). The religion variable can be represented by the dummy variables that follow:

Religion	D_1	D_2	D_3
C	1	0	0
P	0	1	0
J	0	0	1
O	0	0	0

Note that to represent the four categories, we need only three dummy variables. In general, to represent k categories, we need $k - 1$ dummy variables. Here the three variables represent C, P, and J, respectively. For example, $D_1 = 1$ if Catholic and zero otherwise; $D_2 = 1$ if Protestant and zero otherwise; and $D_3 = 1$ if Jewish and zero otherwise. The "other" group has a value of zero on each of the three dummy variables.

The estimated regression equation will have the form

$$Y = A + B_1 D_1 + B_2 D_2 + B_3 D_3$$

The average incomes for the four groups are

$$\begin{aligned} \bar{Y}_C &= A + B_1 \\ \bar{Y}_P &= A + B_2 \\ \bar{Y}_J &= A + B_3 \\ \bar{Y}_O &= A \end{aligned}$$

Therefore,

$$\begin{aligned} B_1 &= \bar{Y}_C - A = \bar{Y}_C - \bar{Y}_O \\ B_2 &= \bar{Y}_P - A = \bar{Y}_P - \bar{Y}_O \\ B_3 &= \bar{Y}_J - A = \bar{Y}_J - \bar{Y}_O \end{aligned}$$

Thus the group "other" is taken as the referent group to which all the others are compared. Although the analysis is independent of which group is chosen as the referent group, the investigator should select the group that makes the interpretation of the B's most meaningful. The mean of the referent group is the constant A in the equation. The chosen referent group should have a fairly large sample size.

If none of the groups represents a natural choice of a referent group, an alternative choice of assigning values to the dummy variables is as follows:

Religion	D_1^*	D_2^*	D_3^*
C	1	0	0
P	0	1	0
J	0	0	1
O	−1	−1	−1

As before, Catholic has a value of 1 on D_1^*, Protestant a value of 1 on D_2^*, and Jewish a value of 1 on D_3^*. However, "other" has a value of −1 on each of the three dummy variables. Note that zero is also a possible value of these dummy variables.

The estimated regression equation is now

$$Y = A^* + B_1^* D_1^* + B_2^* D_2^* + B_3^* D_3^*$$

The group means are

$$\begin{aligned}
\overline{Y}_C &= A^* + B_1^* \\
\overline{Y}_P &= A^* + B_2^* \\
\overline{Y}_J &= A^* + B_3^* \\
\overline{Y}_O &= A^* - B_1^* - B_2^* - B_3^*
\end{aligned}$$

In this case the constant

$$A^* = \frac{1}{4}(\overline{Y}_C + \overline{Y}_P + \overline{Y}_J + \overline{Y}_O)$$

is the unweighted average of the four group means. Thus

$$\begin{aligned}
B_1^* &= \overline{Y}_C - \frac{1}{4}(\overline{Y}_C + \overline{Y}_P + \overline{Y}_J + \overline{Y}_O) \\
B_2^* &= \overline{Y}_P - \frac{1}{4}(\overline{Y}_C + \overline{Y}_P + \overline{Y}_J + \overline{Y}_O)
\end{aligned}$$

and

$$B_3^* = \overline{Y}_J - \frac{1}{4}(\overline{Y}_C + \overline{Y}_P + \overline{Y}_J + \overline{Y}_O)$$

Note that with this choice of dummy variables there is no referent group. Also, there is no slope coefficient corresponding to the "other" group. As before, the choice of the group to receive −1 values is arbitrary. That is, any of the four religious groups could have been selected for this purpose.

One nominal and one interval variable

Another example, which includes one dummy variable and one continuous variable, is the following. Suppose an investigator wants to relate vital capacity (Y) to height (X) for men and women (D) in a restricted age group. One model for the population regression equation is

$$Y = \alpha + \beta X + \delta D + e$$

where $D = 0$ for females and $D = 1$ for males. This equation is a multiple regression equation with $X = X_1$ and $D = X_2$. This equation breaks down to an equation for females,

$$Y = \alpha + \beta X + e$$

and one for males,

$$Y = \alpha + \beta X + \delta + e$$

or

$$Y = (\alpha + \delta) + \beta X + e$$

Figure 9.1a illustrates both equations.

Note that this model forces the equation for males and females to have the same slope β. The only difference between the two equations is in the intercept: α for females and $(\alpha + \delta)$ for males. The effect of height (X) on vital capacity (Y) is also said to be independent of sex since it is the same for males and for females. Also, given the comparison is made for the same height the difference in vital capacity between males and females (δ) is said to be independent of height, since the difference is the same regardless of the height. The intercept terms α for females and $(\alpha + \delta)$ for males are interpreted as the average vital capacity for females and males for zero height. However, unless the value of zero for height represents a meaningful value, the intercept does not have a meaningful interpretation. If the variable height (X) is coded in inches, it might be useful to center this variable around the mean value by subtracting the average height from all individual values to make the interpretation of the intercept meaningful.

Interaction

If we have reason to believe that the effect of height (X) on vital capacity (Y) is different for males and females, a model which includes only height (X) and sex (D) is not sufficient to describe the situation. We need to change the model so that the slope (describing the effect of height on vital capacity) can be different (non-parallel) for males and females.

A model that does not force the lines to be parallel is

$$Y = \alpha + \beta X + \delta D + \gamma(XD) + e$$

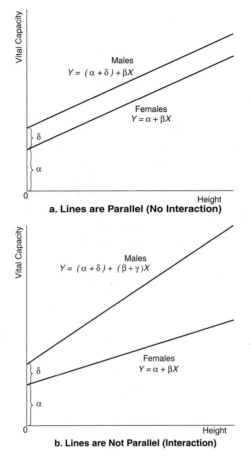

Figure 9.1: *Vital Capacity versus Height for Males and Females*

This equation is a multiple regression equation with $X = X_1, D = X_2$, and $XD = X_3$. The variable X_3 can be generated on the computer as the product of the other two variables. With this model the equation for females $(D = 0)$ is

$$Y = \alpha + \beta X + e$$

and for males it is

$$Y = \alpha + \beta X + \delta + \gamma X + e$$

or

$$Y = (\alpha + \delta) + (\beta + \gamma)X + e$$

The equations for this model are illustrated in Figure 9.1b for the case where γ is a positive quantity.

Thus with this model we allow both the intercept and slope to be different for males and females. The quantity δ is the difference between the two intercepts, and γ is the difference between the two slopes. Note that some investigators would call the product XD the **interaction** between height and sex. With this model we can test the hypothesis of no interaction, i.e., $\gamma = 0$, and thus decide whether parallel lines are appropriate. As discussed in Section 7.6, most statistical programs report a t-statistic for each of the variables in the model, which can be used to formally test the hypothesis that the slope of a specific variable is equal to zero. Under the null hypothesis of no interaction the estimate of the slope divided by its estimated standard error follows a t distribution with $n - p - 1$ degrees of freedom. Thus, if the t-statistic for the slope γ of the interaction XD is significantly different from zero, we would conclude that there is sufficient evidence that the slopes for females and males are different.

If we decide that parallel lines are not appropriate we would keep the interaction between height and sex in the model (in addition to the variables for height (X) and sex (D)). In that case it would be inappropriate to present an **overall** estimate of the effect of height (X) on vital capacity (Y). Since this estimate is different for males and females, we would present an estimate of the effect of height (X) on vital capacity (Y) for males $(\beta + \gamma)$ and for females (β) separately. Furthermore, in the presence of an interaction the difference in vital capacity between males and females cannot be described with one value, since the difference depends on the individual's height. Assuming the interaction could be described by a graph like the one presented in Figure 9.1b, the difference in vital capacity between males and females would increase as height increases. If it is not possible to present the results in a graph (if, e.g., the model contains more variables than height and sex), it might be helpful to choose a few different values for height and present estimates of the difference in vital capacity between males and females for each of those heights.

If the interaction involves one continuous variable (height) and one nominal variable (such as sex), the concept of interactions can be illustrated nicely in graphs, as is done in Figures 9.1a and b. The main concept remains the same for the case where both variables are continuous, both variables are nominal (or ordinal), and for the case of interactions which involve more than two variables. However, the interpretation can become more complex, especially if more than two variables are involved, as discussed in Extensions below.

For further discussion of the effects of interaction in regression analysis see Section 7.8 of this book as well as Kleinbaum *et al.* (2007) or Jaccard *et al.* (1990).

Extensions

We can use the ideas of dummy variables and interactions and apply them to situations where there may be several X variables and several D (or dummy) variables. For example, we can estimate vital capacity by using age and height for males and females and for smokers and nonsmokers. The selection of the model, i.e., whether or not to include interaction, and the interpretation of the resulting equations must be done with caution. The previous methodology for variable selection applies here as well, with appropriate attention given to interpretation. For example, if a nominal variable such as religion is split into three new dummy variables, with "other" used originally as a referent group, a stepwise program may enter only one of these three dummy variables. Suppose the variable D_1 is the only one entered, where $D_1 = 1$ signifies Catholic and $D_1 = 0$ signifies Protestant, Jewish, or other. In this case the referent group becomes Protestant, Jewish, or other. Care must be taken in the interpretation of the results to report the proper referent group. Sometimes, investigators will force in the three dummy variables D_1, D_2, D_3 if any one of them is entered in order to keep the referent group as it was originally chosen.

The investigator is advised to write out the separate equation for each subgroup, as was shown in the previous examples. This technique will help clarify the interpretation of the regression coefficients and the implied referent group (if any). Also, it may be advisable to select a meaningful referent group prior to selecting the model equations rather than rely on the computer program to do it for you.

Another alternative to using dummy variables is to find separate regression equations for each level of the nominal or ordinal variables. For example, in the prediction of FEV1 from age and height, a better prediction can be achieved if an equation is found for males and a separate one for females. Females are not simply "short" men. If the sample size is adequate, it is a good procedure to check whether it is necessary to include an interaction term. If the slopes "seem" equal, no interaction term is necessary; otherwise, such terms should be included. Formal tests that exist for this purpose were given in Section 7.9.

Dummy variables can be created from any nominal, ordinal, or interval data by using the "if–then–" options in the various statistical packages. Alternatively the data can be entered into the data spreadsheet in a form already suitable for use as dummy variables. Some programs, for example SAS and Stata, have options that allow the user to specify that a variable is a categorical variable and then the program will automatically generate dummy variables for the analysis. Each program has its own rules regarding which group is designated to be the referent group so the user needs to make sure that the choice made is the desired one.

At times, investigators cannot justify using certain predictor X variables in a linear fashion. Even transforming X might not produce an approximate linear relationship with Y. In such cases, one could create a nominal or ordinal variable from the original X and insert it in the regression equation using the methods described in this chapter. For example, income may be divided as "low," "medium," and "high." Two dummy variables can then be entered into that equation, producing a regression estimate that does not involve any assumed functional relationship between income and Y.

A problem that arises here is how to select the boundaries between several categories. The underlying situation may suggest natural cut-off points or quantiles can be used to define categories. An empirical method for selecting the cutoff points is called **Regression Trees**, or more generally, classification and regression trees (**CART**). This method is also called "recursive partitioning." In this method, the computer algorithm selects the "best" cutoff point for X to predict Y. Best is defined according to a criterion specified by the user, e.g., minimizing the error of prediction. In typical situations where a number of X variables are used, the algorithm continues producing a "tree" with "branches" defined by cut-off points of the predictor variables. The tree is then refined, "pruned," until a final tree is found that partitions the space of the variables into a number of cells. For each cell, a predicted Y value is computed by a specific equation. For example, if $X_1 = $ age and $X_2 = $ income, then there might be four cells:

1. The first "branch" is defined as age ≤ 25 or age > 25.

2. For those 25 years old or under, another branch could be income $\leq \$30,000$ per year or $> \$30,000$ per year.

3. For those over 25 years old, the branch might be defined as income being $\leq \$50,000$ or $> \$50,000$ per year.

In general, the algorithm produces cells that span the whole space of the X variables. The algorithm then computes an equation to calculate the predicted $\hat{Y}$ for the cases falling in each of those cells separately. The output is in the form of a tree that graphically illustrates this branching process. For each cell, an equation estimates $\hat{Y}$.

Regression trees are available in the program called CART which can be obtained from the web site http://www.salford-systems.com/. Less extensive versions have been implemented in R, S-PLUS, SPSS and STATISTICA. Interested readers can learn more about the subject from Breiman *et al.* (1984) or Zhang and Singer (1999).

9.4 Constraints on parameters

Some packaged programs, known as nonlinear regression programs, offer the user the option of restricting the range of possible values of the parameter estimates. In addition, some programs (e.g., STATISTICA nonlinear estimation module, Stata cnsreg, and SAS REG) offer the option of imposing **linear constraints** on the parameters. These constraints take the form

$$C_1\beta_1 + C_2\beta_2 + \cdots + C_P\beta_P = C$$

where $\beta_1, \beta_2, \ldots, \beta_P$ are the parameters in the regression equation and the $C_1, C_2, \ldots, C_P$ and C are the constants supplied by the user. The program finds estimates of the parameters restricted to satisfy this constraint as well as any other constraint supplied.

Although some of these programs are intended for nonlinear regression, they also provide a convenient method of performing a **linear** regression with constraints on the parameters. For example, suppose that the coefficient of the first variable in the regression equation was demonstrated from previous research to have a specified value, such as $B_1 = 2.0$. Then the constraint would simply be

$$C_1 = 1, \quad C_2 = \ldots = C_P = 0$$

and

$$C = 2.0 \quad \text{or} \quad 1\beta_1 = 2.0$$

Another example of an inequality constraint is the situation when coefficients are required to be nonnegative. For example, if $\beta_2 \geq 0$, this constraint can also be supplied to the program.

The use of linear constraints offers a simple solution to the problem known as **spline regression** or **segmented-curve regression** (see Marsh and Cormier, 2002). For instance, in economic applications we may want to relate the consumption function Y to the level of aggregate disposable income X. A possible nonlinear relationship is a linear function up to some level X_0, i.e., for $X \leq X_0$, and another linear function for $X > X_0$. As illustrated in Figure 9.2, the equation for $X \leq X_0$ is

$$Y = \alpha_1 + \beta_1 X + e$$

and for $X > X_0$ it is

$$Y = \alpha_2 + \beta_2 X + e$$

The two curves must meet at $X = X_0$. This condition produces the linear constraint

$$\alpha_1 + \beta_1 X_0 = \alpha_2 + \beta_2 X_0$$

Equivalently, this constraint can be written as

$$\alpha_1 + X_0\beta_1 - \alpha_2 - X_0\beta_2 = 0$$

To formulate this problem as a multiple linear regression equation, we first define a dummy variable D such that

$$D = 0 \quad \text{if} \quad X \leq X_0$$
$$D = 1 \quad \text{if} \quad X > X_0$$

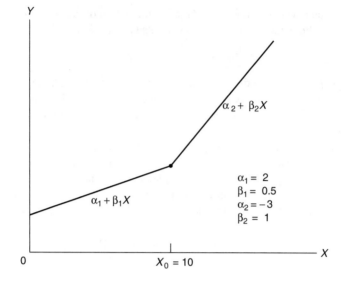

Figure 9.2: *Segmented Regression Curve with $X = X_0$ as the Change Point*

Some programs, for example Stata mkspline and STATISTICA piecewise linear regression, can be used to produce a regression equation that satisfies this condition. If your package does not have such an option, the problem can be solved as follows. The segmented regression equation can be combined into one multiple linear equation as

$$Y = \gamma_1 + \gamma_2 D + \gamma_3 X + \gamma_4 DX + e$$

When $X \leq X_0$, then $D = 0$, and this equation becomes

$$Y = \gamma_1 + \gamma_3 X + e$$

Therefore $\gamma_1 = \alpha_1$ and $\gamma_3 = \beta_1$. When $X > X_0$, then $D = 1$, and the equation becomes

$$Y = (\gamma_1 + \gamma_2) + (\gamma_3 + \gamma_4)X + e$$

Therefore $\gamma_1 + \gamma_2 = \alpha_2$ and $\gamma_3 + \gamma_4 = \beta_2$.

With this model a nonlinear regression program can be employed with the restriction

$$\gamma_1 + X_0\gamma_3 - (\gamma_1 + \gamma_2) - X_0(\gamma_3 + \gamma_4) = 0$$

or

$$-\gamma_2 - X_0\gamma_4 = 0$$

or

$$\gamma_2 + X_0\gamma_4 = 0$$

For example, suppose that X_0 is known to be 10 and that the fitted multiple regression equation is estimated to be

$$Y = 2 - 5D + 0.5X + 0.5DX + e$$

Then we estimate $\alpha_1 = \gamma_1$ as 2, $\beta_1 = \gamma_3$ as 0.5, $\alpha_2 = \gamma_1 + \gamma_2$ as $2 - 5 = -3$, and $\beta_2 = \gamma_3 + \gamma_4$ as $0.5 + 0.5 = 1$. These results are pictured in Figure 9.2. Further examples can be found in Draper and Smith (1998).

Where there are more than two segments with known values of X at which the segments intersect, a similar procedure using dummy variables could be employed. When these points of intersection are unknown, more complicated estimation procedures are necessary. Quandt (1972) presents a method for estimating two regression equations when an unknown number of the points belong to the first and second equation. The method involves numerical maximization techniques and is beyond the scope of this book.

Another kind of restriction occurs when the value of the dependent variable is constrained to be above (or below) a certain limit. For example, it may be physically impossible for Y to be negative. Tobin (1958) derived a procedure for obtaining a multiple linear regression equation satisfying such a constraint.

9.5 Regression analysis with multicollinearity

In Section 7.9, we discussed multicollinearity (i.e., when the independent variables are highly intercorrelated) and showed how it affected the estimate of the standard error of the regression coefficients. Multicollinearity is usually detected by examining tolerance (one minus the squared multiple correlation between X_i and the remaining X variables already entered in the model) or its inverse, the variance inflation factor. A commonly used cutoff value is tolerance less than 0.01 or variance inflation factor greater than 100. Since the tolerance coefficient is itself quite unstable, sometimes reliable slope coefficients can be obtained even if the tolerance is less than 0.01.

Multicollinearity is also detected by examining the size of the **condition index** (or the **condition number**). This quantity is computed in the SAS REG procedure as the square root of the ratio of the largest eigenvalue to the smallest eigenvalue of the covariance matrix of the X variables (see Section 14.5 for a definition of eigenvalues). It can readily be computed from any program that gives the eigenvalues. A large condition index is an indication of multicollinearity. Belsley, Kuh and Welsch (1980) indicate that values greater than 30 can be an indication of serious problems.

A useful method for checking stability is to standardize the data, compute the regular slope coefficients from the standardized data, and then compare them to standardized coefficients obtained from the original regression equation. To standardize the data, new variables are created whereby the sample mean is subtracted from each observation and the resulting difference is divided by the standard deviation. This is done for each independent and dependent variable, causing each variable to have a zero mean and a standard deviation of one. If there are no appreciable differences between the standardized slope coefficients from the original equation and the regular slope coefficients from the standardized data, then there is less chance of a problem due to multicollinearity. Standardizing the X's is also a suggested procedure to stabilize the estimate of the condition index (Chatterjee and Hadi, 1988).

Another indication of lack of stability is a change in the coefficients when the sample size changes. It is possible to take a random sample of the cases using features provided by the statistical packages. If the investigator randomly splits the sample in half and obtains widely different slope coefficients in the two halves, this may indicate a multicollinearity problem. But note that unstable coefficients could also be caused by outliers.

Before shifting to formal techniques for handling multicollinearity, there are straightforward ways to detect the possible cause of the problem. The first is to check for outliers. An outlier that has large leverage can artificially inflate the correlation coefficients between the particular X variable involved and other X variables. Removal of outlier(s) that inflate the correlations may be enough to solve the problem.

The second step is to check for relationships among some of the X variables that will lead to problems. For example, if a researcher included mean arterial blood pressure along with systolic and diastolic blood pressure as X variables, then problems will occur since there is a linear relationship among these three variables. In this case, the simple solution is to remove one or two of the three variables. Also, if two X variables are highly intercorrelated, multicollinearity is likely and again removing one of them could solve the problem. When this solution is used, note that the estimate of the slope coefficient for the X variable that is kept will likely not be the same as it would be if both intercorrelated X variables were included. Hence the investigator has to decide whether this solution creates conceptual problems in interpreting the results of the regression model.

Another technique that may be helpful if multicollinearity is not too severe is to perform the regression analysis on the standardized variables or on so-called centered variables, where the mean is subtracted from each variable (both the X variables and Y variables). Centering reduces the magnitude of the variables and thus tends to lessen rounding errors. When centered variables are used, the intercept is zero, but the estimates of the slope coefficients should remain unchanged. With standardized variables the intercept is zero and the estimates of the coefficients are different from those obtained using the original data (Section 7.7). Some statistical packages always use the correlation matrix in their regression computations to lessen computational problems and then modify them to compute the usual regression coefficients.

If the above methods are insufficient to alleviate the multicollinearity, then the use of ridge regression, which produces biased estimates, can be considered, as discussed in the next section.

9.6   Ridge regression

In the previous section, methods for handling multicollinearity by either excluding variables or modifying them were given, and also in Chapter 8 where variable selection methods were described. However, there are situations where the investigator wishes to use several independent variables that are highly intercorrelated. For example, such will be the case when the independent variables are the prices of commodities or highly related physiological variables on animals. There are two major methods for performing regression analysis when multicollinearity exists. The first method, which will be explained next, uses a technique called ridge regression. The second technique uses a principal components regression and will be described in Section 14.6.

Theoretical background

One solution to the problem of multicollinearity is the so-called **ridge regression procedure** (Marquardt and Snee, 1975; Hoerl and Kennard, 1970; Gunst and Mason, 1980). In effect, ridge regression artificially reduces the correlations among the independent variables in order to obtain more stable estimates of the regression coefficients. Note that although such estimates are biased, they may, in fact, produce a smaller mean square error for the estimates.

To explain the concept of ridge regression, we must follow through a theoretical presentation. Readers not interested in this theory may skip to the discussion of how to perform the analysis, using ordinary least squares regression programs.

We will restrict our presentation to the standardized form of the observations, i.e., where the mean of each variable is subtracted from the observation and the difference is divided by the standard deviation of the variable. The resulting least squares regression equation will automatically have a zero intercept and standardized regression coefficients. These coefficients are functions of only the correlation matrix among the X variables, namely,

$$\begin{bmatrix} 1 & r_{12} & r_{13} & \cdots & r_{1P} \\ r_{12} & 1 & r_{23} & \cdots & \\ r_{13} & r_{23} & 1 & \cdots & \vdots \\ \vdots & \vdots & \vdots & & \\ r_{1P} & r_{2P} & r_{3P} & \cdots & 1 \end{bmatrix}$$

The instability of the least squares estimates stems from the fact that some of the independent variables can be predicted accurately by the other independent variables. (For those familiar with matrix algebra, note that this feature results in the correlation matrix having a nearly zero determinant.) The ridge regression method artificially inflates the diagonal elements of the correlation matrix by adding a positive amount k to each of them. The correlation matrix is modified to

$$
\begin{bmatrix}
1+k & r_{12} & r_{13} & \cdots & r_{1P} \\
r_{12} & 1+k & r_{23} & \cdots & \\
r_{13} & r_{23} & 1+k & \cdots & \vdots \\
\vdots & \vdots & \vdots & & \\
r_{1P} & r_{2P} & r_{3P} & \cdots & 1+k
\end{bmatrix}
$$

The value of k can be any positive number and is usually determined empirically, as will be described later.

Examples of ridge regression

For $P = 2$, i.e., with two independent variables X_1 and X_2, the ordinary least squares **standardized** coefficients are computed as

$$
b_1 = \frac{r_{1Y} - r_{12}r_{2Y}}{1 - r_{12}^2}
$$

and

$$
b_2 = \frac{r_{2Y} - r_{12}r_{1Y}}{1 - r_{12}^2}
$$

The ridge estimators turn out to be

$$
b_1^* = \frac{r_{1Y} - [r_{12}/(1+k)]r_{2Y}}{1 - [r_{12}/(1+k)]^2} \left(\frac{1}{1+k} \right)
$$

and

$$
b_2^* = \frac{r_{2Y} - [r_{12}/(1+k)]r_{1Y}}{1 - [r_{12}/(1+k)]^2} \left(\frac{1}{1+k} \right)
$$

Note that the main difference between the ridge and least squares coefficients is that r_{12} is replaced by $r_{12}/(1+k)$, thus artificially reducing the correlation between X_1 and X_2.

For a numerical example, suppose that $r_{12} = 0.9, r_{1Y} = 0.3$, and $r_{2Y} = 0.5$. Then the standardized least squares estimates are

$$
b_1 = \frac{0.3 - (0.9)(0.5)}{1 - (0.9)^2} = -0.79
$$

and

$$
b_2 = \frac{0.5 - (0.9)(0.3)}{1 - (0.9)^2} = 1.21
$$

For a value of $k = 0.4$ the ridge estimates are

$$b_1^* = \frac{0.3 - [0.9/(1+0.4)](0.5)}{1 - [0.9/(1+0.4)]^2} \left(\frac{1}{1+0.4}\right) = -0.026$$

and

$$b_2^* = \frac{0.5 - [0.9/(1+0.4)](0.3)}{1 - [0.9/(1+0.4)]^2} \left(\frac{1}{1+0.4}\right) = 0.374$$

Note that both coefficients are reduced in magnitude from the original slope coefficients. We computed several other ridge estimates for $k = 0.05$ to 0.5. In practice, these estimates are plotted as a function of k, and the resulting graph, as shown in Figure 9.3, is called the **ridge trace**.

The ridge trace should be supplemented by a plot of the residual mean square of the regression equation. From the ridge trace together with the residual mean square graph, it becomes apparent when the coefficients begin to stabilize. That is, an increasing value of k produces a small effect on the coefficients. The value of k at the point of stabilization (say k^*) is selected to compute the final ridge coefficients. The final ridge estimates are thus a compromise between bias and inflated coefficients.

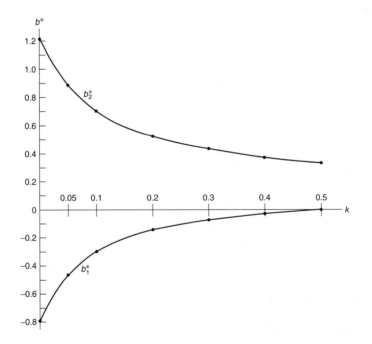

Figure 9.3: *Ridge Regression Coefficients for Various Values of k (Ridge Trace)*

For the above example, as seen from Figure 9.3, the value of $k^* = 0.2$ seems to represent that compromise. Values of k greater than 0.2 do not produce appreciable changes in the coefficients, although this decision is a subjective one on our part. Unfortunately, no objective criterion has been developed to determine k^* from the **sample**. Proponents of ridge regression agree that the ridge estimates will give better predictions in the **population**, although they do not fit the **sample** as well as the least squares estimates. Further discussion of the properties of ridge estimators can be found in Gruber (2010).

Implementation on the computer

Ordinary least squares regression programs such as those discussed in Chapter 7 can be used to obtain ridge estimates. As usual, we denote the dependent variable by Y and the P independent variables by $X_1, X_2, \ldots, X_P$. First, the Y and X variables are standardized by creating new variables in which we subtract the mean and divide the difference by the standard deviation for each variable. Then P additional dummy observations are added to the data set. In these dummy observations the value of Y is always set equal to zero.

For a specified value of k the values of the X variables are defined as follows:

1. Compute $[k(N-1)]^{1/2}$. If $N = 20$ and $k = 0.2$, then $[0.2(19)]^{1/2} = 1.95$.
2. The **first** dummy observation has $X_1 = [k(N-1)]^{1/2}$, $X_2 = 0, \ldots, X_P = 0$ and $Y = 0$.
3. The **second** dummy observation has $X_1 = 0, X_2 = [k(N-1)]^{1/2}$, $X_3 = 0, \ldots, X_P = 0$, $Y = 0$, etc.
4. The Pth dummy observation has $X_1 = 0, X_2 = 0, X_3 = 0, \ldots, X_{P-1} = 0, X_P = [k(N-1)]^{1/2}$ and $Y = 0$.

For a numerical example, suppose $P = 2, N = 20$, and $k = 0.2$. The two additional dummy observations are as follows:

X_1	X_2	Y
1.95	0	0
0	1.95	0

With the N regular observations and these P dummy observations, we use an ordinary least squares regression program in which the intercept is **forced** to be **zero**. The regression coefficients obtained from the output are automatically the ridge coefficients for the specified k. The resulting residual mean square should also be noted. Repeating this process for various values of k will provide the information necessary to plot the ridge trace of the coefficients, similar to Figure 9.3, and the residual mean square. The usual range of interest of k is between zero and one.

Some regression programs allow the user to fit the regression model starting with the correlation matrix. With these programs the user can also perform ridge regression directly by modifying the correlation matrix as shown previously in this section (replace the 1's down the diagonal with $1 + k$). This is recommended for SAS and SPSS. For Stata the method of implementation just given is suggested. The STATISTICA regression program will perform ridge regression directly and provide suitable output from the results.

9.7 Summary

In this chapter we reviewed methods for handling some special problems in regression analysis and introduced some recent advances in the subject. We reviewed the literature on missing values and recommended that when there are few missing values the largest possible complete sample be selected from the data set and then used in the analysis. When there is an appreciable amount of missing values, it may be useful to replace missing values with estimates from the sample. Methods for single and multiple imputation were given and references are provided for obtaining further information. For a discussion of this and other relevant aspects of multiple regression, see Afifi *et al.* (2007).

We gave a detailed discussion of the use of dummy variables in regression analysis. Several situations exist where dummy variables are very useful. These include incorporating nominal or ordinal variables in equations. We also mentioned some extensions of these ideas, including regression trees. The ideas explained here can also be used in other multivariate analyses. One reason we went into detail in this chapter is to enable you to adapt these methods to other multivariate techniques.

Two methods were discussed for producing "biased" parameter estimates, i.e., estimates that do not average out to the true parameter value. The first method is placing constraints on the parameters. This method is useful when you wish to restrict the estimates to a certain range or when natural constraints on some function of the parameters must be satisfied. The second biased regression technique is ridge regression. This method is used only when you must employ variables that are known to be highly intercorrelated.

You should also consider examining the data set for outliers in X and possible redundant variables. The tolerance or variance inflation option provides a warning of which variable is causing the problem when you use a forward selection or forward stepwise selection method.

9.8 Problems

9.1 In the depression data set described in Chapter 3, data on educational level, age, sex, and income are presented for a sample of adults from Los Angeles County. Fit a regression plane with income as the dependent variable and the other variables as independent variables. Use a dummy variable for the variable SEX that was originally coded 1,2 by stating SEX = SEX − 1. Which sex is the referent group?

9.2 Repeat Problem 9.1, but now use a dummy variable for education. Divide the education level into three categories: did not complete high school, completed at least high school, and completed at least a bachelor's degree. Compare the interpretation you would make of the effects of education on income in this problem and in Problem 9.1.

9.3 In the depression data set, determine whether religion has an effect on income when used as an independent variable along with age, sex, and educational level.

9.4 Draw a ridge trace for the accompanying data.

	Variable			
Case	$X1$	$X2$	$X3$	Y
1	0.46	0.96	6.42	3.46
2	0.06	0.53	5.53	2.25
3	1.49	1.87	8.37	5.69
4	1.02	0.27	5.37	2.36
5	1.39	0.04	5.44	2.65
6	0.91	0.37	6.28	3.31
7	1.18	0.70	6.88	3.89
8	1.00	0.43	6.43	3.27
Mean	0.939	0.646	6.340	3.360
Standard deviation	0.475	0.566	0.988	1.100

9.5 Use the lung function data described in Appendix A. For the parents we wish to relate Y = weight to X = height for both men and women in a single equation. Using dummy variables, write an equation for this purpose, including an interaction term. Interpret the parameters. Run a regression analysis, and test whether the rate of change of weight versus height is the same for men and women. Interpret the results with the aid of appropriate graphs.

9.6 Another way to answer the question of interaction between the independent variables in Problem 7.13 is to define a dummy variable that indicates whether an observation is above the median weight, and an equivalent variable for height. Relate FEV1 for the fathers to these dummy variables, including an interaction term. Test the hypothesis that there is no interaction. Compare your results with those of Problem 7.13.

9.7 (Continuation of Problem 9.5.) Do a similar analysis for the first boy and girl. Include age and age squared in the regression equation.

9.8 Unlike the real data used in Problem 9.5, the accompanying data are "ideal" weights published by the Metropolitan Life Insurance Company for American men and women. Compute Y = midpoint of weight range for medium-framed men and women for the various heights shown in the table. Pretending that the results represent a real sample, repeat the analysis requested in Problem 9.5, and compare the results of the two analyses.

Weights of men (lb)[a]

Height	Small frame	Medium frame	Large frame
5 ft 2 in.	128–134	131–141	138–150
5 ft 3 in.	130–136	133–143	140–153
5 ft 4 in.	132–138	135–145	142–156
5 ft 5 in.	134–140	137–148	144–160
5 ft 6 in.	136–142	139–151	146–164
5 ft 7 in.	138–145	142–154	149–168
5 ft 8 in.	140–148	145–157	152–172
5 ft 9 in.	142–151	148–160	155–176
5 ft 10 in.	144–154	151–163	158–180
5 ft 11 in.	146–157	154–166	161–184
6 ft 0 in.	149–160	157–170	164–188
6 ft 1 in.	152–164	160–174	168–192
6 ft 2 in.	155–168	164–178	172–197
6 ft 3 in.	158–172	167–182	176–202
6 ft 4 in.	162–176	171–187	181–207

[a] Including 5 lb of clothing, and shoes with 1 in. heels.

Weights of women (lb)[a]

Height	Small frame	Medium frame	Large frame
4 ft 10 in.	102–111	109–121	118–131
4 ft 11 in.	103–113	111–123	120–134
5 ft 0 in.	104–115	113–126	122–137
5 ft 1 in.	106–118	115–129	125–140
5 ft 2 in.	108–121	118–132	128–143
5 ft 3 in.	111–124	121–135	131–147
5 ft 4 in.	114–127	124–138	134–151
5 ft 5 in.	117–130	127–141	137–155
5 ft 6 in.	120–133	130–144	140–159
5 ft 7 in.	123–136	133–147	143–163
5 ft 8 in.	126–139	136–150	146–167
5 ft 9 in.	129–142	139–153	149–170
5 ft 10 in.	132–145	142–156	152–173
5 ft 11 in.	135–148	145–159	155–176
6 ft 0 in.	138–151	148–162	158–179

[a] Including 3 lb of clothing, and shoes with 1 in. heels.

9.9 Use the data described in Problem 7.7. Since some of the X variables are intercorrelated, it may be useful to do a ridge regression analysis of Y on $X1$ to $X9$. Perform such an analysis, and compare the results to those of Problems 7.10 and 8.7.

9.10 (Continuation of Problem 9.8.) Using the data in the table given in Problem 9.8, compute the midpoints of weight range for all frame sizes for men and women separately. Pretending that the results represent a real sample, so that each height has three Y values associated with it instead of one, repeat the analysis of Problem 9.7. Produce the appropriate plots. Now repeat with indicator variables for the different frame sizes, choosing one as a referent group. Include any necessary interaction terms.

9.11 Take the family lung function data described in Appendix A and delete (label as missing) the height of the middle child for every family with ID divisible by 6, that is, families 6, 12, 18 etc. (To find these, look for those IDs with ID/6=integer part of (ID/6).) Delete the FEV1 of the middle child for families with ID divisible by 10. Assume these data are missing completely at random. Try the various imputation methods suggested in this chapter to fill in the missing data. Find the regression of FEV1 on height and weight using the imputed values. Compare the results using the imputed data with those from the original data set.

9.12 In the depression data set, define $Y = $ the square root of total depression score (CESD), X_1 = log(income), X_2 = Age, X_3 = Health and X_4 = Bed days. Set X_1 = missing whenever $X_3 = 4$ (poor health). Also set X_2 = missing whenever X_2 is between 50 and 59 (inclusive). Are these data missing at random? Try various imputation methods and obtain the regression of Y on $X_1, \ldots, X_4$ with the imputed values as well as using the complete cases only. Compare these results with those obtained from the original data (nothing missing).

9.13 Using the family lung function data, relate FEV1 to height for the oldest child in three ways: simple linear regression (Problem 6.9), regression of FEV1 on height squared, and spline regression (split at HEI = 64). Which method is preferable?

9.14 Perform a ridge regression analysis of the family lung function data using FEV1 of the oldest child as the dependent variable and height, weight and age of the oldest child as the independent variables.

9.15 Using the family lung function data, find the regression of height for the oldest child on mother's and father's height. Include a dummy variable for the sex of the child and any necessary interaction terms.

9.16 Using dummy variables, run a regression analysis that relates CESD as the dependent variable to marital status in the depression data set given in Chapter 3. Do it separately for males and females. Repeat using the combined group, but including a dummy variable for sex and any necessary interaction terms. Compare and interpret the results of the two regressions.

9.17 Using the data from the Parents HIV/AIDS study, for those adolescents who have started to use alcohol, predict the age when they first start their use (AGEALC). Predictive variables should include NGHB11 (drinking in the neighborhood) GENDER, HOWREL (how religious). Choose suitable referent groups for the three predictor variables and perform a regression analysis.

9.18 For the variables describing the average number of cigarettes smoked during the past 3 months (SMOKEP3M) and the variable describing the mother's education (EDUMO) in the Parental HIV data determine the percent with missing values. For each of the variables describe hypothetical scenarios which might have led to these values being a) missing completely at random, b) missing at random, but not completely at random, and c) neither missing completely at random nor missing at random.

Part III

Multivariate Analysis

Chapter 10

Canonical correlation analysis

10.1 Chapter outline

In the previous four regression chapters, there was a single outcome or dependent variable Y. In this chapter, we show how to analyze data where there are several dependent variables as well as several predictor or independent variables. Section 10.2 discusses canonical correlation analysis and gives some examples of its use. Section 10.3 introduces a data example using the depression data set. Section 10.4 presents basic concepts and tests of hypotheses, and Section 10.5 describes a variety of output options. The output available from the six computer programs is summarized in Section 10.6 and precautionary remarks are given in Section 10.7.

10.2 When is canonical correlation analysis used?

The technique of **canonical correlation analysis** is best understood by considering it as an extension of multiple regression and correlation analysis. In multiple regression analysis we find the best linear combination of P variables, $X_1, X_2, \ldots, X_P$, to predict one variable Y. The multiple correlation coefficient is the simple correlation between Y and its predicted value $\hat{Y}$. In multiple regression and correlation analysis our concern was therefore to examine the relationship between the X variables and the single Y variable.

In canonical correlation analysis we examine the linear relationships between a set of X variables and a set of more than one Y variable. The technique consists of finding several linear combinations of the X variables and the same number of linear combinations of the Y variables in such a way that these linear combinations best express the correlations between the two sets. Those linear combinations are called the **canonical variables**, and the correlations between corresponding pairs of canonical variables are called **canonical correlations**.

In a common application of this technique the Y's are interpreted as outcome or dependent variables, while the X's represent independent or predictive variables. The Y variables may be harder to measure than the X variables, as in the calibration situation discussed in Section 6.11.

Canonical correlation analysis applies to situations in which regression techniques are appropriate and where there exists more than one dependent variable. It is especially useful when the dependent or criterion variables are moderately intercorrelated, so it does not make sense to treat them separately, ignoring the interdependence. Another useful application is for testing independence between the sets of Y and X variables. This application will be discussed further in Section 10.4.

An example of an early application is given by Waugh (1942); he studied the relationship between characteristics of certain varieties of wheat and characteristics of the resulting flour. Waugh was able to conclude that desirable wheat was high in texture, density, and protein content, and low on damaged kernels and foreign materials. Similarly, good flour should have high crude protein content and low scores on wheat per barrel of flour and ash in flour. Another early application of canonical correlation is described by Meredith (1964) where he calibrates two sets of intelligence tests given to the same individuals. The literature also includes several health-related applications of canonical correlation analysis.

Canonical correlation analysis is one of the less commonly used multivariate techniques. Its limited use may be due, in part, to the difficulty often encountered in trying to interpret the results.

10.3 Data example

The depression data set presented in previous chapters is used again here to illustrate canonical correlation analysis. We select two dependent variables, CESD and health. The variable CESD is the sum of the scores on the 20 depression scale items; thus a high score indicates likelihood of depression. Likewise, "health" is a rating scale from 1 to 4, where 4 signifies poor health and 1 signifies excellent health. The set of independent variables includes "sex," transformed so that 0 = male and 1 = female; "age" in years; "education," from 1 = less than high school up to 7 = doctorate; and "income" in thousands of dollars per year.

The summary statistics for the data are given in Tables 10.1 and 10.2. Note that the average score on the depression scale (CESD) is 8.9 in a possible range of 0 to 60. The average on the health variables is 1.8, indicating an average perceived health level falling between excellent and good. The average educational level of 3.5 shows that an average person has finished high school and perhaps attended some college.

Examination of the correlation matrix in Table 10.2 shows neither CESD nor health is highly correlated with any of the independent variables. In fact, the highest correlation in this matrix is between education and income. Also, CESD is negatively correlated with age, education, and income (the younger, less-educated and lower-income person tends to be more depressed). The positive correlation between CESD and sex shows that females tend to be more depressed than males. Persons who perceived their health as good are more apt to be high on income and education and low on age.

Table 10.1: *Means and standard deviations for depression data set*

Variable	Mean	Standard deviation
CESD	8.88	8.82
Health	1.77	0.84
Sex	0.62	0.49
Age	44.41	18.09
Education	3.48	1.31
Income	20.57	15.29

Table 10.2: *Correlation matrix for depression data set*

	CESD	Health	Sex	Age	Education	Income
CESD	1.000	0.212	0.124	-0.164	-0.101	-0.158
Health		1.000	0.098	0.304	-0.270	-0.183
Sex			1.000	0.044	-0.106	-0.180
Age				1.000	-0.208	-0.192
Education					1.000	0.492
Income						1.000

In the following sections we will examine the relationships between the dependent variables (perceived health and depression) and the set of independent variables.

10.4 Basic concepts of canonical correlation

Suppose we desire to examine the relationship between a set of variables $X_1, X_2, \ldots, X_P$ and another set $Y_1, Y_2, \ldots, Y_Q$. The X variables can be viewed as the independent or predictor variables, while the Y's are considered dependent or outcome variables. We assume that in any given sample the mean of each variable has been subtracted from the original data so that **the sample means of all X and Y variables are zero**. In this section we discuss how the degree of association between the two sets of variables is assessed and we present some related tests of hypotheses.

First canonical correlation

The basic idea of canonical correlation analysis begins with finding one linear combination of the Y's, say

$$U_1 = a_1 Y_1 + a_2 Y_2 + \cdots + a_Q Y_Q$$

and one linear combination of the X's, say

$$V_1 = b_1 X_1 + b_2 X_2 + \cdots + b_P X_P$$

For any particular choice of the coefficients (a's and b's) we can compute values of U_1 and V_1 for each individual in the sample. From the N individuals in the sample we can then compute the simple correlation between the N pairs of U_1 and V_1 values in the usual manner. The resulting correlation depends on the choice of the a's and b's.

In canonical correlation analysis we select values of a and b coefficients so as to **maximize** the correlation between U_1 and V_1. With this particular choice the resulting linear combination U_1 is called the **first canonical variable** of the Y's and V_1 is called the first canonical variable of the X's. Note that both U_1 and V_1 have a mean of zero. The resulting correlation between U_1 and V_1 is called the **first canonical correlation**. The square of the first canonical correlation is often called the first **eigenvalue**.

The first canonical correlation is thus the highest possible correlation between a linear combination of the X's and a linear combination of the Y's. In this sense it is the maximum linear correlation between the set of X variables and the set of Y variables. The first canonical correlation is analogous to the multiple correlation coefficient between a single Y variable and the set of X variables. The difference is that in canonical correlation analysis we have several Y variables and we must find a linear combination of them also.

The SAS CANCORR program computed the a's and b's, as shown in Table 10.3. Note that the first set of coefficients is used to compute the values of the canonical variables U_1 and V_1.

Table 10.4 shows the process used to compute the canonical correlation. For each individual we compute V_1 from the b coefficients and the individual's X variable values after subtracting the means. We do the same for U_1. These computations are shown for the first three individuals. The correlation coefficient is then computed from the 294 values of U_1 and V_1. Note that the variances of U_1 and V_1 are each equal to 1.

The standardized coefficients are also shown in Table 10.3, and they are to be used with the standardized variables. The standardized coefficients can be obtained by multiplying each unstandardized coefficient by the standard deviation of the corresponding variable. For example, the unstandardized coefficient of y_1 (CESD) is $a_1 = -0.0555$, and from Table 10.1 the standard deviation of y_1 is 8.82. Therefore the standardized coefficient of y_1 is -0.490.

Table 10.3: *Canonical correlation coefficients for first correlation (depression data set)*

Coefficients	Standardized coefficients
$b_1 = 0.051$ (sex)	$b_1 = 0.025$ (sex)
$b_2 = 0.048$ (age)	$b_2 = 0.871$ (age)
$b_3 = -0.29$ (education)	$b_3 = -0.383$ (education)
$b_4 = +0.005$ (income)	$b_4 = 0.082$ (income)
$a_1 = -0.055$ (CESD)	$a_1 = -0.490$ (CESD)
$a_2 = 1.17$ (health)	$a_2 = +0.982$ (health)

In this example the resulting canonical correlation is 0.405. This value represents the highest possible correlation between any linear combination of the independent variables and any linear combination of the dependent variables. In particular, it is larger than any simple correlation between an X variable and a Y variable (Table 10.2). One method for interpreting the linear combination is by examining the standardized coefficients. For the X variables the canonical variable is determined largely by age and education. Thus a person who is relatively old and relatively uneducated would score high on canonical variable V_1. The canonical variable based on the Y's gives a large positive weight to the perceived health variables and a negative weight to CESD. Thus a person with a high health value (perceived poor health) and a low depression score would score high on canonical variable U_1. In contrast, a young person with relatively high education would score low on V_1, and a person in good perceived health but relatively depressed would score low on U_1. Sometimes, because of high intercorrelations between two variables in the same set, one variable may result in another having a small coefficient and thus make the interpretation difficult. No very high correlations within a set existed in the present example.

In summary, we conclude that older but uneducated people tend to be not depressed although they perceive their health as relatively poor. Because the first canonical correlation is the largest possible, this impression is the strongest conclusion we can make from this analysis of the data. However, there may be other important conclusions to be drawn from the data, which will be discussed in the next section.

It should be noted that interpreting canonical coefficients can be difficult, especially when two X variables are highly intercorrelated or if, say, one X variable is almost a linear combination of several other X variables (see Manly, 2004). The same holds true for Y variables. Careful examination of the correlation matrix or of scatter diagrams of each variable with all the other variables is recommended.

Table 10.4: *Computation of U_1 and V_1*

Individual	$V_1 =$	Sex $+b_1(X_1 - \bar{X}_1)$	Age $+b_2(X_2 - \bar{X}_2)$	Education $+b_3(X_3 - \bar{X}_3)$	Income $+b_4(X_4 - \bar{X}_4)$
1	$1.49 =$	$+0.051(1 - 0.62)$	$+0.048(68 - 44.4)$	$-0.29(2 - 3.48)$	$+0.0054(1 - 20.57)$
2	$0.44 =$	$+0.051(0 - 0.62)$	$+0.048(58 - 44.4)$	$-0.29(4 - 3.48)$	$+0.0054(15 - 20.57)$
3	$0.23 =$	$+0.051(1 - 0.62)$	$+0.048(45 - 44.4)$	$-0.29(3 - 3.48)$	$+0.0054(28 - 20.57)$
.					
.					
294					

Individual	$U_1 =$	CESD $a_1(Y_1 - \bar{Y}_1)$	Health $+a_2(Y_2 - \bar{Y}_2)$
1	$0.76 =$	$-0.055(0 - 8.88)$	$+1.17(2 - 1.77)$
2	$-0.64 =$	$-0.055(4 - 8.88)$	$+1.17(1 - 1.77)$
3	$-0.54 =$	$-0.055(4 - 8.88)$	$+1.17(2 - 1.77)$
.			
.			
294			

Other canonical correlations

Additional interpretation of the relationship between the X's and the Y's is obtained by deriving other sets of canonical variables and their corresponding canonical correlations. Specifically, we derive a second canonical variable V_2 (linear combination of the X's) and a corresponding canonical variable U_2 (linear combination of the Y's). The coefficients for these linear combinations are chosen so that the following conditions are met.

1. V_2 is uncorrelated with V_1 and U_1.
2. U_2 is uncorrelated with V_1 and U_1.
3. Subject to conditions 1 and 2, U_2 and V_2 have the maximum possible correlation.

The correlation between U_2 and V_2 is called the **second canonical correlation** and will necessarily be less than or equal to the first canonical correlation.

In our example the second set of canonical variables expressed in terms of the standardized coefficients is

$$V_2 = 0.396(\text{sex}) - 0.443(\text{age}) - 0.448(\text{education}) - 0.555(\text{income})$$

and

$$U_2 = 0.899(\text{CESD}) + 0.288(\text{health})$$

Note that U_2 gives a high positive weight to CESD and a low positive weight to health. In contrast, V_2 gives approximately the same moderate weight to all four variables, with sex having the only positive coefficient. A large value of V_2 is associated with young, poor, uneducated females. A large value of U_2 is associated with a high value of CESD (depressed) and to a lesser degree with a high value of health (poor perceived health). The value of the second canonical correlation is 0.266.

In general, this process can be continued to obtain other sets of canonical variables $U_3, V_3; U_4, V_4$; etc. The maximum number of canonical correlations and their corresponding sets of canonical variables is equal to the minimum of P (the number of X variables) and Q (the number of Y variables). In our data example $P = 4$ and $Q = 2$, so the maximum number of canonical correlations is two.

Tests of hypotheses

Most computer programs print the coefficients for all of the canonical variables, the values of the canonical correlations, and the values of the canonical variables for each individual in the sample (canonical variable scores). Another common feature is a test of the null hypothesis that the k smallest population canonical correlations are zero. Two tests may be found, either Bartlett's chi-square test (Bartlett, 1941; Lawley, 1959) or an approximate F test (Rao, 1973). These tests were derived with the assumption that the X's and Y's are jointly distributed according to a multivariate normal distribution. A large chi-square or a large F is an indication that not all of those k population correlations are zero.

In our data example, the approximate F test that both population canonical correlations are zero was computed by the CANCORR procedure to be $F = 9.68$ with 8 and 576 degrees of freedom and $P = 0.0001$. So we can conclude that at least one population canonical correlation is nonzero and proceed to test that the smallest one is zero. The F value equals 7.31 with 3 and 289 degrees of freedom. The P value is 0.0001 again, and we conclude that both canonical correlations are significantly different from zero. Similar results were obtained from Bartlett's test from STATISTICA.

In data sets with more variables these tests can be a useful guide for selecting the number of significant canonical correlations. The test results are examined to determine at which step the remaining canonical correlations can be considered zero. In this case, as in stepwise regression, the significance level should not be interpreted literally.

10.5   Other topics in canonical correlation

In this section we discuss some useful optional output available from packaged programs.

Plots of canonical variable scores

A useful option available in some programs is a plot of the canonical variable score U_i versus V_i. In Figure 10.1 we show a scatter diagram of U_1 versus V_1 for the depression data. The first individual from Table 10.4 is indicated on the graph. The degree of scatter gives the impression of a somewhat weak but significant canonical correlation (0.405). For multivariate normal data the graph would approximate an ellipse of concentration. Such a plot can be useful in highlighting unusual cases in the sample as possible outliers or blunders. For example, the individual with the lowest value on U_1 is case number 289. This individual is a 19-year-old female with some high school education and with a $28,000 per-year income. These scores produce a value of $V_1 = -0.73$. Also, this woman perceives her health as excellent (1) and is very depressed (CESD = 47), resulting in $U_1 = -3.02$. This individual, then, represents an extreme case in that she is uneducated and young but has a good income. In spite of the fact that she perceives her health as excellent, she is extremely depressed. Although this case gives an unusual combination, it is not necessarily a blunder.

The plot of U_1 versus V_1 does not result in an apparently nonlinear scatter diagram, nor does it look like a bivariate normal distribution (elliptical in

shape). It may be that the skewness present in the CESD distribution has resulted in a somewhat skewed pattern for U_1 even though health has a greater
overall effect on the first canonical variable. If this pattern were more extreme,
it might be worthwhile to consider transformations on some of the variables,
such as CESD.

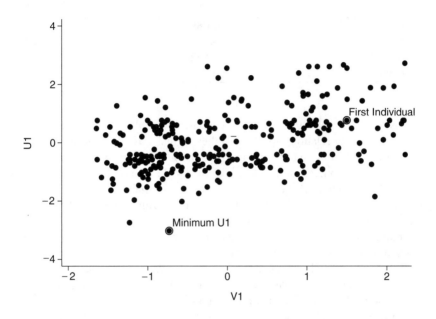

Figure 10.1: *Plot of 294 Pairs of Values of the Canonical Variables U_1 and V_1 for the
Depression Data Set (Canonical Correlation = 0.405)*

Another interpretation of canonical variables

Another useful optional output is the set of correlations between the canonical
variables and the original variables used in deriving them. This output provides
a way of interpreting the canonical variables when some of the variables within
either the set of independent or the set of dependent variables are highly intercorrelated with each other. For the depression data example these correlations
are as shown in Table 10.5. These correlations are sometimes called **canonical
variable loadings**. Other terms are **canonical loadings** and **canonical structural coefficients**.

Since the canonical variable loadings can be interpreted as simple correla-

Table 10.5: *Correlations between canonical variables and corresponding variables (depression data set)*

	U_1	U_2
CESD	-0.281	0.960
Health	0.878	0.478
	V_1	V_2
Sex	0.089	0.525
Age	0.936	-0.225
Education	-0.532	-0.636
Income	-0.254	-0.737

tions between each variable and the canonical variable, they are useful in understanding the relationship between the original variables and the canonical variables. When the set of variables used in one canonical variable are uncorrelated, the canonical variable loadings are equal to the standardized canonical variable coefficients. When some of the original variables are highly intercorrelated, the loadings and the coefficients can be quite different. It is in these cases that some statisticians find it simpler to try to interpret the canonical variable loadings rather than the canonical variable coefficients. For example, suppose that there are two X variables that are highly positively correlated, and that each is positively correlated with the canonical variable. Then it is possible that one canonical variable coefficient will be positive and one negative, while the canonical variable loadings are both positive, the result one expects.

In the present data set, the intercorrelations among the variables are neither zero nor strong. Comparing the results of Tables 10.5 and 10.3 for the first canonical variable shows that the standardized coefficients have the same signs as the canonical variable loadings, but with somewhat different magnitudes.

Redundancy analysis

The average of the squared canonical variable loadings for the first canonical variate, V_1, gives the proportion of the variance in the X variables explained by the first canonical variate. The same is true for U_1 and Y. Similar results hold for each of the other canonical variates. For example, for U_1 we have $[(-0.281)^2 + 0.878^2]/2 = 0.425$, or less than half of the variance in the Y's is explained by the first canonical variate. Sometimes the proportion of variance explained is quite low, even though there is a large canonical correlation. This may be due to only one or two variables having a major influence on the canonical variate.

The above computations provide one aspect of what is known as **redun-**

dancy analysis. In addition, SAS CANCORR, STATISTICA Canonical analysis, and SPSS can compute a quantity called the **redundancy coefficient** which is also useful in evaluating the adequacy of the prediction from the canonical analysis (Muller, 1981). The coefficient is a measure of the average proportion of variance in the Y set that is accounted for by the V set. It is comparable to the squared multiple correlation in multiple linear regression analysis. It is also possible to obtain the proportion of variance in the X variables that is explained by the U variables, but this is usually of less interest.

10.6 Discussion of computer programs

R, S-PLUS, SAS, Stata, and STATISTICA each contains a special purpose canonical correlation program. SPSS MANOVA also performs canonical correlation analysis. In general, the interpretation of the results is simpler from special purpose programs since all the output relates directly to canonical correlation analysis. The various options for these programs are summarized in Table 10.6.

For the data example described in Section 10.3 we want to relate reported health and depression levels to several typical demographic variables. The SAS CANCORR program was used to obtain the results reported in this chapter. The data set "depress1.dat" was located in drive B. The CANCORR procedure is very straightforward to run. We simply call the procedure and specify the following:

```
FILENAME TEST1 ''B:depress1.dat'';
DATA ONE;
   INFILE TEST1;
   INPUT v1 SEX AGE v4 EDUCAT v6 INCOME v8--v28 CESD v30--v31
   HEALTH v33--v37;
PROC CANCORR DATA = ONE;
VAR CESD HEALTH;
WITH SEX AGE EDUCAT INCOME;
RUN;
```

The OUT = Data set produces a new data set that includes the original variables plus the canonical variable scores which can then be plotted.

The output in Stata is less complete but it does include estimates for the standard errors of the canonical coefficients, along with confidence limits, and a test that they are zero. The Stata canon program simply asks for the dependent and independent variables. The original data could be standardized prior to running canon to obtain standardized coefficients. The canonical variable scores could be obtained from the canonical coefficients following the example in Table 10.4 and they could then be plotted.

Table 10.6: *Software commands and output for canonical correlation analysis*

	S-PLUS/R	SAS	SPSS	Stata	STATISTICA
Correlation matrix	cancor	CANCORR	MANOVA	canon	Canonical Analysis
Canonical correlations	cancor	CANCORR	MANOVA	estat correlations	Canonical Analysis
Canonical coefficients	cancor	CANCORR	MANOVA	canon	
Standardized canonical coefficients		CANCORR	MANOVA	canon	Canonical Analysis
Canonical variable loadings	calculate from cancor	CANCORR	MANOVA	estat loadings	Canonical Analysis
Canonical variable scores		CANCORR	MANOVA	predict	Canonical Analysis
Plot of canonical variable score	plot	PLOT	DISCRIMINANT	graph	Canonical Analysis
Bartlett's test and P value			ONEWAY		Canonical Analysis
Wilk's lambda and F approx.	cancor	CANCORR	MANOVA	canon	Canonical Analysis
Redundancy analysis	cancor	CANCORR	MANOVA		Canonical Analysis

The STATISTICA canonical correlation options are quite complete. In STATISTICA the results are obtained by first telling the program which variables you wish to use and then pointing and clicking on the options you desire. In STATISTICA the standardized coefficients are called canonical weights. Since STATISTICA prints the standardized coefficients, the first step is to obtain the unstandardized coefficients by dividing by the standard deviation of the corresponding variable (Section 10.4). Then the computation given in Table 10.4 is performed using transformation options. The results labeled "factors structure" are the correlations between the canonical variables and the corresponding variables given in Table 10.5. The numerical value of the second canonical correlation is given with the chi-square test results.

With SPSS MANOVA the user specifies the variables and output. An example is included in the manual.

10.7   What to watch out for

Because canonical correlation analysis can be viewed as an extension of multiple linear regression analysis, many of the precautionary remarks made at the ends of Chapters 6–8 apply here. The user should be aware of the following points:

1. The sample should be representative of the population to which the investigator wishes to make inferences. A simple random sample has this property. If this is not attainable, the investigator should at least make sure that the cases are selected in such a way that the full range of observations occurring in the population can occur in the sample. If the range is artificially restricted, the estimates of the correlations will be affected.

2. Poor reliability of the measurements can result in lower estimates of the correlations among the X's and among the Y's.

3. A search for outliers should be made by obtaining histograms and scatter diagrams of pairs of variables.

4. Stepwise procedures are not available in the programs described in this chapter. Variables that contribute little and are not needed for theoretical models should be candidates for removal. It may be necessary to run the programs several times to arrive at a reasonable choice of variables.

5. The investigator should check that the canonical correlation is large enough to make examination of the coefficients worthwhile. In particular, it is important that the correlation not be due to just one dependent variable and the independent variable. The proportion of variance should be examined, and if it is small then it may be sensible to reduce the number of variables in the model.

6. If the sample size is large enough, it is advisable to split it, run a canonical analysis on both halves, and compare the results to see if they are similar.

7. If the canonical coefficients and the canonical variable loadings differ considerably (i.e., if they have different signs), then both should be examined carefully to aid in interpreting the results. Problems of interpretation are often more difficult in the second or third canonical variates than in the first. The condition that subsequent linear combinations of the variables be independent of those already obtained places restrictions on the results that may be difficult to understand.

8. Tests of hypotheses regarding canonical correlations assume that the joint distribution of the X's and Y's is multivariate normal. This assumption should be checked if such tests are to be reported.

9. Since canonical correlation uses both a set of Y variables and a set of X variables, the total number of variables included in the analysis may be quite large. This can increase the problem of many cases being not used because of missing values. Either careful choice of variables or imputation techniques may be required.

10.8   Summary

In this chapter we presented the basic concepts of canonical correlation analysis, an extension of multiple regression and correlation analysis. The extension is that the dependent variable is replaced by two or more dependent variables. If Q, the number of dependent variables, equals 1, then canonical correlation reduces to multiple regression analysis.

In general, the resulting canonical correlations quantify the strength of the association between the dependent and independent sets of variables. The derived canonical variables show which combinations of the original variables best exhibit this association. The canonical variables can be interpreted in a manner similar to the interpretation of principal components or factors, as will be explained in Chapters 14 and 15.

10.9   Problems

10.1 For the depression data set, perform a canonical correlation analysis between the following.

- Set 1: AGE, MARITAL (married versus other), EDUCAT (high school or less versus other), EMPLOY (full-time versus other), and INCOME.
- Set 2: the last seven variables.

Perform separate analyses for men and women. Interpret the results.

10.2 For the data set given in Appendix A, perform a canonical correlation analysis on height, weight, FVC, and FEV1 for fathers versus the same variables for mothers. Interpret the results.

10.3 For the chemical companies' data given in Table 8.1, perform a canonical correlation analysis using P/E and EPS5 as dependent variables and the remaining variables as independent variables. Write the interpretations of the significant canonical correlations in terms of their variables.

10.4 Using the data described in Appendix A, perform a canonical correlation analysis using height, weight, FVC, and FEV1 of the oldest child as the dependent variables and the same measurements for the parents as the independent variables. Interpret the results.

10.5 Generate the sample data for $X1, X2,\ldots, X9, Y$ as in Problem 7.7. For each of the following pairs of sets of variables, perform a canonical correlation analysis between the variables in Set 1 and those in Set 2.

Set 1	Set 2
a. $X1, X2, X3$	$X4–X9$
b. $X4, X5, X6$	$X7, X8, X9$
c. $Y, X9$	$X1, X2, X3$

Interpret the results in light of what you know about the population.

10.6 For the depression data set run a canonical analysis using HEALTH, ACUTEILL, CHRONILL, and BEDDAYS as independent variables and AGE, SEX (transformed to 0, 1), EDUCAT, and INCOME as independent variables. Interpret the results.

10.7 For the Parental HIV data set, run a canonical correlation analysis using LIVWITH, JOBMO, and EDUMO as predictor variables and HOWREL and SCHOOL as dependent variables. Interpret the results.

Chapter 11

Discriminant analysis

11.1 Chapter outline

Discriminant analysis is used to classify a case into one of two or more populations. Two methods are given in this book for classification into populations or groups, discriminant analysis and logistic regression analysis (Chapter 12). In both of these analyses, you must know which population the individual belongs to for the sample initially being analyzed (also called the training sample). Either analysis will yield information that can be used to classify future individuals whose population membership is unknown into one of two or more populations. If population membership is unknown in the training sample, the cluster analysis given in Chapter 16 could be used.

Sections 11.2 and 11.3 provide further discussion on when discriminant function analysis is useful and introduce an example that will show how to classify individuals as depressed or not depressed. Section 11.4 presents the basic concepts used in classification as given by Fisher (1936) and shows how they apply to the example. The interpretation of the effects of each variable based on the discriminant function coefficients is also presented. The assumptions made in making inferences from discriminant analysis are discussed in Section 11.5.

Interpretation of various output options from the statistical programs for two populations is given in Section 11.6. Often in practice, the number of cases in the two populations are unequal; in Section 11.7 a method of adjusting for this is given. Also, if the seriousness of misclassification into one group is higher than in the other, a technique for incorporating costs of misclassification into discriminant function analysis is given in the same section. Section 11.8 gives additional methods for determining how well the discriminant function will classify cases in the overall population. Section 11.9 presents formal tests to determine if the discriminant function results in better classification of cases than chance alone. In Section 11.10, the use of stepwise procedures in discriminant function analysis is discussed.

Section 11.11 summarizes the output from the various computer programs and Section 11.12 lists what to watch out for in doing discriminant function analysis.

11.2   When is discriminant analysis used?

Discriminant analysis techniques are used to classify individuals into one of two or more alternative groups (or populations) on the basis of a set of measurements. The populations are known to be distinct, and each individual belongs to one of them. These techniques can also be used to identify which variables contribute to making the classification. Thus, as in regression analysis, we have two uses, prediction and description.

As an example, consider an archeologist who wishes to determine which of two possible tribes created a particular statue found in a dig. The archeologist takes measurements on several characteristics of the statue and must decide whether these measurements are more likely to have come from the distribution characterizing the statues of one tribe or from the other tribe's distribution. The distributions are based on data from statues known to have been created by members of one tribe or the other. The problem of classification is therefore to guess who made the newly found statue on the basis of measurements obtained from statues whose identities are certain.

The measurements on the new statue may consist of a single observation, such as its height. However, we would then expect a low degree of accuracy in classifying the new statue since there may be quite a bit of overlap in the distribution of heights of statues from the two tribes. If, on the other hand, the classification is based on several characteristics, we would have more confidence in the prediction. The discriminant analysis methods described in this chapter are multivariate techniques in the sense that they employ several measurements.

As another example, consider a loan officer at a bank who wishes to decide whether to approve an applicant's automobile loan. This decision is made by determining whether the applicant's characteristics are more similar to those persons who in the past repaid loans successfully or to those persons who defaulted. Information on these two groups, available from past records, would include factors such as age, income, marital status, outstanding debt, and home ownership.

A third example, which is described in detail in the next section, comes from the depression data set (Chapters 1 and 3). We wish to predict whether an individual living in the community is more or less likely to be depressed on the basis of readily available information on the individual.

The examples just mentioned could also be analyzed using logistic regression, as will be discussed in Chapter 12.

11.3   Data example

As described in Chapter 1, the depression data set was collected for individuals residing in Los Angeles County. To illustrate the ideas described in this chapter, we will develop a method for estimating whether an individual is likely to be depressed. For the purposes of this example "depression" is defined by a score of 16 or greater on the CESD scale (see the codebook given in Table 3.4). This information is given in the variable called "cases." We will base the estimation on demographic and other characteristics of the individual. The variables used are education and income. We may also wish to determine whether we can improve our prediction by including information on illness, sex, or age. Additional variables are an overall health rating, number of bed days in the past two months (0 if less than eight days, 1 if eight or more), acute illness (1 if yes in the past two months, 0 if no), and chronic illness (0 if none, 1 if one or more).

The first step in examining the data is to obtain descriptive measures of each of the groups. Table 11.1 lists the means and standard deviations for each variable in both groups. Here an [a] indicates where a significant difference exists between the means at a $P = .01$ level when a normal distribution is assumed. Note that in the depressed group, group II, we have a significantly higher percentage of females and lower incomes while the age and education is somewhat lower. The standard deviations in the two groups are similar except for income, where they are slightly different. The variances for income are 255.4 for the nondepressed group and 96.8 for the depressed group. The ratio of the variance for the nondepressed group to the depressed is 2.64 for income, which supports the impression of differences in variation between the two groups as well as mean values. The variances for the other variables are more similar.

Note also that the health characteristics of the depressed group are generally worse than those of the nondepressed, even though the members of the depressed group tend to be younger on the average. Because sex is coded males = 1 and females = 2, the average sex of 1.80 indicates that 80% of the depressed group are females. Similarly, 59% of the nondepressed individuals are female.

Suppose that we wish to predict whether or not individuals are depressed, on the basis of their incomes. Examination of Table 11.1 shows that the mean value for depressed individuals is significantly lower than that for the nondepressed. Thus, intuitively, we would classify those with lower incomes as depressed and those with higher incomes as nondepressed. Similarly, we may classify the individuals on the basis of age alone, or sex alone, etc. However, as in the case of regression analysis, the use of several variables simultaneously can be superior to the use of any one variable. The methodology for achieving this result will be explained in the next sections.

Table 11.1: *Means and standard deviations for nondepressed and depressed adults in Los Angeles County*

Variable	Grp I, nondepressed ($N = 244$)		Grp II, depressed ($N = 50$)	
	Mean	Standard deviation	Mean	Standard deviation
Sex (male = 1, female = 2)	1.59[a]	0.49	1.80[a]	0.40
Age (in years)	45.20	18.10	40.40	17.40
Education (1 to 7, 7 high)	3.55	1.33	3.16	1.17
Income (thousands of dollars per year)	21.68[a]	15.98	15.20[a]	9.84
Health index (1 to 4, 1 = excellent)	1.71[a]	0.80	2.06[a]	0.98
Bed days (0 = less than 8 days per year, 1 = 8 or more days)	0.17[a]	0.38	0.42[a]	0.50
Acute conditions (0 = no, 1 = yes)	0.28	0.45	0.38	0.49
Chronic conditions (0 = none, 1 = one or more)	0.48	0.50	0.62	0.49

11.4 Basic concepts of classification

In this section, we present the underlying concepts of classification as given by Fisher (1936) and give an example illustrating its use. We also briefly discuss the coefficients from the Fisher discriminant function.

Statisticians have formulated different ways of performing and evaluating discriminant function analysis. One method of evaluating the results uses what are called **classification functions**. This approach will be described next. Necessary computations are given in Section 11.6. In addition, discriminant function analysis can be viewed as a special case of canonical correlation analysis, presented in Chapter 10. Many of the programs print out canonical coefficients and graphical output based upon evaluation of canonical variates.

In general, when discriminant function analysis is used to discriminate between **two** groups, Fisher's method and classification functions are used. Although discriminant analysis has been generalized to cover three or more groups, many investigators find the two group comparisons easier to interpret. In addition, when there are more than two groups, it still is sometimes sensible to compare the groups two at a time. For example, one group can be used as a referent or control group so that the investigator may want to compare each group to the control group. In this chapter we will present the two group case; for information on how to use discriminant function analysis for more than two groups, see, e.g., Rencher (2002) or Timm (2002) and programs such as SAS for computations. Alternatively, you can use the methods for nominal or ordinal logistic regression given in Section 12.9. The choice between using discriminant function analysis and the nominal or ordinal logistic analysis depends on which analysis is more appropriate for the data being analyzed. As noted in Section 11.5, if the data follow a multivariate normal distribution and the variances and covariances in the groups are equal, then discriminant function analysis is recommended. If the data follow the less restrictive assumptions given in Section 12.4, then logistic regression is recommended.

Principal ideas

Suppose that an individual may belong to one of two populations. We begin by considering how an individual can be classified into one of these populations on the basis of a measurement of one characteristic, say X. Suppose that we have a representative sample from each population, enabling us to estimate the distributions of X and their means. Typically, these distributions can be represented as in Figure 11.1.

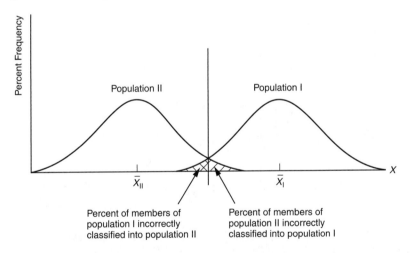

Figure 11.1: *Hypothetical Frequency Distributions of Two Populations Showing Percentage of Cases Incorrectly Classified*

From the figure it is intuitively obvious that a low value of X would lead us to classify an individual into population II and a high value would lead us to classify an individual into population I. To define what is meant by **low** or **high**, we must select a dividing point. If we denote this dividing point by C, then we would classify an individual into population I if $X \geq C$. For any given value of C we would be incurring a certain percentage of error. If the individual came from population I but the measured X were less than C, we would incorrectly classify the individual into population II, and *vice versa*. These two types of errors are illustrated in Figure 11.1. If we can assume that the two populations have the same variance, then the usual value of C is

$$C = \frac{\overline{X}_\mathrm{I} + \overline{X}_\mathrm{II}}{2}$$

This value ensures that the two probabilities of error are equal.

The idealized situation illustrated in Figure 11.1 is rarely found in practice. In real-life situations the degree of overlap of the two distributions is frequently large, and the variances are rarely precisely equal. For example, in the depression data the income distributions for the depressed and nondepressed individuals do overlap to a large degree, as illustrated in Figure 11.2. The usual dividing point is

$$C = \frac{15.20 + 21.68}{2} = 18.44$$

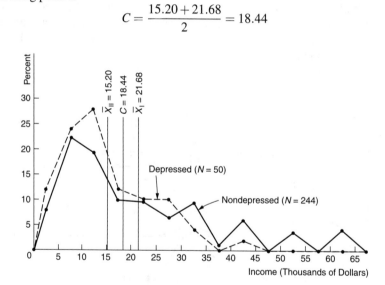

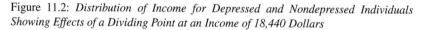

Figure 11.2: *Distribution of Income for Depressed and Nondepressed Individuals Showing Effects of a Dividing Point at an Income of 18,440 Dollars*

As can be seen from Figure 11.2, the percentage of errors is rather large. The exact data on the errors are shown in Table 11.2. These numbers were obtained by first checking whether each individual's income was greater than or equal to 18.44×10^3 and then determining whether the individual was correctly classified. For example, of the 244 nondepressed individuals, 121 had incomes greater than or equal to $\$18.44 \times 10^3$ and were therefore correctly classified as not depressed (Table 11.2). Similarly, of the 50 depressed individuals, 31 were correctly classified. The total number of correctly classified individuals is $121 + 31 = 152$, amounting to 51.7% of the total sample of 294 individuals, as shown in Table 11.2. Thus, although the mean incomes were significantly different from each other $(P < 0.01)$, income alone is not very successful in identifying whether an individual is depressed.

Table 11.2: *Classification of individuals as depressed or not depressed on the basis of income alone*

	Classified as		
Actual status	Not depressed	Depressed	% correct
Not depressed ($N = 244$)	121	123	49.6
Depressed ($N = 50$)	19	31	62.0
Table $N = 294$	140	154	51.7

Combining two or more variables may provide better classification. Note that the number of variables used must be less than $N_I + N_{II} - 1$. For two variables X_1 and X_2 concentration ellipses may be illustrated, as shown in Figure 11.3 (see Section 7.5 for an explanation of concentration ellipses). Figure 11.3 also illustrates the univariate distributions of X_1 and X_2 separately. The univariate distribution of X_1 is what is obtained if the values of X_2 are ignored. On the basis of X_1 alone, and its corresponding dividing point C_1, a relatively large amount of error of misclassification would be encountered. Similar results occur for X_2 and its corresponding dividing point C_2. To use both variables simultaneously, we need to divide the plane of X_1 and X_2 into two regions, each corresponding to one population, and classify the individuals accordingly. A simple way of defining the two regions is to draw a straight line through the points of intersection of the two concentration ellipses, as shown in Figure 11.3.

The percentage of individuals from population II incorrectly classified is shown in the striped areas. The cross-hatched areas show the percentage of individuals from population I who are misclassified. The errors incurred by using two variables are often much smaller than those incurred by using either variable alone. In the illustration in Figure 11.3 this result is, in fact, the case.

The dividing line was represented by Fisher (1936) as an equation $Z = C$, where Z is a linear combination of X_1 and X_2 and C is a constant defined as

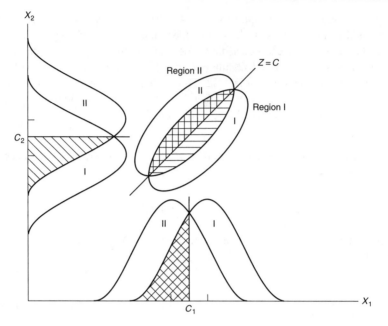

Figure 11.3: *Classification into Two Groups on the Basis of Two Variables*

follows:

$$C = \frac{\overline{Z}_{\mathrm{I}} + \overline{Z}_{\mathrm{II}}}{2}$$

where $\overline{Z}_{\mathrm{I}}$ is the average value of Z in population I and $\overline{Z}_{\mathrm{II}}$ is the average value of Z for population II.

In this book we will call Z the **Fisher discriminant function**, written as

$$Z = a_1 X_1 + a_2 X_2$$

for the two-variable case. The formulas for computing the coefficients a_1 and a_2 were derived by Fisher (1936) in order to maximize the "distance" between the two groups (see Section 11.5).

For each individual from each population, the value of Z is calculated. When the frequency distributions of Z are plotted separately for each population, the result is as illustrated in Figure 11.4. In this case, the bivariate classification problem with X_1 and X_2 is reduced to a univariate situation using the single variable Z.

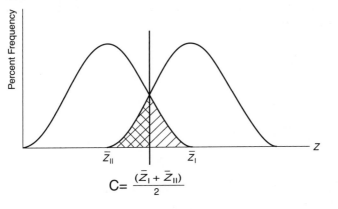

Figure 11.4: *Frequency Distributions of Z for Populations I and II*

Example

As an example of this technique for the depression data, it may be better to use both income and age to classify depressed individuals. Unfortunately, this equation cannot be obtained directly from the output, and some intermediate computations must be made, as explained in Section 11.6. The result is

$$Z = 0.0209(\text{age}) + 0.0336(\text{income})$$

The mean Z value for each group can be obtained as follows, using the means from Table 11.1:

$$\text{mean } Z = 0.0209(\text{mean age}) + 0.0336(\text{mean income})$$

Thus

$$\bar{Z}_{\text{not depressed}} = 0.0209(45.2) + 0.0336(21.68) = 1.67$$

and

$$\bar{Z}_{\text{depressed}} = 0.0209(40.4) + 0.0336(15.20) = 1.36$$

The dividing point is therefore

$$C = \frac{1.67 + 1.36}{2} = 1.515$$

An individual is then classified as depressed if his or her Z value is less than 1.52.

For two variables it is possible to illustrate the classification procedure as shown in Figure 11.5. The dividing line is a graph of the equation $Z = C$, i.e.,

$$0.0209(\text{age}) + 0.0336(\text{income}) = 1.515$$

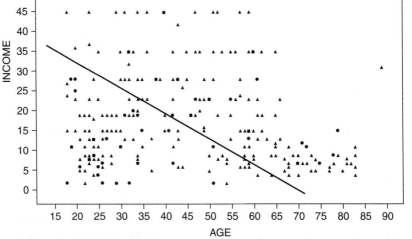

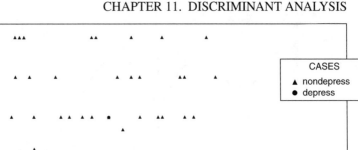

Figure 11.5: *Classification of Individuals as Depressed or Not Depressed on the Basis of Income and Age*

An individual falling in the region above the dividing line is classified as not depressed. Note that, indeed, very few depressed persons fall far above the dividing line.

To measure the degree of success of the classification procedure for this sample, we must count how many of each group are correctly classified. The computer program produces these counts automatically; they are shown in Table 11.3. Note that 63.1% of the nondepressed are correctly classified. This value is compared with 50.4%, which results when income alone is used (Table 11.2). The percentage of depressed correctly classified is comparable in both tables. Combining age with income improved the overall percentage of correct classification from 52.4% to 62.6%.

Interpretation of coefficients

In addition to its use for classification, the Fisher discriminant function is helpful in indicating the direction and degree to which each variable contributes to the classification. The first thing to examine is the sign of each coefficient: if it is positive, the individuals with larger values of the corresponding variable

Table 11.3: *Classification of individuals as depressed or not depressed on the basis of income and age*

Actual status	Classified as		
	Not depressed	Depressed	% correct
Not depressed ($N = 244$)	154	90	63.1
Depressed ($N = 50$)	20	30	60.0
Table $N = 294$	174	120	62.6

tend to belong to population I, and *vice versa*. In the depression data example, both coefficients are positive, indicating that large values of both variables are associated with a lack of depression. In more complex examples, comparisons of those variables having positive coefficients with those having negative coefficients can be revealing. To quantify the magnitude of the distribution, the investigator may find standardized coefficients helpful, as explained in Section 11.6.

The concept of discriminant functions applies as well to situations where there are more than two variables, say $X_1, X_2, \ldots, X_P$. As in multiple linear regression, it is often sufficient to select a small number of variables. Variable selection will be discussed in Section 11.10. In the next section we present some necessary theoretical background for discriminant function analysis.

11.5 Theoretical background

In deriving his linear discriminant function, Fisher (1936) did not have to make any distributional assumptions for the variables used in classification. Fisher denoted the discriminant function by

$$Z = a_1 X_1 + a_2 X_2 + \cdots + a_P X_P$$

As in the previous section, we denote the two mean values of Z by $\overline{Z}_{\mathrm{I}}$ and $\overline{Z}_{\mathrm{II}}$. We also denote the pooled sample variance of Z by S_Z^2 (this statistic is similar to the pooled variance used in the standard two-sample t test). To measure how "far apart" the two groups are in terms of values of Z, we compute

$$D^2 = \frac{(\overline{Z}_{\mathrm{I}} - \overline{Z}_{\mathrm{II}})^2}{S_Z^2}$$

Fisher selected the coefficients $a_1, a_2, \ldots, a_P$ so that D^2 has the maximum possible value.

The term D^2 can be interpreted as the squared distance between the means of the standardized values of the Zs. A larger value of D^2 indicates that it is easier to discriminate between the two groups. The quantity D^2 is called the **Mahalanobis distance**. Both a_i and D^2 are functions of the group means and the pooled variances and covariances of the variables. The manuals for the statistical package programs make frequent use of these formulas, and you can find them in standard multivariate textbooks, e.g., Rencher (2002).

Some distributional assumptions make it possible to develop further statistical procedures relating to the problem of classification. These procedures include tests of hypotheses for the usefulness of some or all of the variables and methods for estimating errors of classification.

The variables used for classification are denoted by $X_1, X_2, \ldots, X_P$. The standard model makes the assumption that for each of the two populations these variables have a multivariate normal distribution. It further assumes that the covariance matrix is the same in both populations. However, the mean values for a given variable may be different in the two populations. Figure 11.3 illustrates these assumptions. Note that both X_1 and X_2 are univariately normally distributed. Also, the bivariate distribution of X_1 and X_2 is depicted as an ellipse, indicating that the data come from a bivariate normal distribution. Multivariate normality can be viewed as a direct extension of bivariate normality. We further assume that we have a random sample from each of the populations. The sample sizes are denoted by N_I and N_{II}.

Alternatively, we may think of the two populations as subpopulations of a single population. For example, in the depression data the original population consists of all adults over 18 years old in Los Angeles County. Its two subpopulations are the depressed and nondepressed. A single sample was collected and later diagnosed to form two subsamples.

If we examine the variables age and income to see if they meet the theoretical assumptions made in performing discriminant function analysis, it becomes clear that they do not. We had examined the distribution of income in Section 4.3, where it was found that income had quite a skewed distribution and a logarithmic transformation could be used to reduce that skewness. Theoretically, if a set of data follows a multivariate normal distribution, then each variable in the set will be univariately normally distributed. But this does not work *vice versa*. Each variable can be univariately normally distributed without the set of variables being multivariately normally distributed. But in that case, in practice, it is very likely that the multivariate distribution is approximately normal, so that most investigators are satisfied with checking the variables one at a time.

In addition, the variances for income in the depressed and nondepressed groups appear to be different. A formal test for equal covariance matrices (see Section 7.5 for a description and example of a covariance matrix) in the depressed and nondepressed groups was rejected at a $P = 0.00054$ level. This data set clearly does not meet the theoretical assumptions for performing tests of hypotheses and making inferences based on the assumptions since one of the variables is not normally distributed and the covariance matrices are not equal.

However, it is not clear that formal tests of hypotheses of underlying assumptions are always the best way of determining when a statistical analysis, such as discriminant function analysis, should be used. Numerous statistical researchers have studied how discriminant function analysis performs when different distributions or sample sizes are tried. This research has been done using simulations. Here, the researcher may take a multivariate normal distribution with all the variances equal to one and another distribution with all variances equal to, say, four. The distributions are also designed to have means that are a fixed distance apart. Then, samples of a given size are taken from each distribution and the researcher sees how well discriminant function analysis performs (what percentage of the two samples is correctly classified). This process is repeated for numerous sets of two samples. The same process is then repeated again for some other method of classification such as logistic regression analysis (Chapter 12) or a more complex type of discriminant function analysis. The method of choice is then the one which empirically has the highest proportion of cases correctly classified.

If an investigator has data that are similar to what has been studied using simulations, then this is often used to guide the choice of analysis since it shows what works in practice. The difficulty with this method is that the simulations seldom precisely fit the data set that an investigator has, so usually some judgment is needed to decide what to trust.

In the example using age and income, even though the covariance matrices were not equal, they were close enough to justify using regular discriminant function analysis for classification assuming the multivariate data were normally distributed (Marks and Dunn, 1974).

11.6 Interpretation

In this section we present various methods for interpreting discriminant functions. Specifically, we discuss the regression analogy, computations of the coefficients, standardized coefficients, and posterior probabilities.

Regression analogy

Connections exist between regression and discriminant analyses. For the regression interpretation we think of the classification variables $X_1, X_2, \ldots, X_P$ as the independent variables. The dependent variable is a dummy variable indicating the population from which each observation comes. Specifically,

$$Y = \frac{N_{II}}{N_I + N_{II}}$$

if the observation comes from population I, and

$$Y = -\frac{N_{\mathrm{I}}}{N_{\mathrm{I}} + N_{\mathrm{II}}}$$

if the observation comes from population II. For instance, for the depression data $Y = 50/(244 + 50)$ if the individual is not depressed and $Y = -244/(244 + 50)$ if the individual is depressed.

When the usual multiple regression analysis is performed, the resulting regression coefficients are proportional to the discriminant function coefficients $a_1, a_2, \ldots, a_P$. The value of the resulting multiple correlation coefficient R is related to the Mahalanobis D^2 by the following formula:

$$D^2 = \frac{R^2}{1 - R^2} \frac{(N_{\mathrm{I}} + N_{\mathrm{II}})(N_{\mathrm{I}} + N_{\mathrm{II}} - 2)}{N_{\mathrm{I}} N_{\mathrm{II}}}$$

Hence from a multiple regression program it is possible to obtain the coefficients of the discriminant function and the value of D^2. The $\bar{Z}$'s for each group can be obtained by multiplying each coefficient by the corresponding variable's sample mean and adding the results. The dividing point C can then be computed as

$$C = \frac{\bar{Z}_{\mathrm{I}} + \bar{Z}_{\mathrm{II}}}{2}$$

As in regression analysis, some of the independent variables (or classification variables) may be dummy variables (Section 9.3). In the depression example we may, for instance, use sex as one of the classification variables by treating it as a dummy variable. Research has shown that even though such variables do not follow a normal distribution, their use in linear discriminant analysis can still help improve the classification.

Computing the Fisher discriminant function

In the discriminant analysis programs to be discussed in Section 11.11, some computations must be performed to obtain the values of the discriminant coefficients. The programs print what is called a "classification function" for each group. SPSS DISCRIMINANT calls these "classification function coefficients" and SAS DISCRIM procedure calls them the "linearized discriminant functions." For each population the coefficients are printed for each variable. The discriminant function coefficients $a_1, a_2, \ldots, a_P$ are then obtained by subtraction.

As an example, we again consider the depression data using age and income. The classification functions are shown in Table 11.4.

The coefficient a_1 for age is $0.1634 - 0.1425 = 0.0209$. For income, $a_2 = 0.1360 - 0.1024 = 0.0336$. The dividing point C is also obtained by subtraction, but in **reverse order**. Thus $C = -4.3483 - (-5.8641) = 1.5158$. (This

Table 11.4: *Classification function and discriminant coefficients for age and income*

| | Classification function | | |
| | Group I, | Group II, | |
Variables	not depressed	depressed	Discriminant function
Age	0.1634	0.1425	$0.0209 = a_1$
Income	0.1360	0.1024	$0.0336 = a_2$
Constant	−5.8641	−4.3483	$1.5158 = C$

agrees closely with the previously computed value $C = 1.515$, which we use throughout this chapter.) For more than two variables the same procedure is used to obtain $a_1, a_2, \ldots, a_p$ and C.

In practice each case is assigned to the group for which it has the largest classification function score. For example, the first respondent in the depression data set is 68 years old and has an income of \$4000. The classification function for not depressed is $0.1634(68) + 0.1360(4) = 5.7911$ and for depressed it is 5.7513, so she would be assigned to the not depressed group.

Renaming the groups

If we wish to put the depressed persons into group I and the nondepressed into group II, we can do so by reversing the zero and one values for the "cases" variable. In the data used for our example, "cases" equals 1 if a person is depressed and 0 if not depressed. However, we can make cases equal 0 if a person is depressed and 1 if a person is not depressed. Note that this reversal does not change the classification functions but simply changes their order so that **all** the signs in the linear discriminant function are changed. The new constant and discriminant functions are

$$-1.515 \text{ and } -0.0209(\text{age}) - 0.0336(\text{income}), \text{ respectively}$$

The ability to discriminate is exactly the same, and the number of individuals correctly classified is the same.

Standardized coefficients

As in the case of regression analysis, the values of $a_1, a_2, \ldots, a_p$ are not directly comparable. However, an impression of the relative effect of each variable on the discriminant function can be obtained from the **standardized discriminant coefficients**. This technique involves the use of the pooled (or within-group) covariance matrix from the computer output. In the original example this co-variance matrix is as follows:

	Age	Income
Age	327.08	−53.01
Income	−53.01	233.79

Thus the pooled standard deviations are $(327.08)^{1/2} = 18.08$ for age and $(233.79)^{1/2} = 15.29$ for income. The standardized coefficients are obtained by multiplying the a_i's by the corresponding pooled standard deviations. Hence the standardized discriminant coefficients are

$$(0.0209)(18.09) = 0.378 \text{ for age}$$

and

$$(0.0336)(15.29) = 0.514 \text{ for income}$$

It is therefore seen that income has a slightly larger effect on the discriminant function than age.

Posterior probabilities

Thus far the classification procedure assigned an individual to either group I or group II. Since there is always a possibility of making the wrong classification, we may wish to compute the probability that the individual has come from one group or the other. We can compute such a probability under the multivariate normal model discussed in Section 11.5. The formula is

$$\text{probability of belonging to population I} = \frac{1}{1 + exp(-Z + C)}$$

where $exp(-Z + C)$ indicates e raised to the power $(-Z + C)$. The probability of belonging to population II is one minus the probability of belonging to population I.

For example, suppose that an individual from the depression study is 42 years old and earns 24×10^3 income per year. For that individual the value of the discriminant function is

$$Z = 0.0209(42) + 0.0336(24) = 1.718$$

Since $C = 1.515$, and therefore Z is greater than C, we classify the individual as not depressed (in population I). To determine how likely this person is to be not depressed, we compute the probability

$$\frac{1}{1 + exp(-1.718 + 1.515)} = 0.55$$

The probability of being depressed is $1 - 0.55 = 0.45$. Thus this individual is only slightly more likely to be not depressed than to be depressed.

Several packaged programs compute the probabilities of belonging to both groups for each individual in the sample. In some programs these probabilities are called the **posterior probabilities** since they express the probability of belonging to a particular population posterior to (i.e., after) performing the analysis.

Posterior probabilities offer a valuable method of interpreting classification results. The investigator may wish to classify only those individuals whose probabilities clearly favor one group over the other. Judgment could be withheld for individuals whose posterior probabilities are close to 0.5. In the next section another type of probability, called prior probability, will be defined and used to modify the dividing point.

Finally, we note that the discriminant function presented here is a sample estimate of the population discriminant function. We would compute the latter if we had the actual values of the population parameters. If the populations were both multivariate normal with equal covariance matrices, then the population discriminant classification procedure would be optimal, i.e., no other classification procedure would produce a smaller total classification error.

11.7 Adjusting the dividing point

In this section we indicate how prior probabilities and costs of misclassification can be incorporated into the choice of the dividing point C.

Incorporating prior probabilities into the choice of C

Thus far, the dividing point C has been used as the point producing an equal percentage of errors of both types, i.e., the probability of misclassifying an individual from population I into population II, or *vice versa*. This use can be seen in Figure 11.4. But the choice of the value of C can be made to produce any desired ratio of these probabilities of errors. To explain how this choice is made, we must introduce the concept of **prior probability**. Since the two populations constitute an overall population, it is of interest to examine their relative size. The prior probability of population I is the probability that an individual selected at random actually comes from population I. In other words, it is the proportion of individuals in the overall population who fall in population I. This proportion is denoted by q_I.

In the depression data the definition of a depressed person was originally designed so that 20% of the population would be designated as depressed and 80% nondepressed. Therefore the prior probability of not being depressed (population I) is $q_I = 0.8$. Likewise, $q_{II} = 1 - q_I = 0.2$. Without knowing any of the characteristics of a given individual, we would thus be inclined to classify him or her as nondepressed, since 80% were in that group. In this case we would be correct 80% of the time. This example offers an intuitive interpretation of prior probabilities. Note, however, that we would always be wrong in identifying depressed individuals.

The theoretical choice of the dividing point C is made so that the total probability of misclassification is minimized. This total probability is defined as q_I times the probability of misclassifying an individual from population I into population II plus q_{II} times the probability of misclassifying an individual from population II into population I:

$$q_I \cdot \text{Prob(II given I)} + q_{II} \cdot \text{Prob(I given II)}$$

Under the multivariate normal model mentioned in Section 11.5 the optimal choice of the dividing point C is

$$C = \frac{\overline{Z}_I + \overline{Z}_{II}}{2} + \ln \frac{q_{II}}{q_I}$$

where ln stands for the natural logarithm. Note that if $q_I = q_{II} = \frac{1}{2}$, then $q_{II}/q_I = 1$ and $\ln q_{II}/q_I = 0$. In this case C is

$$C = \frac{\overline{Z}_I + \overline{Z}_{II}}{2} \quad \text{if} \quad q_I = q_{II}$$

Thus in the previous sections we have been implicitly assuming that $q_I = q_{II} = \frac{1}{2}$.

For the depression data we have seen that $q_I = 0.8$, and therefore the theoretical dividing point should be

$$C = 1.515 + \ln(0.25) = 1.515 - 1.386 = 0.129$$

In examining the data, we see that using this dividing point classifies all of the nondepressed individuals correctly and all of the depressed individuals incorrectly. Therefore the probability of classifying a nondepressed individual (population I) as depressed (population II) is zero. On the other hand, the probability of classifying a depressed individual (population II) as nondepressed (population II) is 1. Therefore the total probability of misclassification is $(0.8)(0) + (0.2)(1) = 0.2$. When $C = 1.515$ was used, the two probabilities of misclassification were 0.369 and 0.400, respectively (Table 11.3). In that case the total probability of misclassification is $(0.8)(0.379) + (0.2)(0.400) = 0.383$. This result verifies that the theoretical dividing point did produce a smaller value of this total probability of misclassification.

In practice, however, it is not appealing to identify none of the depressed individuals. If the purpose of classification were preliminary screening, we would be willing to incorrectly label some individuals as depressed in order to avoid missing too many of those who are truly depressed. In practice, we would choose various values of C and for each value determine the two probabilities of misclassification. The desired choice of C would be made when some balance of these two is achieved.

Incorporating costs into the choice of C

One method of weighting the errors is to determine the relative costs of the two types of misclassification. For example, suppose that it is four times as serious to falsely label a depressed individual as nondepressed as it is to label a nondepressed individual as depressed. These costs can be denoted as

$$\text{cost(II given I)} = 1$$

and

$$\text{cost(I given II)} = 4$$

The dividing point C can then be chosen to minimize the total cost of misclassification, namely

$$q_1 \cdot \text{Prob(II given I)} \cdot \text{cost(II given I)} + q_{\text{II}} \cdot \text{Prob(I given II)} \cdot \text{cost(I given II)}$$

The choice of C that achieves this minimization is

$$C = \frac{\overline{Z}_{\text{I}} + \overline{Z}_{\text{II}}}{2} + K$$

where

$$K = \ln \frac{q_{\text{II}} \cdot \text{cost(I given II)}}{q_1 \cdot \text{cost(II given I)}}$$

In the depression example the value of K is

$$K = \ln \frac{0.2(4)}{0.8(1)} = \ln 1 = 0$$

In other words, this numerical choice of cost of misclassification and the use of prior probabilities counteract each other so that $C = 1.515$, the same value obtained without incorporating costs and prior probabilities.

Many of the computer programs allow the user to adjust the prior probabilities but not the costs. We can trick a program into incorporating costs into the prior probabilities, as the following example illustrates. Suppose $q_{\text{I}} = 0.4, q_{\text{II}} = 0.6$, cost(II given I) $= 5$, and cost(I given II) $= 1$. Then

$$\text{adjusted } q_{\text{I}} = q_{\text{I}} \cdot \text{cost(II given I)} = (0.4)(5) = 2$$

and

$$\text{adjusted } q_{\text{II}} = q_{\text{II}} \cdot \text{cost(I given II)} = (0.6)(1) = 0.6$$

Since the prior probabilities must add up to one (not $2 + 0.6 = 2.6$), we further adjust the computed prior probabilities such that their sum is one, i.e.,

$$\text{adjusted } q_{\text{I}} = \frac{2}{2.6} = 0.77 \quad \text{and} \quad \text{adjusted } q_{\text{II}} = \frac{0.6}{2.6} = 0.23$$

Finally, it is important to note that incorporating the prior probabilities and costs of misclassification alters only the choice of the dividing point C. It does not affect the computation of the coefficients $a_1, a_2, \ldots, a_P$ in the discriminant function. If the computer program does not allow the option of incorporating those quantities, you can easily modify the dividing point, as was done in the above example.

11.8   How good is the discrimination?

A **measure of goodness** for the classification procedure consists of the two probabilities of misclassification, probability (II given I) and probability (I given II). Various methods exist for estimating these probabilities. One method, called the **empirical method**, was used in the previous examples. That is, we applied the discriminant function to the same samples used for deriving it and computed the proportion incorrectly classified from each group (Tables 11.2 and 11.3). This process is a form of validation of the discriminant function. Although this method is intuitively appealing, it does produce biased estimates. In fact, the resulting proportions underestimate the true probabilities of misclassification, because the same sample is used for deriving and validating the discriminant function.

Ideally, we would like to derive the function from one sample and apply it to another sample to estimate the proportion misclassified. This procedure is called **cross-validation**, and it produces unbiased estimates. The investigator can achieve cross-validation by randomly splitting the original sample from each group into two subsamples: one for deriving the discriminant function and one for cross-validating it.

The investigator may be hesitant to split the sample if it is small. An alternative method sometimes used in this case, which imitates splitting the samples, is called the **jackknife procedure**. In this method we exclude one observation from the first group and compute the discriminant function on the basis of the remaining observations. We then classify the excluded observation. This procedure is repeated for each observation in the first sample. The proportion of misclassified individuals is the jackknife estimate of Prob(II given I). A similar procedure is used to estimate Prob(I given II). This method produces nearly unbiased estimators. SAS offers this option.

If we accept the multivariate normal model with equal covariance matrices, theoretical estimates of the probabilities are also available and require only an estimate of D^2. The formulas are

$$\text{estimated Prob(II given I)} = \text{area to left of} \left[\frac{K - \frac{1}{2}D^2}{D} \right]$$

under standard normal curve

and

$$\text{estimated Prob(I given II)} = \text{area to left of} \left[\frac{-K - \frac{1}{2}D^2}{D} \right]$$

under standard normal curve

where

$$K = \ln \frac{q_{II} \cdot \text{cost(I given II)}}{q_I \cdot \text{cost(II given I)}}$$

If $K = 0$, these two estimates are each equal to the area to the left of $(-D/2)$ under the standard normal curve. For example, in the depression example $D^2 = 0.319$ and $K = 0$. Therefore $D/2 = 0.282$, and the area to the left of -0.282 is 0.389. From this method we estimate both Prob(II given I) and Prob(I given II) as 0.39. This method is particularly useful if the discriminant function is derived from a regression program, since D^2 can be easily computed from R^2 (Section 11.6).

Unfortunately, this last method also underestimates the true probabilities of misclassification. An **unbiased estimator** of the population Mahalanobis D^2 is

$$\text{unbiased } D^2 = \frac{N_I + N_{II} - P - 3}{N_I + N_{II} - 2} D^2 - P \left(\frac{1}{N_I} + \frac{1}{N_{II}} \right)$$

In the depression example we have

$$\begin{aligned} \text{unbiased } D^2 &= \frac{50 + 244 - 2 - 3}{50 + 244 - 2} (0.319) - 2 \left(\frac{1}{50} + \frac{1}{244} \right) \\ &= 0.316 - 0.048 \\ &= 0.268 \end{aligned}$$

The resulting area is computed in a similar fashion to the last method. Since unbiased $D/2 = 0.259$, the resulting area to the left of $-D/2$ is 0.398. In comparing this result with the estimate based on the biased D^2, we note that the difference is small because (1) only two variables are used and (2) the sample sizes are fairly large. On the other hand, if the number of variables P were close to the total sample size $(N_I + N_{II})$, the two estimates could be very different from each other.

Whenever possible, it is recommended that the investigator obtain at least some of the above estimates of errors of misclassification and the corresponding probabilities of correct prediction.

To evaluate how well a particular discriminant function is performing, the investigator may also find it useful to compute the probability of correct prediction based on pure **guessing**. The procedure is as follows. Suppose that the prior probability of belonging to population I is known to be q_I. Then $q_{II} = 1 - q_I$. One way to classify individuals using these probabilities alone is to imagine a coin that comes up heads with probability q_I and tails with probability q_{II}. Every time an individual is to be classified, the coin is tossed. The individual is classified into population I if the coin comes up heads and into population II if it is tails. Overall, a proportion q_I of all individuals will be classified into population I.

Next, the investigator computes the total probability of correct classification. Recall that the probability that a person comes from population I is q_I, and the probability that any individual is classified into population I is q_I. Therefore the probability that a person comes from population I and is **correctly** classified into population I is q_I^2. Similarly, q_{II}^2 is the probability that an individual comes from population II and is correctly classified into population II. Thus the total probability of correct classification using only knowledge of the prior probabilities is $q_I^2 + q_{II}^2$. Note that the lowest possible value of this probability occurs when $q_I = q_{II} = 0.5$, i.e., when the individual is equally likely to come from either population. In that case $q_I^2 + q_{II}^2 = 0.5$.

Using this method for the depression example, with $q_I = 0.8$ *and* $q_{II} = 0.2$, gives us $q_I^2 + q_{II}^2 = 0.68$. Thus we would expect more than two-thirds of the individuals to be correctly classified if we simply flipped a coin that comes up heads 80% of the time. Note, however, that we would be wrong on 80% of the depressed individuals, a situation we may not be willing to tolerate. (In this context you might recall the role of costs of misclassification discussed in Section 11.7.)

11.9   Testing variable contributions

Can we classify individuals by using variables available to us better than we can by chance alone? One answer to this question assumes the multivariate normal model presented in Section 11.4. The question can be formulated as a hypothesis-testing problem. The null hypothesis being tested is that none of the variables improves the classification based on chance alone. Equivalent null hypotheses are that the two population means for each variable are identical, or that the population D^2 is zero. The test statistic for the null hypothesis is

$$F = \frac{N_I + N_{II} - P - 1}{P(N_I + N_{II} - 2)} \times \frac{N_I N_{II}}{N_I + N_{II}} \times D^2$$

with degrees of freedom of P and $N_I + N_{II} - P - 1$ (Rao, 1973). The P value is the tail area to the right of the computed test statistic. We point out that $N_I N_{II} D^2 / (N_I + N_{II})$ is known as the two-sample Hotelling T^2, derived originally for testing the equality of two sets of means.

For the depression example using age and income, the computed F value is

$$F = \frac{244 + 50 - 2 - 1}{2(244 + 50 - 2)} \times \frac{244 \times 50}{244 + 50} \times 0.319 = 6.60$$

with 2 and 291 degrees of freedom. The P value for this test is less than 0.005. Thus these two variables together significantly improve the prediction based on chance alone. Equivalently, there is statistical evidence that the population means are not identical in both groups. It should be noted that most existing computer programs do not print the value of D^2. However, from the printed value of the above F statistic we can compute D^2 as follows:

$$D^2 = \frac{P(N_I + N_{II})(N_I + N_{II} - 2)}{(N_I N_{II})(N_I + N_{II} - P - 1)} F$$

Another useful test is whether one additional variable improves the discrimination. Suppose that the population D^2 based on $X_1, X_2, \ldots, X_P$ variables is denoted by pop D_P^2. We wish to test whether an additional variable X_{P+1} will significantly increase the pop D^2, i.e., we test the hypothesis that pop $D_{P+1}^2 = $ pop D_P^2. The test statistic under the multivariate normal model is also an F statistic and is given by

$$F = \frac{(N_I + N_{II} - P - 2)(N_I N_{II})(D_{P+1}^2 - D_P^2)}{(N_I + N_{II})(N_I + N_{II} - 2) + N_I N_{II} D_P^2}$$

with 1 and $(N_I + N_{II} - P - 2)$ degrees of freedom (Rao, 1965).

For example, in the depression data we wish to test the hypothesis that age improves the discriminant function when combined with income. The D_1^2 for income alone is 0.183, and D_2^2 for income and age is 0.319. Thus with $P = 1$,

$$F = \frac{(50 + 244 - 1 - 2)(50 \times 244)(0.319 - 0.183)}{(294)(292) + 50 \times 244 \times 0.183} = 5.48$$

with 1 and 291 degrees of freedom. The P value for this test is equal to 0.02. Thus, age significantly improves the classification when combined with income.

A generalization of this last test allows for the checking of the contribution of several additional variables simultaneously. Specifically, if we start with $X_1, X_2, \ldots, X_P$ variables, we can test whether $X_{P+1}, \ldots, X_{P+Q}$ variables improve the prediction. We test the hypothesis that pop $D_{P+Q}^2 = $ pop D_P^2. For the multivariate normal model the test statistic is

$$F = \frac{(N_I + N_{II} - P - Q - 1)}{Q} \times \frac{N_I N_{II}(D_{P+Q}^2 - D_P^2)}{(N_I + N_{II})(N_I + N_{II} - 2) + N_I N_{II} D_P^2}$$

with Q and $(N_I + N_{II} - P - Q - 1)$ degrees of freedom.

The last two formulas for F are useful in variable selection, as will be shown in the next section.

11.10   Variable selection

Recall that there is an analogy between regression analysis and discriminant function analysis. Therefore much of the discussion of variable selection given in Chapter 8 applies to selecting variables for classification into two groups. In fact, the computer programs discussed in Chapter 8 may be used here as well. These include, in particular, forward selection, stepwise regression programs, and subset regression programs. In addition, some computer programs are available for performing stepwise discriminant analysis. They employ the same concepts discussed in connection with stepwise regression analysis.

In discriminant function analysis, instead of testing whether the value of multiple R^2 is altered by adding (or deleting) a variable, we test whether the value of pop D^2 is altered by adding or deleting variables. The F statistic given in Section 11.9 is used for this purpose. As before, the user may specify a value for the F-to-enter and F-to-remove values. For F-to-enter, Costanza and Afifi (1979) recommend using a value corresponding to a P of 0.15. No recommended value from research can be given for the F-to-remove value, but a reasonable choice may be a P of 0.30.

11.11   Discussion of computer programs

The computer output for the six packages in this book is summarized in Table 11.5.

The D^2 given in Table 11.5 is a measure of the distance between each point representing an individual and the point representing the estimated population mean. A small value of D^2 indicates that the individual is like a typical individual in that population and hence probably belongs to that population.

The discriminant function analyses in this chapter have been restricted to "linear" discriminant analyses analogous to linear regression. Quadratic discriminant function analysis uses the separate group covariance matrices instead of the pooled covariance matrix for computing the various output. For further discussion of quadratic discriminant function analysis see McLachlan (2004).

Quadratic discriminant analysis is theoretically appropriate when the data are multivariately normally distributed but the covariance matrices are quite unequal. In the depression example that was given in this chapter, the null hypothesis of equal covariance matrices was rejected (Section 11.5). The variable "age" had a similar variance in the two groups but the variance of income was higher in the nondepressed group. Also, the covariance between age and income was a small positive number for the depressed respondents but negative and larger in magnitude for the nondepressed (older people and smaller income). We did not test if the data were multivariate normal. When we tried quadratic discrimination on this data set, the percentage correctly classified went down somewhat, not up.

We also tried making a transformation on income to see if that would improve the percentage correctly classified. Using the suggested transformation given in Section 4.3, we created a new variable log_inc = log(income +2) and used that with age. The variances now were similar but the covariances were still of opposite sign as before. The linear discriminant function came out with the same percentage classified correctly as given in Section 11.4 without the transformation, although slightly more respondents were classified as normal than before the transformation.

Marks and Dunn (1974) showed that the decision to use a quadratic discriminant function depended not only on the ratio of the variances but also on the sample size, separation between the groups, and the number of variables. They did not consider cases with equal variances but unequal covariances. With small sample sizes and variances not too different, linear was preferred to quadratic discrimination. When the ratio of the variances was two or less, little difference was seen and even ratios up to four resulted in not too much difference under most conditions. Multivariate normality was assumed in all their simulations.

SAS has three discriminant function procedures, DISCRIM, CANDISC, and STEPDISC. DISCRIM is a comprehensive procedure that does linear, quadratic, and also nonparametric (not discussed here) discriminant analysis. STEPDISC performs stepwise discriminant function analysis and CANDISC performs canonical discriminant function analysis.

The SPSS DISCRIMINANT program is a general purpose program that performs stepwise discriminant analysis and provides clearly labelled output. It provides a test for equality of covariance matrices but not correlation matrices by groups.

The STATISTICA discriminant program is quite complete and is quick to run. With STATISTICA and using the equal prior option, we got the same output as with the other packages with the exception of the constants in the classification function. They came out somewhat different and the resulting classification table was also different.

11.12 What to watch out for

Partly because there is more than one group, discriminant function analysis has additional aspects to watch out for beyond those discussed in regression analysis. A list of important trouble areas is as follows:

1. Theoretically, a simple random sample from each population is assumed. As

Table 11.5: *Software commands and output for discriminant function analysis*

	S-PLUS/R	SAS	SPSS	Stata	STATISTICA
Pooled covariances and correlations	discrim*b	All[a]	DISCRIMINANT	corr	Discr. Anal.
Covariances and correlations by group	discrim*b	All[a]	DISCRIMINANT[b]	corr, by	
Classification function	discrim*	DISCRIM	DISCRIMINANT	estat classification	Discr. Anal.
D^2	mahalanobis	DISCRIM STEPDISC	DISCRIMINANT	estat grdistances	Discr. Anal.
F statistics		All[a]	DISCRIMINANT	candisc;estat canontest	Discr. Anal.
Wilks' lambda		All[a]	DISCRIMINANT	estat manova	Discr. Anal.
Stepwise options	summary.discrim*	STEPDISC	DISCRIMINANT		Discr. Anal.
Classified tables		DISCRIM	DISCRIMINANT	candisc;discrim	Discr. Anal.
Jackknife classification table		DISCRIM	DISCRIMINANT	candisc;estat classtable	
Cross-validation with subsample	crossvalidate.discrim*,lda	DISCRIM	DISCRIMINANT	predict	Discr. Anal.
Canonical coefficients	discrim*	CANDISC	DISCRIMINANT	estat loadings	Discr. Anal.
Standardized canonical coefficients		CANDISC	DISCRIMINANT	candisc;estat loadings	Discr. Anal.
Canonical loadings		CANDISC	DISCRIMINANT	candisc;estat structure	Discr. Anal.
Canonical plots		CANDISC	DISCRIMINANT	scoreplot	Discr. Anal.
Prior probabilities	discrim*,lda	DISCRIM	DISCRIMINANT	candisc;estat classtable	Discr. Anal.
Posterior probabilities	predict.discrim*,lda	DISCRIM	DISCRIMINANT	predict	Discr. Anal.
Quadratic discriminant analysis	discrim*qda	DISCRIM	DISCRIMINANT	discrim, qda	Discr. Anal.
Nonparametric discriminant analysis		DISCRIM	kNN	discrim, knn	
Test of equal covariance matrices	summary.discrim*	DISCRIM	DISCRIMINANT	mvtest covariances	

[a] STEPDISC, DISCRIM and CANDISC.
[b] Covariance matrices by groups.
*discrim is in S-PLUS only; lda and qda are in both S-PLUS and R.

this is often not feasible, the sample taken should be examined for possible biasing factors.

2. It is critical that the correct group identification be made. For example, in the example in this chapter it was assumed that all persons had a correct score on the CESD scale so that they could be correctly identified as normal or depressed. Had we identified some individuals incorrectly, this would increase the reported error rates given in the computer output. It is particularly troublesome if one group contains more misidentifications than another.

3. The choice of variables is also important. Analogies can be made here to regression analysis. Similar to regression analysis, it is important to remove outliers, make necessary transformations on the variables, and check independence of the cases. It is also a matter of concern if there are considerably more missing values in one group than another.

4. The assumption of multivariate normality is made in discriminant function analysis when computing posterior probabilities or performing statistical tests. The use of dummy variables has been shown not to cause undue problems, but very skewed or long-tailed distributions for some variables can increase the total error rate.

5. Another assumption is equal covariance matrices in the groups. If one covariance matrix is very different from the other, then quadratic discriminant function analysis should be considered for two groups or transformations should be made on the variables that have the greatest difference in their variances.

6. If the sample size is sufficient, researchers sometimes obtain a discriminant function from one-half or two-thirds of their data and apply it to the remaining data to see if the same proportion of cases are classified correctly in the two subsamples. Often when the results are applied to a different sample, the proportion classified correctly is smaller. If the discriminant function is to be used to classify individuals, it is important that the original sample come from a population that is similar to the one it will be applied to in the future.

7. If some of the variables are dichotomous or not normally distributed, then logistic regression analysis (discussed in the next chapter) should be considered. Logistic regression does not require the assumption of multivariate normality or equal covariances for inferences and should be considered as an alternative to discriminant analysis when these assumptions are not met or approximated.

8. As in regression analysis, significance levels of F should not be taken as valid if variable selection methods such as forward or backward stepwise methods are used. Also, caution should be taken when using small samples with numerous variables (Rencher and Larson, 1980). If a nominal or dis-

crete variable is transformed into a dummy variable, we recommend that only two binary outcomes per variable be considered.

11.13   Summary

In this chapter we discussed discriminant function analysis, a technique dating back to at least 1936. However, its popularity began with the introduction of large scale computers in the 1960s. The method's original concern was to classify an individual into one of several populations. It is also used for explanatory purposes to identify the relative contributions of a single variable or a group of variables to the classification.

In this chapter we presented discriminant analysis for two populations since that is the case that is most commonly used. We gave some theoretical background and presented an example of the use of a packaged program for this situation.

Interested readers may pursue the subject of classification further by consulting McLachlan (2004), Johnson and Wichern (2007), or Rencher (2002).

11.14   Problems

11.1 Using the depression data set, perform a stepwise discriminant function analysis with age, sex, log(income), bed days, and health as possible variables. Compare the results with those given in Section 11.13.

11.2 For the data shown in Table 8.1, divide the chemical companies into two groups: group I consists of those companies with a P/E less than 9, and group II consists of those companies with a P/E greater than or equal to 9. Group I should be considered mature or troubled firms, and group II should be considered growth firms. Perform a discriminant function analysis, using ROR5, D/E, SALESGR5, EPS5, NPM1, and PAYOUTR1. Assume equal prior probabilities and costs of misclassification. Test the hypothesis that the population $D^2 = 0$. Produce a graph of the posterior probability of belonging to group I versus the value of the discriminant function. Estimate the probabilities of misclassification by several methods.

11.3 (Continuation of Problem 11.2.) Test whether D/E alone does as good a classification job as all six variables.

11.4 (Continuation of Problem 11.2.) Choose a different set of prior probabilities and costs of misclassification that seems reasonable and repeat the analysis.

11.5 (Continuation of Problem 11.2.) Perform a variable selection analysis, using stepwise and best-subset programs. Compare the results with those of the variable selection analysis given in Chapter 8.

11.6 (Continuation of Problem 11.2.) Now divide the companies into three groups: group I consists of those companies with a P/E of 7 or less, group

II consists of those companies with a P/E of 8 to 10, and group III consists of those companies with a P/E greater than or equal to 11. Perform a stepwise discriminant function analysis, using these three groups and the same variables as in Problem 11.2. Comment.

11.7 In this problem you will modify the data set created in Problem 7.7 to make it suitable for the theoretical exercises in discriminant analysis. Generate the sample data for $X1, X2,\ldots,X9$ as in Problem 7.7 (Y is not used here). Then for the first 50 cases, add 6 to $X1$, add 3 to $X2$, add 5 to $X3$, and leave the values for $X4$ to $X9$ as they are. For the last 50 cases, leave all the data as they are. Thus the first 50 cases represent a random sample from a multivariate normal population called population I with the following means: 6 for $X1$, 3 for $X2$, 5 for $X3$, and zero for $X4$ to $X9$. The last 50 observations represent a random sample from a multivariate normal population (called population II) whose mean is zero for each variable. The population Mahalanobis D^2's for each variable separately are as follows: 1.44 for $X1$, 1 for $X2$, 0.5 for $X3$, and zero for each of $X4$ to $X9$. It can be shown that the population D^2 is as follows: 3.44 for $X1$ to $X9$; 3.44 for $X1$, $X2$ and $X3$; and zero for $X4$ to $X9$. For all nine variables the population discriminant function has the following coefficients; 0.49 for $X1$, 0.5833 for $X2$, -0.25 for $X3$, and zero for each each of $X4$ to $X9$. The population errors of misclassification are

$$\text{Prob(I given II)} = \text{Prob(II given I)} = 0.177$$

Now perform a discriminant function analysis on the data you constructed, using all nine variables. Compare the results of the sample with what you know about the populations.

11.8 (Continuation of Problem 11.7.) Perform a similar analysis, using only $X1$, $X2$, and $X3$. Test the hypothesis that these three variables do as well as all nine classifying the observations. Comment.

11.9 (Continuation of Problem 11.7.) Do a variable selection analysis for all nine variables. Comment.

11.10 (Continuation of Problem 11.7.) Do a variable selection analysis, using variables $X4$ to $X9$ only. Comment.

11.11 From the family lung function data in Appendix A create a data set containing only those families from Burbank and Long Beach (AREA = 1 or 3). The observations now belong to one of two AREA-defined groups.
(a) Assuming equal prior probabilities and costs, perform a discriminant function analysis for the fathers using FEV1 and FVC.
(b) Now choose prior probabilities based on the population of each of these two cities (at the time the study was conducted, the population of Burbank was 84,625 and that of Long Beach was 361,334), and assume the cost of Long Beach given Burbank is twice that of Burbank given Long Beach. Repeat the analysis of part (a) with these assumptions. Compare and comment.

11.12 At the time the study was conducted, the population of Lancaster was 48,027 while Glendora had 38,654 residents. Using prior probabilities based on these population figures and those given in Problem 11.11, and the entire lung function data set (with four AREA-defined groups), perform a discriminant function analysis on the fathers. Use FEV1 and FVC as classifying variables. Do you think the lung function measurements are useful in distinguishing among the four areas?

11.13 (a) In the family lung function data in Appendix A divide the fathers into two groups: group I with FEV1 less than or equal to 4.09, and group II with FEV1 greater than 4.09. Assuming equal prior probabilities and costs, perform a stepwise discriminant function analysis using the variables height, weight, age, and FVC. What are the probabilities of misclassification?
(b) Use the classification function you found in part (a) to classify first the mothers and then the oldest children. What are the probabilities of misclassification? Why is the assumption of equal prior probabilities not realistic?

11.14 Divide the oldest children in the family lung function data set into two groups based on weight: less than or equal to 101 versus greater than 101. Perform a stepwise discriminant function analysis using the variables OCHEIGHT, OCAGE, MHEIGHT, MWEIGHT, FHEIGHT, and FWEIGHT. Now temporarily remove from the data set all observations with OCWEIGHT between 87 and 115 inclusive. Repeat the analysis and compare the results.

11.15 Is it possible to distinguish between men and women in the depression data set on the basis of income and level of depression? What is the classification function? What are your prior probabilities? Test whether the following variables help discriminate: EDUCAT, EMPLOY, HEALTH.

11.16 Refer to the table of ideal weights given in Problem 9.8 and calculate the midpoint of each weight range for men and women. Pretending these represent a real sample, perform a discriminant function analysis to classify observations as male or female on the basis of height and weight. How could you include frame size in the analysis? What happens if you do?

11.17 Calculate the Parental Bonding Overprotection and Parental Bonding Care score for the Parental HIV data (see Appendix A and the codebook). Perform a discriminant function analysis to classify adolescents into a group who has been absent from school without a reason (HOOKEY) and a group who has not on the basis of the Parental Bonding scores and the adolescents' age. What are the probabilities of misclassification? Are the assumptions underlying discriminant function analysis met? If not, what approaches could be used to meet the assumptions?

Chapter 12

Logistic regression

12.1   Chapter outline

In Chapter 11 we presented discriminant analysis, a method originally developed to classify individuals into one of two possible populations using continuous predictor variables. Historically, discriminant analysis was the predominant method used for classification. From this chapter, you will learn another classification method, logistic regression analysis, which applies to discrete or continuous predictor variables. Particularly in the health sciences, this method has become commonly used to perform regression analysis on an outcome variable that represents two or more groups (e.g., disease or no disease; or no, mild, moderate, or severe disease).

Section 12.2 presents examples of the use of logistic regression analysis. Section 12.3 derives the basic formula for logistic regression models and illustrates the appearance of the logistic function. In Section 12.4, a definition of odds is given along with the basic model assumed in logistic regression. Sections 12.5, 12.6, and 12.7 discuss the interpretation of categorical, continuous and interacting predictor variables, respectively. Methods of variable selection, checking the fit of the model, and evaluating how well logistic regression predicts outcomes are given in Section 12.8. Logistic regression analysis techniques available for more than two outcome levels are discussed in Section 12.9. Section 12.10 discusses the use of logistic regression in cross-sectional, case-control, matched samples and imputation. A brief introduction to Poisson regression analysis is included in Section 12.11. A summary of the output of the six statistical packages for logistic regression is given in Section 12.12 and what to watch out for when interpreting logistic regression is discussed in Section 12.13.

12.2   When is logistic regression used?

Logistic regression can be used whenever an individual is to be classified into one of **two** populations. Thus, it is an alternative to the discriminant analysis presented in Chapter 11. When there are more than two groups, what is called polychotomous or generalized logistic regression analysis can be used.

In the past, most of the applications of logistic regression were in the medical field, but it is also frequently used in epidemiologic research. It has been used, for example, to calculate the risk of developing heart disease as a function of certain personal and behavioral characteristics such as age, weight, blood pressure, cholesterol level, and smoking history. Similarly, logistic regression can be used to decide which characteristics are predictive of teenage pregnancy. Different variables such as grade point average in elementary school, religion, or family size could be used to predict whether or not a teenager will become pregnant. In industry, operational units could be classified as successful or not according to some objective criteria. Then several characteristics of the units could be measured and logistic regression could be used to determine which characteristics best predict success. The use of this technique for regression analysis is widespread since very often the outcome variable only takes on two values. We also describe how to use logistic regression analysis when there are more than two outcomes. These outcomes can be either nominal or ordinal.

Logistic regression also represents an alternative method of classification when the multivariate normal model is not justified. As we discuss in this chapter, logistic regression analysis is applicable for any combination of discrete and continuous predictor variables. (If the multivariate normal model with equal covariance matrices is applicable, the methods discussed in Chapter 11 would result in a better classification procedure.)

Logistic regression analysis requires knowledge of both the dependent (or outcome) variable and the independent (or predictor) variables in the sample being analyzed. The results can be used in future classification when only the predictor variables are known, similar to the results in discriminant function analysis.

12.3   Data example

The same depression data set described in Chapter 11 will be used in conjunction with discriminant analysis to derive a formula for the probability of being depressed. This formula is the basis for logistic regression analysis. Readers who have not studied the material in Chapter 11 may skip to the next section.

In Section 11.6 we estimated the probability of belonging to the first group (not depressed). In this chapter the first group will consist of people who are depressed rather than not depressed. Based on the discussion given in Section 11.6, this redefinition of the groups implies that the discriminant function based on age and income is

$$Z = -0.0209(\text{age}) - 0.0336(\text{income})$$

with a dividing point $C = -1.515$. Assuming equal prior probabilities, the posterior probability of being depressed is

$$\text{Prob(depressed)} = \frac{1}{1 + \exp[-1.515 + 0.0209(\text{age}) + 0.0336(\text{income})]}$$

For a given individual with a discriminant function value of Z, we can write this posterior probability as

$$P_Z = \frac{1}{1 + e^{C-Z}}$$

As a function of Z, the probability P_Z has the logistic form shown in Figure 12.1. Note that P_Z is always positive; in fact, it must lie between 0 and 1 because it is a probability. The minimum age is 18 years, and the minimum income is $\$2 \times 10^3$. These minimums result in a Z value of -0.443 and a probability $P_Z = 0.745$. When Z is equal to the dividing point C ($Z = -1.515$), then $P_Z = 0.5$. Larger values of Z occur when age is younger and/or income is lower. For an older person with a higher income, the probability of being depressed is low.

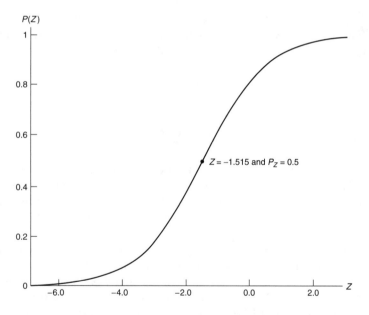

Figure 12.1: *Logistic Function for the Depression Data Set*

Figure 12.1 is an example of the cumulative distribution function for the logistic distribution.

Recall from Chapter 11 that $Z = a_1X_1 + a_2X_2 + \cdots + a_PX_P$. If we rewrite

$C - Z$ as $-(a + b_1 X_1 + b_2 X_2 + \cdots + b_P X_P)$, thus $a = -C$ and $b_i = a_i$ for $i = 1$ to P, the equation for the posterior probability can be written as

$$P_Z = \frac{1}{1 + e^{-(a + b_1 X_1 + b_2 X_2 + \cdots + b_P X_P)}}$$

which is mathematically equivalent to

$$P_Z = \frac{e^{a + b_1 X_1 + b_2 X_2 + \cdots + b_P X_P}}{1 + e^{a + b_1 X_1 + b_2 X_2 + \cdots + b_P X_P}}$$

This expression for the probability is the basis for logistic regression analysis.

12.4   Basic concepts of logistic regression

The **logistic function** has the form

$$P_Z = \frac{e^{\alpha + \beta_1 X_1 + \beta_2 X_2 + \cdots + \beta_P X_P}}{1 + e^{\alpha + \beta_1 X_1 + \beta_2 X_2 + \cdots + \beta_P X_P}}$$

This equation is called the **logistic regression equation**, where Z is the linear function $\alpha + \beta_1 X_1 + \cdots + \beta_P X_P$. It may be transformed to produce a new interpretation. Specifically, we define the **odds** as the following ratio:

$$\text{odds} = \frac{P_Z}{1 - P_Z}$$

or in terms of P_Z,

$$P_Z = \frac{\text{odds}}{1 + \text{odds}}$$

Computing the odds is a commonly used technique of interpreting probabilities (Fleiss *et al.*, 2003). For example, in sports we may say that the odds are 3 to 1 that one team will defeat another in a game. This statement means that the favored team has a probability of $3/(3 + 1)$ of winning or 0.75.

Note that as the value of P_Z varies from 0 to 1, the odds vary from 0 to ∞. When $P_Z = 0.5$, the odds are 1. On the odds scale the values from 0 to 1 correspond to values of P_Z from 0 to 0.5. On the other hand, values of P_Z from 0.5 to 1.0 result in odds of 1 to ∞. Taking the natural logarithm of the odds will cure this asymmetry. When $P_Z = 0$, $\ln(\text{odds}) = -\infty$; when $P_Z = 0.5$, $\ln(\text{odds}) = 0.0$; and when $P_Z = 1.0$, $\ln(\text{odds}) = +\infty$. The term **logit** is sometimes used instead of $\ln(\text{odds})$.

By performing some algebraic manipulation and taking the natural logarithm of the odds, we obtain

$$\text{odds} = \left(\frac{P_Z}{1 - P_Z} \right) = e^{\alpha + \beta_1 X_1 + \beta_2 X_2 + \cdots + \beta_P X_P}$$

Table 12.1: *Classification of individuals by depression level and sex*

	Depression		
Sex	Yes	No	Total
Female (1)	40	143	183
Male (0)	10	101	111
Total	50	244	294

$$\ln(\text{odds}) = \ln\left(\frac{P_Z}{1 - P_Z}\right) = \alpha + \beta_1 X_1 + \beta_2 X_2 + \cdots + \beta_P X_P = Z$$

In words, the logarithm of the odds is linear in the independent variables $X_1, \cdots, X_P$. This equation is in the same form as the multiple linear regression equation (Chapter 7). For this reason the logistic function has been called the **multiple logistic regression equation**, and the coefficients in the equation can be interpreted as regression coefficients. The right hand side of the equation is called the linear predictor. As described in Section 12.11 on Poisson regression, the function which relates the mean, in this case the mean of $\frac{P_Z}{1 - P_Z}$, to the linear predictor is called the **link function**. Thus, for logistic regression the ln(odds) is the link function.

The fundamental assumption in logistic regression analysis is that ln(odds) is linearly related to the independent variables. No assumptions are made regarding the distributions of the X variables. In fact, one of the major advantages of this method is that the X variables may be discrete or continuous.

As mentioned earlier, the technique of linear discriminant analysis can be used to compute estimates of the parameters $\alpha, \beta_1, \beta_2, \ldots, \beta_P$. These discriminant function estimates can be obtained by minimizing the Mahalanobis distance function (Section 11.5). However, the method of **maximum likelihood** produces estimates that depend only on the logistic model. The maximum likelihood estimates should, therefore, be more robust than the linear discriminant function estimates.

Most logistic regression programs use the method of maximum likelihood to compute estimates of the parameters. The procedure is iterative and a theoretical discussion of it is beyond the scope of this book.

The maximum likelihood estimates of the probabilities of belonging to one population or the other are preferred to the discriminant function estimates when the X's are nonnormal.

12.5 Interpretation: categorical variables

One of the most important benefits of the logistic model is that it facilitates the understanding of associations between a dichotomous outcome variable and categorical independent variables. Any number of categorical independent variables can be used in the model. The simplest situation is one in which we have a single X variable with two possible values. For example, for the depression data we can attempt to predict who is depressed on the basis of the individual's sex. Table 12.1 shows the individuals classified by depression and sex.

If the individual is female, then the odds of being depressed are $40/143$. Similarly, for males the odds of being depressed are $10/101$. The **ratio** of these odds is

$$\text{odds ratio} = \frac{40/143}{10/101} = \frac{40 \times 101}{10 \times 143} = 2.825$$

The odds of a female being depressed are 2.825 times that of a male. Note that we could just as well compute the odds ratio of **not** being depressed. In this case we have

$$\text{odds ratio} = \frac{143/40}{101/10} = 0.354$$

Odds ratios are used extensively in biomedical applications (Fleiss *et al.*, 2003). They are a measure of association of a binary variable (risk factor) with the occurrence of a given event such as disease.

To represent a variable such as sex, we customarily use a dummy variable: $X = 0$ if male and $X = 1$ if female. This makes males the referent group (see Section 9.3). (Note that in the depression data set, sex is coded as a 1, 2 variable. To produce a 0, 1 variable, we transform the original variable by subtracting 1 from each sex value.) The logistic regression equation can then be written as

$$\text{Prob(depressed)} = \frac{e^{\alpha+\beta X}}{1 + e^{\alpha+\beta X}}$$

The sample estimates of the parameters are

$$a = \text{estimate of } \alpha = -2.313$$
$$b = \text{estimate of } \beta = 1.039$$

We note that the estimate of β is the natural logarithm of the odds ratio of females to males, or

$$1.039 = \ln 2.825$$

Equivalently,

$$\text{odds ratio} = e^b = e^{1.039} = 2.825$$

Also, the estimate of α is the natural logarithm of the odds for males, the referent group, or

$$-2.313 = \ln\frac{10}{101}$$

When there is only a single dichotomous variable, it is not worthwhile to perform a logistic regression analysis. However, in a multivariate logistic equation the value of the coefficient of a dichotomous variable can be related to the odds ratio in a manner similar to that outlined above. For example, for the depression data, if we include age, sex, and income in the same logistic model, the estimated equation is

Prob(depressed)

$$= \frac{\exp\{-0.676 - 0.021(\text{age}) - 0.037(\text{income}) + 0.929(\text{sex})\}}{1 + \exp\{-0.676 - 0.021(\text{age}) - 0.037(\text{income}) + 0.929(\text{sex})\}}$$

Since sex is a 0, 1 variable, its coefficient can be given an interesting interpretation. The quantity $e^{0.929} = 2.582$ may be interpreted as the odds ratio of being depressed for females compared to males after adjusting for the linear effects of age and income. It is important to note that such an interpretation is valid only when we do **not** include the interaction of the dichotomous variable with any of the other variables (see Section 12.7).

Approximate confidence intervals for the odds ratio for a binary variable can be computed using the slope coefficient b and its standard error listed in the output. For example, 95% approximate confidence limits for the odds ratio can be computed as $\exp(b \pm 1.96 \times \text{standard error of } b)$.

The odds ratio provides a directly understandable statistic for the relationship between the outcome variable and a specified predictor variable (given all the other predictor variables in the model are fixed). The odds ratio of 2.582 for sex can thus be interpreted as indicated previously. In contrast, the estimated β coefficient is not linearly related to the probability of the occurrence, P_Z, since $P_Z = e^{\alpha + \beta X}/(1 + e^{\alpha + \beta X})$, a nonlinear equation (Demaris, 1990). Since the slope coefficient for the variable "sex" is not linearly related to the probability of being depressed, it is difficult to interpret it in an intuitive manner.

12.6 Interpretation: continuous variables

We perform a logistic regression on age and income in the depression data set. The estimates obtained from the logistic regression analysis are as follows:

Term	Coefficient	Standard error
Age	−0.020	0.009
Income	−0.041	0.014
Constant	0.028	0.487

So the estimate of α is 0.028, of β_1 is −0.020, and β_2 is −0.041. The odds of being depressed are less if the respondent has a higher income and is older. The equation for the ln(odds), or logit, is estimated by $0.028 - 0.020(\text{age})$ $- 0.041(\text{income})$. The coefficients −0.020 and −0.041 are interpreted in the same manner as they were in a multiple linear regression equation, where the dependent variable is $\ln[P_Z/(1-P_Z)]$ and P_Z is the logistic regression equation, estimated as

$$\text{Prob(depressed)} = \frac{\exp\{0.028 - 0.020(\text{age}) - 0.041(\text{income})\}}{1 + \exp\{0.028 - 0.020(\text{age}) - 0.041(\text{income})\}}$$

We compare these results to what we obtained using discriminant analysis (Section 11.6). Recall that the coefficients for age and income in the discriminant function are -0.0209 and -0.0336, respectively, when the depressed are in group I. These estimates are within one standard error of -0.020 and -0.041, respectively, the estimates obtained for the logistic regression. The constant 0.028 corresponds to a dividing point of -0.028. This value is different from the dividing point of -1.515 obtained by the discriminant analysis. The explanation for this discrepancy is that the logistic regression program implicitly uses prior probability estimates of being in the first or second group, obtained from the sample. In this example these prior probabilities are $q_{II} = 244/294$ and $q_I = 50/294$, respectively. When these prior probabilities are used, the discriminant function dividing point is $-1.515 + \ln q_{II}/q_I = -1.515 + 1.585 = 0.070$. This value is closer to the value 0.028 obtained by the logistic regression program than is -1.515.

For a continuous variable X with slope coefficient b, the quantity $\exp(b)$ is interpreted as the ratio of the odds for a person with value $(X+1)$ relative to the odds for a person with value X. Therefore, $\exp(b)$ is the incremental odds ratio corresponding to an increase of one unit in the variable X, assuming that the values of all other X variables remain unchanged. The incremental odds ratio corresponding to the change of k units in X is $\exp(kb)$. For example, for the depression data, the odds of depression can be written as

$$P_Z/(1-P_Z) = \exp[0.028 - 0.020(\text{age}) - 0.041(\text{income})]$$

For the variable age, a reasonable increment is $k = 10$ years. The incremental odds ratio corresponding to an increase of 10 years in age can be computed as $\exp(10b)$ or $\exp(10 \times -0.020) = \exp(-0.200) = 0.819$. In other words, the odds of depression are estimated as 0.819 of what they would be if a person were ten years younger. An odds ratio of one signifies no effect, but in this case increasing the age has the effect of lowering the odds of depression. If the computed value of $\exp(10b)$ had been two, then the odds of depression would have doubled for that incremental change in a variable. This statistic can be called the ten-year odds ratio for age or, more generally, the k incremental odds ratio for variable X (Hosmer and Lemeshow, 2000).

Ninety-five percent confidence intervals for the above statistic can be computed from $\exp[kb \pm k(1.96)\text{se}(b)]$, where se stands for standard error. For example, for age

$$\exp(10 \times -0.020 \pm 10 \times 1.96 \times 0.009) \quad \text{or}$$

$$\exp(-0.376 \text{ to} - 0.024) \quad \text{or}$$
$$0.687 < \text{odds ratio} < 0.976$$

Note that for each coefficient, many of the programs will supply an **asymptotic standard error** of the coefficient. These standard errors can be used to test hypotheses and obtain confidence intervals. For example, to test the hypothesis that the coefficient for age is zero, we compute the test statistic

$$Z = \frac{-0.020 - 0}{0.009} = -2.22$$

For large samples an approximate P value can be obtained by comparing this Z statistic to percentiles of the standard normal distribution.

12.7   Interpretation: interactions

The interpretation of the coefficients of a logistic regression model as described in the preceding sections needs to be revised if there is evidence that two (or more) variables are interacting. We described in Section 9.3 the meaning and modelling of interactions for multiple linear regression. To summarize: the inclusion of an interaction is necessary if the effect of an independent variable depends on the level of another independent variable. For linear regression this also means that if an interaction is present, the combined effect of two (or more) interacting variables cannot be determined by adding the individual effects. For logistic regression the first part stays the same, in that the inclusion of an interaction is necessary if the effect of an independent variable depends on the level of another independent variable. But the difference for the logistic regression setting is that if we determine that the inclusion of an interaction is necessary, then the combined effect of two (or more) interacting variables cannot be determined by multiplying the individual effects. This is because the logistic regression model is a multiplicative model. We will use the depression data to illustrate why the logistic regression model is multiplicative.

Suppose we would like to investigate the relationship between income and depression, and employment status and depression. For the purpose of this example we will dichotomize income into low and high by creating a dummy variable which has the value one for individuals with income <10 (in units of $\$1,000$ per year) and a value of zero for individuals with income ≥ 10. We will create a dummy variable with two values $(0, 1)$. If income is less than 10, we will call it low income and assign it a value of one. A value of zero is assigned for "higher" income. In addition, we might suspect that individuals who are unemployed or part-time employed might be at a greater risk of depression than individuals who are full-time employed, retired, in school, or stay at home. Thus we would create a dummy variable which is one for part-time or unemployed individuals and is zero otherwise. For ease of describing the different groups, in this section we use "underemployed" for part-time or unemployed and "full-time," etc. for full-time employed, retired, in school, or stay at home. For this analysis, we exclude individuals who report "other" as their employment status by setting the dummy variable to missing for these individuals.

Estimates from a logistic regression analysis using the dummy variables for income and employment status but no interaction variable are as follows:

Variable	Coefficient	Standard error
Indicator: Low income (1)	0.272	0.338
Indicator: Underemployed (1)	1.028	0.349
Constant	−1.935	0.226

The corresponding odds ratios for being depressed for low income versus high income and for comparing underemployed versus full-time, etc. are $\exp(0.272) = 1.31$ and $\exp(1.028) = 2.80$, respectively. The interpretation of these individual odds ratios is similar to what is described in Section 12.5. Note that we assume for this model that income and employment status do not interact. In other words, the effect of income is assumed independent of employment status and, conversely, we assume that the effect of employment status is independent of income. Thus, the odds ratio of 2.80 comparing underemployed to full-time, etc. would be assumed to be the same whether we compare individuals who earn less than $10,000$ per year or whether we compare individuals who earn low or high income, as long as we compare individuals from the same income group.

Under the assumption of no interaction, an estimate of the combined effect of income and employment status would be calculated as follows. We would compare the odds of being depressed for individuals who report low income and who are underemployed to the odds of being depressed for individuals who report high income and whose employment status is full-time, etc. First,

$$\text{odds(income = low and employment status = underemployed)}$$

$$= \exp(-1.935 + 0.272 \times 1 + 1.028 \times 1)$$

and second,

$$\text{odds(income = high and employment status = full-time, etc.)}$$

$$= \exp(-1.935 + 0.272 \times 0 + 1.028 \times 0)$$
$$= \exp(-1.935)$$

Thus the odds **ratio** becomes

$$\text{odds ratio} = \frac{\exp\left(-1.935 + 0.272 \times 1 + 1.028 \times 1\right)}{\exp\left(-1.935\right)} = \exp(0.272 + 1.028)$$

Recall that $\exp(a + b) = \exp(a) \times \exp(b)$ so that

$$\begin{aligned}
\text{odds ratio} &= \exp\left(0.272 + 1.028\right) \\
&= \exp\left(0.272\right) \times \exp\left(1.028\right) \\
&= 1.31 \times 2.80 \\
&= 3.67
\end{aligned}$$

Therefore, assuming that there is no interaction, the combined effect of low income and underemployment can be calculated by multiplying the individual effects.

An important next question is: how do we assess whether the assumption of "no interaction" is adequate? We can do this similarly to what we described for multiple linear regression (see Section 9.3). To formally test whether an interaction term is necessary we can add such an interaction term to the model and assess whether the coefficient for the interaction term is significantly different from zero. The interaction term in our example represents a dummy variable which is created by multiplying the individual dummy variables for income and employment status. Estimates from a model including the interaction term are as follows:

Variable	Coefficient	Standard error
Low income (1)	−0.376	0.435
Underemployed (1)	0.318	0.452
Low income (1) and underemployed (1)	2.198	0.789
Constant	−1.735	0.221

Obviously, the coefficients for "Low income" and "Underemployed" when the interaction term is included are very different from the corresponding coefficients if it is not included. This is not a contradiction. Even though the labelling of the included variables stays the same, the interpretation of the coefficients is different in the presence of an interaction term. We will return to the interpretation shortly, but first we describe two approaches to determining whether the interaction term is necessary.

In Section 7.6 we describe for a multiple linear regression model how we can test the hypothesis that a coefficient is equal to zero. We can use a similar test for the logistic regression setting to assess whether an interaction term is necessary. For each variable in the model, most statistical software packages report a so-called **Wald statistic**. This statistic is calculated as

$$\frac{b^2}{SE(b)^2}$$

where b represents the estimated coefficient β and $SE(b)$ is its standard error.

Under the null hypothesis of a zero slope (i.e., no significant association between the corresponding variable and the outcome) and based on asymptotic theory, this quantity follows a chi-square distribution with one degree of freedom. The test statistic puts the value of the slope in perspective relative to the estimated variability of the slope. Roughly speaking, if the estimated value of the slope is small and its estimated variability is large, we do not have enough evidence to conclude that the slope is significantly different from zero and vice versa. Some statistical software packages report a "Z" score, which is the square root of the Wald test statistic. In this case the test statistic value would be compared to a standard normal distribution and would yield exactly the same P value.

The Wald test is an approximate test in the sense that it might not be "right on target," but hopefully it is close. Unfortunately, there is no way to measure exactly how close it is. One other approximate test to assess whether a coefficient is zero is the **likelihood ratio test**. The general idea of this test is to compare a global measure of fit for the data without the interaction term to a global measure of fit for the data with the interaction term. If the inclusion of the interaction term improves the fit of the model significantly, then we would conclude that there is sufficient evidence that the interaction term is necessary. The test statistic is calculated as two times the difference of the logarithms of the likelihood ratio test statistic between the model excluding, and the model including, the interaction term. The test statistic is assumed to follow a chi-square distribution with one degree of freedom. Most of the statistical software packages provide this test statistic as part of the default output. Note that some software packages (like Stata) provide the test statistic value as the logarithm of the likelihood ratio test, whereas others provide -2 times the logarithm of the likelihood ratio test statistic (e.g., SAS). Some statistical software packages provide automated commands to calculate the test statistic value and the corresponding P value (e.g., Stata), whereas others (e.g., SAS) only provide -2(loglikelihood) for each model, and the user has to calculate the actual value of the test statistic and the corresponding P value.

In our example, the value of the Wald test statistic for the difference in slopes is 7.78, which corresponds to a P value of 0.005, whereas the value of the likelihood ratio test statistic is 8.23 with a corresponding P value of 0.004. Thus, based on either test, we would conclude that it is necessary to include the interaction term. We note that these two tests rarely lead to different conclusions, but if they do, the likelihood ratio test is preferable.

Now, how do we interpret the results in the presence of an interaction? To illustrate, we calculate the odds ratio for the combined effect of income and employment status. Similarly to the calculations of the same odds ratio in the absence of an interaction, we first calculate the odds of being depressed for low income and underemployment. Secondly, we calculate the odds of being depressed for high income and full-time, etc. employment. Finally we combine both to calculate the odds ratio. In general the odds for being depressed under the model which includes the interaction is odds =

$$\exp(-1.735 - 0.376(\text{Income}) + 0.318(\text{Employ}) + 2.198(\text{Income} \times \text{Employ}))$$

Note that in the previous equation Income and Employ are dummy variables derived from the original variables.

It follows that, first,

odds(income = low and employment status = underemployed)

$$= \exp(-1.735 - 0.376 \times 1 + 0.318 \times 1 + 2.198 \times 1 \times 1)$$

and second,

odds(income = high and employment status = full-time, etc.)

$$
\begin{aligned}
&= \exp(-1.735 - 0.376 \times 0 + 0.318 \times 0 + 2.198 \times 0 \times 0) \\
&= \exp(-1.735)
\end{aligned}
$$

Thus, the odds **ratio** becomes

$$
\begin{aligned}
\text{odds ratio} &= \frac{\exp(-1.735 - 0.376 + 0.318 + 2.198)}{\exp(-1.735)} \\
&= \exp(-0.376 + 0.318 + 2.198) \\
&= 8.50
\end{aligned}
$$

Recall that, without inclusion of the interaction term, we would have estimated the combined effect to be 3.67. With the interaction term we estimate the same odds ratio to be 8.50. Because the difference between these estimates is statistically significant, we do have sufficient evidence that we need to include the interaction term.

For our example, we calculate all odds ratios (OR) that might be of interest.

1) OR(Income = low versus high
          given Employ = full-time, etc.)

$$\text{odds ratio} \;=\; \frac{\exp\left(-1.735 - 0.376 \times 1 + 0.318 \times 0 + 2.198 \times (1 \times 0)\right)}{\exp\left(-1.735 - 0.376 \times 0 + 0.318 \times 0 + 2.198 \times (0 \times 0)\right)}$$
$$=\; \exp(-0.376)$$
$$=\; 0.687$$

Thus, in the model including the interaction term, the exponentiated coefficient for Income represents a *conditional* odds ratio. It is conditional on the employment status being full-time employed, etc., i.e., Employ $= 0$.

2) OR(Income = low versus high
    given Employ = underemployed)

$$\text{odds ratio} \;=\; \frac{\exp\left(-1.735 - 0.376 \times 1 + 0.318 \times 1 + 2.198 \times (1 \times 1)\right)}{\exp\left(-1.735 - 0.376 \times 0 + 0.318 \times 1 + 2.198 \times (0 \times 1)\right)}$$
$$=\; \exp(-0.376 + 2.198)$$
$$=\; 6.18$$

3) OR(Employ = underemployed versus full-time, etc.
    given Income = high)

$$\text{odds ratio} \;=\; \frac{\exp\left(-1.735 - 0.376 \times 0 + 0.318 \times 1 + 2.198 \times (0 \times 1)\right)}{\exp\left(-1.735 - 0.376 \times 0 + 0.318 \times 0 + 2.198 \times (0 \times 0)\right)}$$
$$=\; \exp(0.318)$$
$$=\; 1.37$$

Similar to 1), the exponentiated coefficient for employment status represents an odds ratio of Employed conditional on Income $= 0$.

4) OR(Employ = underemployed versus full-time, etc.
    given Income = low)

$$\text{odds ratio} \;=\; \frac{\exp\left(-1.735 - 0.376 \times 1 + 0.318 \times 1 + 2.198 \times (1 \times 1)\right)}{\exp\left(-1.735 - 0.376 \times 1 + 0.318 \times 0 + 2.198 \times (1 \times 0)\right)}$$
$$=\; \exp(0.318 + 2.198)$$
$$=\; 12.4$$

   5) OR(Employ = underemployed and Income = low
                    versus Employ = full-time, etc. and Income = high)
                    = 8.50

as calculated above.

   Only the odds ratios for 1) and 3) are directly available from the usual output from statistical packages. They represent conditional odds ratios, but are not labelled as such. Odds ratios for 2), 4), and 5) need to be obtained manually as done above.

   A common mistake is to interpret the coefficients for the income and employment variables as if an interaction term was not included. Thus, one might incorrectly interpret the odds ratio of 0.69 (under 1) as the odds ratio for low versus high income. It would be correctly interpreted as the odds ratio for low versus high income **given** full-time, etc. employment.

   Another common mistake is to interpret the exponentiated coefficient for the interaction term as an odds ratio. The exponentiated coefficient for the interaction term does not represent an odds ratio. This mistake is easily made since statistical software packages typically report odds ratios as the exponentiated coefficients whether or not they represent a true odds ratio. However, the exponentiated interaction coefficient *can* be interpreted as the ratio of two odds ratios. For example, the ratio of the two odds ratios computed in parts 2) and 1) above is $6.18/0.687 = 9.00$. This value is equivalent to exp(interaction coefficient) $= \exp(2.198) = 9.00$. This value is also the ratio of the two odds ratios computed in parts 4) and 3) above. Another interpretation of the exponentiated interaction coefficient is that it is the multiplier of the odds ratio obtained in part 1) to get the one in part 2), and similarly for parts 3) and 4). But we do not recommend reporting the exponentiated interaction coefficient because it could be confusing, as explained above. In the presence of an interaction it is best to determine which odds ratios are of interest and then calculate them manually.

   The results in our example are not surprising. One can easily imagine that having little money and being part-time or unemployed (underemployed) has a synergistic effect on the risk of being depressed. It is generally advisable to check any results for feasibility.

   It is also possible to include an interaction term between a categorical and a continuous variable. We discuss an example of such an interaction in Section 13.8 in the context of survival analysis.

   Calculation of appropriate confidence intervals for the odds ratios is more complicated as well. In fact, the calculation of any confidence interval for an OR that includes more than one estimated coefficient (as is the case for 2), 4), and 5) above) requires incorporating the estimated covariance of the involved parameters. These covariances are typically not available as part of the default output accompanying logistic regression analysis results, but can be requested through separate options or commands. A description of calculating confidence intervals for odds ratios in the presence of interactions can be found in Hosmer and Lemeshow (2000) and in Kleinbaum and Klein (2010).

Note that in the presence of an interaction, it is usually not sufficient and potentially incorrect to present the odds ratios obtained from the statistical software package. It is important to decide which ORs are of interest and then present them with the appropriate confidence intervals. These confidence intervals unfortunately are often not available from the output. Some software programs, e.g., SAS and Stata, do provide such confidence intervals but the user must specify additional code.

Further discussion of this subject may be found in Breslow and Day (1993, 1994), Schlesselman (1982), Jaccard (2001), and Hosmer and Lemeshow (2000). In particular, the case of several dummy variables is discussed in some detail in these books.

Confounding and effect modification

In many logistic regression analyses, we examine the association between a disease and a risk factor, e.g., between lung cancer (yes/no: the outcome) and smoking (yes/no: a risk factor). Another covariate, e.g., gender (F/M) may be taken into account. Such a covariate would be called a **confounder** if it is associated with both the outcome and the risk factor. If we perform a logistic regression analysis that includes both smoking and gender, then the resulting odds ratio for smoking is adjusted for gender. It would have the same value for males and females, that is, it compares a female smoker to a female nonsmoker or a male smoker to a male nonsmoker. On the other hand, if the odds ratio depends on gender, then gender would be called an **effect modifier** for smoking and we must include the interaction term, as discussed in this section. The odds ratios resulting from the logistic regression would be different for males and females. That is, gender modifies the effect of smoking on lung cancer.

To check for whether a covariate (e.g., gender) is a confounder or an effect modifier for the risk factor (e.g., smoking), we first perform a logistic regression analysis that includes the covariate and the risk factor as well as their interaction. If the interaction term is statistically significant, then the covariate may be considered an effect modifier for the risk factor. If the interaction is not statistically significant, then we may consider it not to be an effect modifier and we run another analysis that includes the covariate and the risk factor but not their interaction. If the covariate is statistically significant then it may be considered a confounder for the risk factor.

Detailed discussion of confounding and effect modification may be found in Hosmer and Lemeshow (2000), Jewell (2004) or Rothman *et al.* (2008).

12.8   Refining and evaluating logistic regression

Many of the same strategies used in linear regression analysis or discriminant function analysis to obtain a good model or to decide if the computed model is useful in practice also apply to logistic regression analysis. In this section we describe typical steps taken in logistic regression analysis in the order that they are performed.

Variable selection

As with linear regression analysis, often the investigator has in mind a prede-termined set of predictor variables to predict group membership. For example, in Section 12.2 we discussed using logistic regression to predict future heart at-tacks in adults. In this case the data came from longitudinal studies of initially healthy adults who were followed over time to obtain periodic measurements. Some individuals eventually had heart attacks and some did not. The investiga-tors had a set of variables which they thought would predict who would have a heart attack. These variables were used in logistic regression models. Thus, the choice of the predictor variables was determined in advance.

In other examples of logistic regression applications, one of the variables indicates which of two treatments each individual received and other variables describe quantities or conditions that may be related to the success or failure of the treatments. The analysis is then performed using treatment (say experi-mental and control) and the other variables that affect outcome, such as age or gender, as predictor variables. Again, in this example the investigator is usually quite sure of which variables to include in the logistic regression equation. The investigator wishes to find out if the effect of the treatment is significant after adjusting for the other predictor variables.

But in exploratory situations, the investigator may not be certain about which variables to include in the analysis. Often the variables are screened one at a time to see which ones are associated with the dichotomous outcome variable. This screening can be done by using the common chi-square test of association described in Section 17.5 for discrete predictor variables or a t test of equal group means for continuous predictor variables. A fairly large signif-icance level is usually chosen for this screening so that useful predictors will not be missed (perhaps $\alpha = 0.15$). Then, any variables with P values less than say $P = 0.15$ are kept for further analyses. Such a screening will often reduce the number of variables to ten or less.

Another reason for reducing the number of variables used relates to a sam-ple size problem that often occurs in logistic regression analysis. Suppose that the outcome event that is being measured does not occur very often. In general, let n_0 be the frequency of the event that occurs less often. For example, in the depression study we had 50 people who were depressed and 244 who were not. Thus, the event that occurs less often is that of being depressed and $n_0 = 50$.

Let P equal the number of parameters being estimated excluding the intercept. For example, if we include two predictor variables $P = 2$, since we would estimate two βs and one α (intercept). The rule given by Hosmer and Lemeshow (2000) is that $P + 1$ should be less than $n_0/10$, which in our example is equal to $50/10 = 5$. In the depression data based on this rule we would be safe using up to four predictor variables. Thus, the recommended maximum number of variables to be used is not limited by the total number of observations (294), but rather by the number of events that occur less often (50). The issue of minimum sample size for logistic regression analysis is complex and it may depend on more than simply the number of events in the two outcome groups.

After this initial screening, **stepwise** logistic regression analysis can be performed. The test for whether a *single* variable improves the prediction forms the basis for the stepwise logistic regression procedure. The principle is the same as that used in linear regression or stepwise discriminant function analysis. Forward and backward procedures as well as different criteria for deciding how many and which variables to select are available. These tests are typically based on chi-square statistics. A large chi-square or a small P value indicates that the variable should be added in forward stepwise selection. In some programs this statistic is called the "improvement chi-square." As in regression analysis, any levels of significance should be viewed with caution. The test should be viewed as a screening device, not a test of significance.

Several simulation studies have shown that logistic regression provides better estimates of the coefficients and better prediction than discriminant function analysis when the distribution of the predictor variables is not multivariate normal. However, a simulation study found that discriminant function analysis often did better in correctly selecting variables than logistic regression when the data were log-normally distributed or were a mixture of dichotomous and log-normal variables. Even with all dichotomous predictor variables, the discriminant function did as well as logistic regression analysis. These simulations were performed for sample sizes of 50 and 100 in each group (O'Gorman and Woolson, 1991).

Many of the logistic regression programs also include a chi-square statistic that tests whether or not the total set of variables in the logistic regression equation is useless in classifying individuals. A large value of this statistic (sometimes called the model chi-square) is an indication that the variables are useful in classification. We note that SAS also offers the option to perform a best subset logistic regression that is used in a manner similar to multiple linear regression.

Checking the fit of the model

For logistic regression there is no need to require that the independent variables follow a specific distribution. But what are the assumptions underlying the logistic regression model? We are assuming that the logistic regression model holds as described in Section 12.4. If the model includes a continuous variable, we also assume that the relationship between the continuous variable and the ln(odds) is correctly modelled. Typically, a linear relationship between any continuous variable and the ln(odds) is assumed, at least initially. Another aspect of model fit is whether there are individuals with "excessive" influence on the estimated parameters. We will describe methods to assess each one of these aspects of fit separately.

First, we focus on the functional form of a continuous variable. Consider again the depression data using age, income, and sex as predictors. In the preceding sections we modelled age as being linearly related to the ln(odds) of being depressed. A relatively easy way to check this assumption graphically is the following. First the continuous variable is divided into quantiles (see Section 4.3), e.g., quartiles. We then generate dummy variables for these quantiles and refit the model using the dummy variables, while specifying one of the quantile groups as the referent group. A graph of the coefficients for the dummy variables versus the midpoints of the groups forming the quantiles should exhibit a roughly linear shape. In our example we choose quartiles of age. Estimated coefficients for the dummy variables for age when using the youngest age quartile as a referent group are:

Term	Coefficient	Standard error
Referent: Age group <28	–	–
Age group 28 – 42	0.075	0.432
Age group 43 – 58	−0.571	0.474
Age group 59 – 89	−0.885	0.456
Income	−0.038	0.015
Sex (1 = Female)	0.924	0.386
Constant	−2.159	0.783

If we now graph the coefficients for the dummy variables versus the midpoint of the quartiles we can graphically assess whether the assumption of a linear relationship between age and ln(odds) of being depressed seems justified (see Figure 12.2). Note that the referent group has an estimated coefficient of zero.

Even though graphs usually cannot give a definite answer, it seems that the linear model might be reasonable. We could formally assess whether, e.g., a quadratic function of age would be a significant improvement compared to the linear model. To do so, we would fit a model which includes the variables age and age^2 for the quadratic model. We could then assess whether these represent a significant improvement by using either the Wald test or the likelihood ratio test to assess whether the coefficient of the variable age^2 is zero.

The above-described graphical method for assessing the functional form is

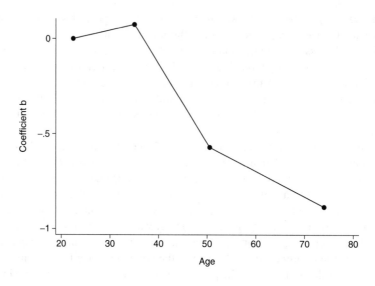

Figure 12.2: *Estimated Coefficients for Age Quartiles by Midpoint of the Quartile*

not efficient. A lot of information is lost by forming quantiles of a continuous variable. Also the choice of the number of groups used is likely to influence the conclusions drawn from the graph. Nevertheless, this method represents a quick and easy approach to obtaining a visual impression of the functional form of the relationship. Other, more precise methods for estimating the most appropriate functional form of a continuous variable are available, but are not discussed here. For further details see, e.g., Hosmer and Lemeshow (2000) and Demaris (1992).

Another way of checking the model is to look at the overall fit. Several approaches have been proposed for testing how well the logistic regression model fits the data overall. Most approaches rely on the idea of comparing an observed number of individuals to the number expected if the fitted model were valid. These observed (O) and expected (E) numbers are combined to form a chi-square (χ^2) statistic called the **goodness-of-fit** χ^2. Large values of the test statistic indicate a poor fit of the model. Equivalently, small P values are an indication of poor fit. Here we discuss two different approaches to the goodness-of-fit test.

The **classical approach** begins with identifying different combinations of values of the variables used in the equation, called *patterns*. For example, if there are two dichotomous variables, sex and employment status (employed/unemployed), there are four distinct pattern: male employed, male unemployed, female employed, and female unemployed. For each of these combinations we count the observed number of individuals (O) in the groups with and without the outcome. In the depression data example, these groups might represent the individuals who are depressed and the individuals who are not depressed, respectively. For each of these individuals we compute the probability of being depressed and not depressed from the fitted logistic regression equation. The sum of these probabilities for a given pattern is denoted by E. The goodness-of-fit test statistic is computed as

$$\text{goodness-of-fit } \chi^2 = \sum 2O \left(\ln \frac{O}{E} \right)$$

where the summation is extended over all the distinct patterns. If the number of these distinct patterns is J, then the number of degrees of freedom of the above statistic is $(J - P - 1)$, where P is the number of covariates, including powers or interactions if any. The reader is warned, though, that this statistic may give misleading results if the number of distinct patterns is large since this situation may lead to small values of E. A large number of distinct patterns occurs when continuous variables are used. Thus, for the depression data the classical approach would not be recommended if age and/or income are included in the model as continuous variables.

Another goodness-of-fit approach was developed by Hosmer and Lemeshow (1980) and Lemeshow and Hosmer (1982). In this approach the probability of belonging to population I (depressed) is calculated for every individual in the sample, and the resulting numbers are arranged in increasing order. The range of probability values is then divided into subgroups (usually deciles). For each decile (subgroup) the observed number of individuals in population I (depressed) is computed (O). Also, the expected numbers (E) are calculated by adding the logistic probabilities for all individuals in each decile. The goodness-of-fit statistic is calculated from the Pearson chi-square statistic as:

$$\text{goodness-of-fit } \chi^2 = \sum \frac{(O-E)^2}{E}$$

where the summation extends over the two groups (depressed and not depressed) and the ten deciles. The degrees of freedom for this statistic is the number of subgroups minus two.

A similar statistic and its P value are given in the Stata logistic lfit option. The procedure calculates the observed and expected numbers for all individuals, or in each category defined by the predictor variables if they are all categorical. The above summation then extends over all individuals or over all categories, respectively. Note that Hosmer and Lemeshow (2000) indicate that when six or fewer subgroups are used, instead of deciles, then the test statistic is likely to indicate that the model fits, whether it really does or not. Thus, this test should only be used when the number of patterns is sufficiently large to justify the use of seven or more subgroups. They also indicate that different implementations of the statistic in different software packages might lead to different conclusions if the number of observations with the same pattern is large.

The classical approach as well as the Hosmer-Lemeshow test are approximate and require large sample sizes. A large goodness-of-fit statistic (or a small P value) indicates that the fit may not be good. Unfortunately, if they indicate that a poor fit exists, they do not say what is causing it or how to modify the variables to improve the fit.

As a third aspect of checking the fit of the model, we consider the influence of individual patterns. One should first check that there are no gross outliers in the predictor variables by visually checking graphical output (box plots or histograms) for continuous variables and frequencies for the discrete variables. Note that the variables do not have to be normally distributed so that formal tests for outliers may not apply.

Recall from Chapter 6 that a measure of the distance between the observed value Y and the model-predicted value $\hat{Y}$ is called a residual. For logistic regression, the observed value Y is either a zero or a one, whereas the predicted value $\hat{Y}$ is an estimated probability, i.e., a value from zero to one (inclusive). Thus, the largest discrepancies between observed and predicted values in logistic regression will occur either when the predicted value is low (close to zero) and the observed value is one or when the predicted value is large (close to one) and the observed value is zero. Most software packages can produce such residuals for each observation. A residual value close to ± 1 is then a preliminary indication that the corresponding pattern may be an outlier, or that it may have excessive influence on the estimated logistic regression.

A more sophisticated measure to assess the influence that individual patterns have on the estimated parameters is the difference between the estimated parameters when all patterns are included and the estimated parameters when a **specific pattern is excluded** from the analysis. This measure is only meaningful if at least one continuous variable is included in the model. In a model with only categorical variables, excluding a pattern is equivalent to excluding a comparison group and might lead to elimination of a parameter.

Typically, the measure is standardized and represents the collective influence of the pattern on all the estimated β's simultaneously. An intuitive way of obtaining this difference would be to rerun the model as many times as there are patterns and for each run calculate the difference in estimated parameters. To avoid having to rerun the model multiple times, approximations to the measure have been developed and can be obtained by most statistical packages as the so-called **delta beta statistic** ($\Delta\beta$, where delta stands for the difference). Thus $\Delta\beta$ measures for each pattern the change in estimated coefficients if we were to exclude that pattern. When plotting the values of $\Delta\beta$ for each pattern against the predicted probabilities, one can identify individuals with relatively large influence on the estimated coefficients. For the depression data when including age, income, and sex as independent variables the resulting $\Delta\beta$ for each covariate pattern is plotted in Figure 12.3 on the vertical axis while the estimated probability of being depressed is plotted on the horizontal axis.

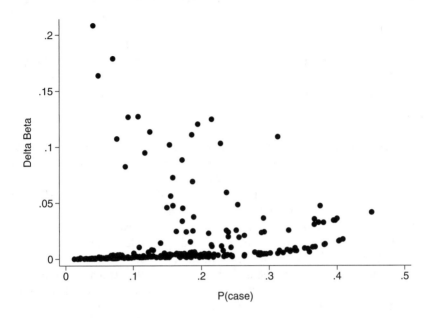

Figure 12.3: *Delta Beta Measures to Assess the Influence of Individual Patterns on Estimated Coefficients*

It might be surprising at first that the patterns with the lowest probability of being depressed have the largest value of $\Delta\beta$. The reason for this becomes clear when we look at the patterns that we would expect to have the largest influence. We would expect those patterns to have the largest influence on parameter estimates for which the discrepancy between the observed and the predicted value is the greatest. For logistic regression these are either the patterns for which the outcome is present (value one), but for which the probability of the outcome is estimated to be low (close to zero) or the patterns for which the outcome is not present (value zero), but for which the probability of the outcome is estimated to be high (close to one). The latter case is not present in these data, since none of the estimated probabilities is larger than 0.46.

That none of the estimated probabilities is larger than 0.46 might be surprising as well. The maximum estimated probability depends mostly on the prevalence of the outcome in the data. Based on our model, we would expect the prevalence to be largest among young females with low income. If we, e.g., calculate the prevalence of depression among females who are 27 years or

Table 12.2: *Percent change in estimated parameters when including and excluding influential patterns*

Model	Age	Change	Income	Change	Sex	Change
Overall	−0.0210		−0.0366		0.929	
1st $\Delta\beta$	−0.0215	2%	−0.0417	14%	1.023	10%
2nd $\Delta\beta$	−0.0234	11%	−0.0358	2%	1.051	13%
3rd $\Delta\beta$	−0.0229	9%	−0.0389	6%	1.042	12%
All 3 $\Delta\beta$	−0.0263	25%	−0.0440	20%	1.303	40%

younger who have a yearly income less than \$15,000 we obtain a prevalence of depression of 44%. This estimate comes very close to the maximum estimated probability.

From the graph we can identify patterns with large values for $\Delta\beta$. One could then sort all observations by $\Delta\beta$ value and print the identification numbers of the observations with the largest values. As a next step we could re-run the logistic regression model and individually exclude the patterns with the highest values of $\Delta\beta$ by excluding the observations that exhibit these patterns. Of most interest would be the degree of change of the estimated coefficients when excluding each of the patterns. How many patterns to exclude is a subjective decision. If there are points in the graph that are clearly outliers, then these should be examined. The number of patterns examined in more detail also depends on the total number of observations. For the depression data we decide to exclude the patterns with the three largest values of $\Delta\beta$. The pattern with the largest value of $\Delta\beta$ represents three individuals. The patterns with the second and third largest $\Delta\beta$ values represent one individual each. Table 12.2 summarizes the estimated coefficients for age, income, and sex for the model excluding no pattern, excluding the pattern with the largest, second largest, and third largest $\Delta\beta$ individually (called the 1st, 2nd, and 3rd $\Delta\beta$), and finally excluding the patterns with the three largest $\Delta\beta$ (all 3 $\Delta\beta$).

Thus, when excluding these three patterns, representing five individuals from the logistic regression (1.7% of the sample), the coefficients for age, income, and sex change by 25%, 20%, and 40%, respectively. If we examine the pattern, outcome, and predicted probability for these individuals, it becomes clear why these individuals have a relatively large impact on the estimated coefficients (see Table 12.3).

All of the individuals have a very low estimated probability of being depressed. In general, the model predicts that the odds of being depressed decrease with increasing age, but individuals with ID numbers 288 and 99 are depressed despite their advanced age. On the other hand, the model predicts that the odds of being depressed decrease as income increases, but one of the

Table 12.3: *Estimated probability of being a case ($\hat{p}$) for five influential observations*

ID	age	income	sex	cases	$\hat{p}$
288	61	28	M	1	0.05
99	72	11	M	1	0.07
143	40	45	M	0	0.04
232	40	45	M	0	0.04
68	40	45	M	1	0.04

individuals (ID number 68) with a relatively large income (at least in this data set) is depressed. But how is it that the coefficient for sex changes by 40% when excluding the individuals with the three largest values of $\Delta\beta$? Recall from Table 12.1 that the sample includes 50 depressed individuals. Of these 10 are males. When excluding the above patterns, we exclude 3 (out of the 10) males who are depressed. Since percentage-wise this is a large change it is not surprising that the coefficient for sex changes dramatically.

An interesting question is: What constitutes "excessive" influence? This is subjective; typically a large percent change in coefficients would be considered "excessive." Some researchers might consider 25% large, whereas others might consider 15% or 30% large. This judgment depends on the research question and specific area of research. In general, if excluding a pattern would change any conclusion substantially then the influence of that pattern would be called "excessive."

What should we do with observations identified as having a large influence on the estimated parameters? First, one should check whether some of the values for the individuals might represent data entry errors. Unless that is the case, we would usually not exclude these individuals, but rather describe them and the associated findings carefully. We would expect that some individuals exhibit outcomes which are contrary to the general findings of the model. This is not a sufficient reason to exclude them. An important aspect to include in any description of the findings is whether our conclusions would change substantively when excluding the influential patterns. In our example, the estimated odds ratios for age, income, and sex would have all changed in the direction of making the estimated effects stronger. Thus, our conclusions would not have changed substantively but only the estimated degree of association.

In addition to measuring the effect of patterns on the estimated parameters, one can also measure the effect of patterns on the fit in general. As described above, the fit can be summarized in an overall chi-square statistic. Changes to this overall chi-square statistic when excluding patterns (called **delta chi-square**, $\Delta\chi^2$ by most programs) can be measured and examined as was done with changes in the estimated coefficients. One could repeat the same steps as described above in order to identify patterns with excessive influence on the overall chi-square value. These procedures ($\Delta\beta$ and $\Delta\chi^2$) will not necessarily identify the same influential patterns. One can see this by obtaining a graph of the estimated probability of being a case (here depressed) on the horizontal axis versus $\Delta\chi^2$ on the vertical axis while allowing the size of the symbol to depend on the corresponding $\Delta\beta$ value.

For the depression data we obtained $\Delta\chi^2$ for all patterns and plotted these versus the estimated probabilities. Each point on the graph is magnified by the corresponding $\Delta\beta$ value. The result can be seen in Figure 12.4.

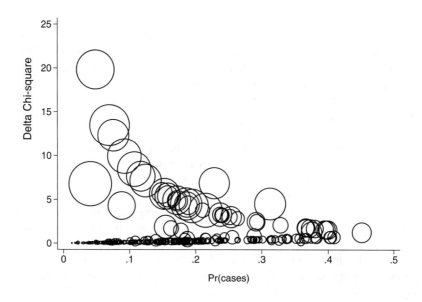

Figure 12.4: *Delta Chi-square Measure to Assess Influence of Pattern on Overall Fit with Symbol Size Proportional to Delta Beta*

Even though the patterns with the largest values of $\Delta\chi^2$ also have large values of $\Delta\beta$, there are some patterns with only intermediate values of $\Delta\chi^2$ but with large values of $\Delta\beta$.

Evaluating how well logistic regression predicts outcomes

In Section 7.6, the coefficient of determination was defined and used as an overall indicator of how well the regression equation predicted the outcomes. A similar coefficient has been proposed for logistic regression analysis (see Menard, 2001). Here

$$R^2 = \frac{\ln L_o - \ln L_m}{\ln L_o}$$

where L_o is the numerical value of the likelihood function for the logistic model when only the intercept term is included and L_m is the numerical value of the likelihood function when all the desired variables are included. The numerical values of these likelihood functions are included in the default output of some statistical programs.

Stata prints the result of this formula as "Pseudo R^2". SAS uses a different formula:

$$\text{Pseudo } R^2 = 1 - (L_m/L_0)^{2/N}$$

We note that pseudo R^2 should not be interpreted as R^2 in multiple regression. In particular, it does not represent the proportion of variation explained by the covariates and thus many statisticians consider it to be of limited value.

If the purpose of performing the logistic regression analysis is to obtain an equation to predict into which of two groups a person or thing could be classified, then the investigator would want to know how much to trust such predictions. In other words, can the equation predict correctly a high proportion of the time? This question is different from asking about the statistical significance, as it is possible to obtain statistically significant results that do not predict very well. Here we will discuss what information is available from logistic regression programs that can be used to evaluate prediction.

If the investigator wishes to classify cases, a cutoff point on the probability of being depressed, for example, must be found. This cutoff point is denoted by P_c. We would classify a person as depressed if the probability of depression is greater than or equal to P_c. The percent of individuals who are correctly classified for various cutpoints can be graphed. These percentages are shown in Figure 12.5. For example, for a cutoff point of about 0.175 each group has approximately 60% correctly classified. This may be a good choice for a cutoff point because it treats both groups equally. In contrast, a cutoff point of 0.002 would result in 98% of the depressed group but only 7% of the normals classified correctly.

Some statistical packages provide a great deal of information to help the user choose a cutoff point and some do not. Others treat logistic regression as a special case of nonlinear regression and provide output similar to that of multiple linear regression (e.g., STATISTICA). The STATISTICA output includes the predicted probabilities for each individual. These probabilities can be saved and separated by whether the person was actually depressed or not. Two histograms (one for depressed and one for nondepressed) can then be formed to see where a reasonable cutoff point would fall. Once a cutoff point is chosen then a two-way table program can be used to make a classification

table similar to Table 11.3. One variable will be depressed or not depressed
and the other the classification according to the cutoff point. In R and S-PLUS,
logistic regression is done using either the glm or gam function and setting
the family argument to binomial. Both functions include an option to save the
predicted probabilities by saving the "fitted values."

SAS LOGISTIC computes the predicted probabilities along with residual
diagnostic output. The value of the cutpoint can be adjusted using the PPROB
option and a classification table is obtained from the CTABLE option. LO-
GISTIC regression in SPSS can also save the predicted probabilities for each
person and make CLASSPLOTS of them. A classification table is printed similar to Table 11.3.

Stata lstat provides considerable output to assist in finding a good cutoff
point. Stata lstat can print a classification table for any cutoff point you supply.
In addition to a two-way table similar to Table 11.3, other statistics are printed.
The user can quickly try a set of cutoff points and zero in on a good one.

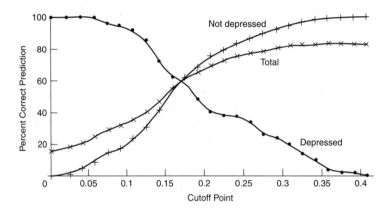

Figure 12.5: *Percentage of Individuals Correctly Classified by Logistic Regression*

Most programs can also plot an ROC (**receiver operating characteristic**)
curve. ROC was originally proposed for signal-detection work when the sig-
nals were not always correctly received. In an ROC curve, the proportion of
depressed persons correctly classified as depressed is plotted on the vertical
axis for the various cutoff points. This is often called the **true positive fraction**, or **sensitivity** in medical research. On the horizontal axis is the proportion
of nondepressed persons classified as depressed (called **false positive fraction**
or one minus **specificity** in medical studies). Swets (1973) gives a summary of
its use in psychology and Metz (1978) presents a discussion of its use in medical studies (see also Kraemer, 1988). SAS also gives the sensitivity, specificity,
and false negative rates.

Three ROC curves are drawn in Figure 12.6. The top one represents a hy-

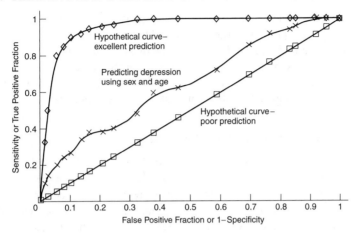

Figure 12.6: *ROC Curve from Logistic Regression for the Depression Data Set*

pothetical curve which would be obtained from a logistic regression equation that resulted in excellent prediction. Even for small values of the proportion of normal persons incorrectly classified as depressed, it would be possible to get a large proportion of depressed persons correctly classified as depressed. The middle curve is what was actually obtained from the prediction of depression using just age and sex. Here we can see that to obtain a high proportion, say 0.80, of depressed persons classified as depressed results in a proportion of about 0.65 of the normal persons classified as depressed, an unacceptable level. The lower hypothetical curve (straight line) represents the chance-alone assignment (i.e., flipping a coin). The closeness of the middle curve to the lower curve shows that we perhaps need other predictor variables in order to obtain a better logistic regression model. This is the case even though our model was significant at the $P = 0.009$ level.

If we believe that the prevalence of depression is low, then a cutoff point on the lower part of the ROC curve is chosen since most of the population is normal and we do not want too many normals classified as depressed. A case would be called depressed only if we were quite sure the person were depressed. This is called a **strict threshold**. The downside of this approach is that many depressed persons would be missed. If we thought we were taking persons from a population with a high rate of depression, then a cutoff point higher up the curve should be chosen. Here a person is called depressed if there is any indication that the person might fall in that group. Very few depressed persons would be missed, but many normals would be called depressed. This type of cutpoint is called a **lax threshold**.

Note that the curve must pass through the points (0,0) and (1,1). The maximum area under the curve is one. The numerical value of this area would be close to one if the prediction is excellent and close to one-half if it is poor. The ROC curve can be useful in deciding which of two logistic regression models to use. All else being equal, the one with the greater area would be chosen. Alternatively, you might wish to choose the one with the greatest height to the ROC curve at the desired cutoff point. Stata gives the area under the curve.

12.9 Nominal and ordinal logistic regression

In the following we use the term **binary logistic regression** to refer to a logistic regression with two groups or outcome categories. When there are more than two groups, extensions of binary logistic regression can be used to model the probability of belonging to a specific group. An important characteristic of these groups is whether they can be ordered or not. If the groups *cannot* be ordered, **nominal logistic regression** is the appropriate analysis. An example of groups or outcome categories which cannot be ordered is the political parties that are voted for in an election. Nominal logistic regression is also called **polytomous**, **polychotomous**, or **multinomial** logistic regression.

If the categories can be ordered, an **ordinal** logistic regression model can be used to model the probability of belonging to a specific group. Several approaches have been suggested to extend binary logistic regression to ordinal logistic regression. We will discuss the most commonly used one, **proportional odds logistic regression**. We note that nominal logistic regression can also be used when the outcome is ordinal.

In general, the more groups or outcome categories are examined, the more effort the analysis takes. Since the basic concepts remain the same, we restrict the presentation in this section to three groups. Extension to four or more groups is straightforward. Nominal logistic regression will be described first and then ordinal logistic regression.

Nominal logistic regression

Recall from binary logistic regression that the odds are defined as

$$\text{odds} = \frac{P_Z}{1 - P_Z}$$

where Z is an appropriate linear function of the independent variables. In the depression data example P_Z represents the probability of being depressed. The denominator $(1 - P_Z)$ represents the complement, the probability of not being depressed. We will again use the depression data but will now form three groups according to the CESD score. We group individuals as either not depressed (0–9), borderline depressed (10–15), or clinically depressed (≥ 16).

The cutoff point for the borderline depressed group is chosen rather arbitrarily, mainly for illustrative purposes.

Parallel to binary logistic regression, we could specify the probability P_Z of belonging to a specific outcome group, e.g., Prob(borderline depression). The denominator in the definition of the odds would be [1-Prob(borderline depression)], which is the same as Prob(no depression or clinical depression). However, this probability is not of interest in and of itself. Of interest is the probability of belonging to the referent group, i.e., we would like to compare borderline depression to no depression. In general, the odds for nominal logistic regression are defined as:

$$\frac{\text{Prob(belonging to specific group)}}{\text{Prob(belonging to referent group)}}$$

Thus, in the depression data, the odds for nominal logistic regression are

$$\frac{\text{Prob(borderline depression)}}{\text{Prob(no depression)}}$$

and

$$\frac{\text{Prob(clinical depression)}}{\text{Prob(no depression)}}$$

The above expressions do not represent odds in a strict mathematical sense; nevertheless, they are typically called odds. We chose no depression as the referent group, but any of the groups can serve as the referent group.

For binary logistic regression, the ln(odds) are defined as

$$\ln(\text{odds}) = \alpha + \beta_1 X_1 + \beta_2 X_2 + \cdots + \beta_P X_P$$

where P represents the number of independent variables and the right hand side is what we called Z. For nominal logistic regression we need a separate set of parameters for each of the odds. We will index the parameters with numbers 1 and 2 to distinguish between the different comparisons.

$$\begin{array}{l} \ln(\text{odds) of} \\ \text{borderline depression} \end{array} = \alpha_1 + \beta_{11} X_1 + \beta_{12} X_2 + \cdots + \beta_{1P} X_P$$

$$\begin{array}{l} \ln(\text{odds) of} \\ \text{clinical depression} \end{array} = \alpha_2 + \beta_{21} X_1 + \beta_{22} X_2 + \cdots + \beta_{2P} X_P$$

Hence, the number of parameters to be estimated in a nominal logistic regression with three groups is $2 \times (P+1)$.

When estimating the parameters for the depression data using sex, age, and income using Stata mlogit, we obtain the results presented in Table 12.4.

Table 12.4: *Estimated coefficients from nominal logistic regression*

Term	Coefficient	Standard error	P value
Borderline depressed			
Sex (1=Female)	−0.017	0.332	0.96
Age (years)	−0.016	0.009	0.08
Income ($1,000)	−0.017	0.012	0.14
Constant	−0.296	0.755	0.70
Clinically depressed			
Sex (1=Female)	0.925	0.393	0.02
Age (years)	−0.024	0.009	0.01
Income ($1,000)	−0.040	0.014	0.01
Constant	−1.136	0.867	0.19

How do we interpret the results? Recall from binary logistic regression that the odds ratio represents the ratio of two odds: for example, the ratio of the odds of being depressed (versus no depression) for females to the odds of being depressed (versus no depression) for males. With three groups there are two analogous odds ratios, the ratio of the odds of being borderline depressed (versus no depression) for females to the odds of being borderline depressed (versus no depression) for males and the ratio of the odds of being clinically depressed (versus no depression) for females to the odds of being clinically depressed (versus no depression) for males. Hence, in nominal logistic regression we need to not only specify who is compared to whom (females versus males), but also the outcome category or group (borderline or clinical depression). The calculation of the odds ratio then proceeds as for binary logistic regression.

The odds ratio for borderline depression for females versus males can be calculated as:

$$\text{odds ratio(borderline depression, females versus males)} = e^{-0.017} = 0.98$$

and the odds ratio for being clinically depressed for females versus males can be calculated as

$$\text{odds ratio(clinical depression, females versus males)} = e^{0.925} = 2.52$$

Similarly, confidence intervals for these odds ratios can also be obtained as in binary logistic regression (Section 12.5) as $\exp(b \pm 1.96 \times \text{standard error}$ of $b)$. Thus, the 95% confidence interval for the odds ratio of being clinically depressed for females versus males is

$$\exp(0.925 \pm 1.96 \times 0.393) \quad \text{or}$$

$$\exp(0.155 \text{ to } 1.695) \quad \text{or}$$
$$1.17 < \text{odds ratio} < 5.45$$

Since the confidence interval does not include 1, we would conclude at the 5% significance level that the odds of clinical depression are significantly larger (about 2.5 times) for females compared to males. In contrast, the odds of borderline depression do not appear to be significantly different for females and males.

How should we proceed with a variable which shows a significant association in one part of the model but not in another? It is not possible to include a variable only for one part of the model. To aid in the modelling decision of including or excluding a specific variable we might want to test whether the variable has an effect across the different outcomes. The Wald or likelihood ratio test can be used for testing such hypotheses. The test statistic values for hypotheses which concern effects across outcomes are typically not part of the default output of statistical packages, but can be obtained via separate commands or options.

When calculating the likelihood ratio statistic to test, e.g., whether each of the variables sex, age, and income is statistically significant across both outcome categories, the resulting test statistic values are 6.48, 8.72, and 10.63. Based on a chi-square distribution with 2 degrees of freedom, the associated P values are 0.04, 0.01, and 0.005. Hence, we would conclude that each of the variables shows a significant association across the outcomes. Note that the degrees of freedom for a given test are the number of parameters that are set to zero under the null hypothesis. To test whether any one variable has no effect, this is the number of outcome categories minus one, or $3 - 1 = 2$ in our example.

We might also be interested in testing whether the coefficients for a particular variable are the same for the different outcomes. When performing Wald tests to check whether the coefficients for each of the variables sex, age, and income are the same for borderline depression and clinical depression, we obtain test statistic values of 4.13, 0.53, and 1.90 which based on a chi-square distribution with one degree of freedom are associated with P values of 0.04, 0.47, and 0.17, respectively. (Here we test the null hypothesis that the parameter for borderline depression minus that for clinical depression is zero. This difference is only one parameter, giving us one degree of freedom.) Thus only for sex would we conclude that the coefficients (-0.017 and 0.925) are significantly different. For specific commands or options to obtain such test statistics the reader is referred to the specific software manual.

In general, if we would not observe any significant differences between the coefficients for the different outcome categories, we might consider combining categories and potentially performing a binary logistic regression. In the depression data example we do observe a significant difference for sex, but

more importantly none of the variables seems to be significant in their relationship to borderline depression. This, together with the knowledge that the
CESD scale was designed to assess clinical depression (not our rather arbitrarily defined group of borderline depression), would lead us to combine the not
depressed and the borderline depressed group and perform binary regression
as was done in previous sections.

Also, if we would not want to combine outcome categories and a variable
shows a significant effect with respect to one but not the other outcome category, this might be of interest in and of itself. We would keep the variable in
the model and describe and discuss the discrepancy between the effects.

Strategies and methods for selecting variables for nominal logistic regression are conceptually the same as for binary logistic regression. They can require considerably more effort, though, due to the multiple outcome categories.

How would the estimated coefficients differ if we were to run two separate binary logistic regression models instead of a nominal logistic regression
model? The main difference would be that for each of the individual models
we would ignore the fact that there is a possible different outcome. For example, when comparing borderline depression to no depression we would ignore
the fact that an individual could also have been clinically depressed. Estimates
and corresponding standard errors based on individual binary logistic regression models can be but are typically not very different from a nominal logistic
regression model.

Ordinal logistic regression

Analyzing ordinal outcome categories by nominal logistic regression represents the most general approach; any information regarding the order would
not be used. The basic idea of ordinal logistic regression is to take advantage
of the ordered nature of the outcome categories.

The most commonly used ordinal logistic regression model is the so-called
proportional odds model. It is also referred to as the **cumulative logit model**.
Assume for now that the outcome categories are coded as values 0, 1, 2, ... In
the proportional odds model, the probability of belonging to outcome category
k or a lower category is compared to belonging to a category higher than k, i.e.,
we consider:

$$\frac{\text{Prob(belonging to category } k \text{ or lower)}}{\text{Prob(belonging to category higher than } k)}$$

This ratio represents the odds of belonging to category k or lower. Suppose that
no depression, borderline depression, and clinical depression are coded as 0, 1,
and 2 for the depression data. Then the following are the possible odds to be
modeled.

Table 12.5: *Estimated coefficients from ordinal logistic regression*

Term	Coefficient	Standard error	P value
Sex	0.520	0.266	0.050
Age	−0.021	0.007	0.003
Income	−0.028	0.009	0.003
Intercept 1	0.040	0.594	
Intercept 2	1.001	0.598	

$$\frac{\text{Prob}(Y \leq 0)}{\text{Prob}(Y > 0)} = \frac{\text{Prob}(\text{no depression})}{\text{Prob}(\text{borderline or clinical depression})}$$

and

$$\frac{\text{Prob}(Y \leq 1)}{\text{Prob}(Y > 1)} = \frac{\text{Prob}(\text{no or borderline depression})}{\text{Prob}(\text{clinical depression})}$$

In the proportional odds model it is assumed that the effect of any variable is the same across all odds, i.e., regardless of which groups are compared. Thus, if we look at the expression for ln(odds)

$$\ln\left(\frac{\text{Prob}(Y \leq 0)}{\text{Prob}(Y > 0)}\right) = \alpha_1 + \beta_1 X_1 + \beta_2 X_2 + \cdots + \beta_P X_P$$

and

$$\ln\left(\frac{\text{Prob}(Y \leq 1)}{\text{Prob}(Y > 1)}\right) = \alpha_2 + \beta_1 X_1 + \beta_2 X_2 + \cdots + \beta_P X_P$$

there is only one parameter for each variable in the model. The intercept is allowed to be different for each of the odds. Thus, there are $p + $ (number of groups−1) parameters to be estimated.

If we run the proportional odds model using Stata ologit for the depression data with sex, age, and income as independent variables, we obtain the estimates presented in Table 12.5. Note that, according to the above models, the odds are in the opposite direction to those for binary and nominal logistic regression, i.e., we are comparing lower values for the outcome categories to higher values for the outcome categories. As a result, we would have expected the coefficients to be in the opposite direction. However, they are not in the opposite direction, since we estimated the model in Stata, and the coefficients in Stata are parameterized as:

$$\ln\left(\frac{\text{Prob}(Y \le 0)}{\text{Prob}(Y > 0)}\right) = \alpha_1 - \beta_1 X_1 - \beta_2 X_2 - \cdots - \beta_P X_P$$

and

$$\ln\left(\frac{\text{Prob}(Y \le 1)}{\text{Prob}(Y > 1)}\right) = \alpha_2 - \beta_1 X_1 - \beta_2 X_2 - \cdots - \beta_P X_P$$

This observation illustrates the fact that, particularly for logistic regression analyses, one should make sure that the statistical program being used parameterizes the model as expected.

For the odds ratio comparing females versus males and comparing the odds of less depression versus more depression using the proportional odds logistic regression model we calculate

$$\frac{\frac{P(y \le j, \text{female})}{P(y > j, \text{female})}}{\frac{P(y \le j, \text{male})}{P(y > j, \text{male})}} = \frac{\exp(-0.520 \times 1)}{\exp(-0.520 \times 0)} = 0.59, \quad \text{for } j = 0 \text{ or } 1$$

(Here we used 1 = female and 0 = male.) Thus, the odds for females to have less depression are estimated to be 41% less than the odds for males. Note that a minus sign was inserted in front of the estimated parameter of 0.520 in order to reverse the parameterization Stata uses for the proportional odds model.

Similarly, we can calculate the odds ratio for an increase of 10 years in age when comparing the odds of less depression to more depression as

$$
\begin{aligned}
\frac{\frac{P(y \le j, \text{age}+10)}{P(y > j, \text{age}+10)}}{\frac{P(y \le j, \text{age})}{P(y > j, \text{age})}} &= \frac{\exp(-(-0.021) \times (\text{age}+10))}{\exp(-(-0.021) \times (\text{age}))} \\
&= \exp(0.021 \times 10) \\
&= 1.23, \quad \text{for } j = 0 \text{ or } 1
\end{aligned}
$$

As a result, an increase of 10 years in age is estimated to increase the odds of being less depressed by 23%. These estimated odds ratios are assumed to be the same whether we compare no depression to (borderline or clinical depression) or whether we compare (borderline or no depression) to clinical depression. This is a relatively stringent assumption. It is possible though to formally test this assumption, and SAS, e.g., provides a test statistic, called the "score" statistic, as part of the default output of the proportional odds model in PROC LOGISTIC. The idea behind the test is to compare the fit of the proportional odds model to the fit of a nominal logistic regression model. If the unrestricted nominal model fits the data significantly better than the restricted one there is evidence that we should not base our conclusions on the restricted proportional odds model. It should be noted that this test is only suggestive, since the models are not nested, i.e., one model is not a subset of the other.

If the program does not provide this test directly, we can compute it as follows. The proportional odds and the nominal model each has an associated log-likelihood value. The likelihood ratio statistic is computed as two times the difference of these log-likelihood values. Under the null hypothesis, it is assumed to be distributed as chi-square with degrees of freedom equal to the difference in the number of estimated parameters. For the depression example, the log-likelihood values are -243.880 and -245.706 for the two models, respectively. Taking the difference and multiplying by 2, we obtain the likelihood ratio test statistic as 3.65, which is associated with a P value of 0.30 when compared to a chi-square distribution with three degrees of freedom. Thus, we fail to reject the hypothesis that the ordinal logistic regression model fits the data adequately when compared with the nominal model. Readers interested in other ordinal models or more details are referred to Agresti (2002) or Hosmer and Lemeshow (2000).

Issues of statistical modeling such as variable selection and goodness-of-fit analysis for ordinal logistic regression are similar to those for binary logistic regression. Statistical tests and procedures for these are mostly not yet available in software packages. The interested reader is referred to Agresti (2002), Hosmer and Lemeshow (2000), and Kleinbaum and Klein (2010) for further discussion of this subject.

12.10 Applications of logistic regression

Multiple logistic regression equations are often used to estimate the probability of a certain event occurring to a given individual. Examples of such events are failure to repay a loan, the occurrence of a heart attack, or death from lung cancer. In such applications the period of time in which such an event is to occur must be specified. For example, the event might be a heart attack occurring within ten years from the start of observation.

Adjusting for different types of samples

For estimation of the equation a sample is needed in which each individual has been observed for the specified period of time and values of a set of relevant variables have been obtained at or up to the start of the observation. Such a sample can be selected and used in the following two ways.

1. A sample is selected in a random manner and observed for the specified period of time. This sample is called a **cross-sectional sample**. From this single sample two subsamples result, namely, those who experience the event

and those who do not. The methods described in this chapter are then used
to obtain the estimated logistic regression equation. This equation can be
applied directly to a new member of the population from which the original
sample was obtained. This application assumes that the population is in a
steady state, i.e., no major changes occur that alter the relationship between
the variables and the occurrence of the event. Use of the equation with a
different population may require an adjustment, as described in the next
paragraph.

2. The second way of obtaining a sample is to select two random samples, one
for which the event occurred and one for which the event did not occur.
This sample is called a **case-control sample**. Values of the predictive variables must be obtained in a retrospective fashion, i.e., from past records or
recollection. The data can be used to estimate the logistic regression equation. This method has the advantage of enabling us to specify the number of
individuals with or without the event. In the application of the equation the
regression coefficients $b_1, b_2, \ldots, b_p$ are valid. However, the constant a must
be adjusted to reflect the true population proportion of individuals with the
event. To make the adjustment, we need an estimate of the probability P of
the event in the target population. The adjusted constant a^* is then computed
as follows:

$$a^* = a + \ln[\hat{P}n_0/(1-\hat{P})n_1]$$

where
a = the constant obtained from the program
$\hat{P}$ = estimate of P
n_0 = the number of individuals in the sample for whom the event did
not occur
n_1 = the number of individuals in the sample for whom the event did
occur

Further details regarding this subject can be found in Breslow (1996).

For example, in the depression data suppose that we wish to estimate the
probability of being depressed on the basis of sex, age, and income. From the
depression data we obtain an equation for the probability of being depressed
as follows:

Prob(depressed)

$$= \frac{\exp\{-0.676 - 0.021(\text{age}) - 0.037(\text{income}) + 0.929(\text{sex})\}}{1 + \exp\{-0.676 - 0.021(\text{age}) - 0.037(\text{income}) + 0.929(\text{sex})\}}$$

This equation is derived from data on an urban population with a depression
rate of approximately 20%. Since the sample has 50/294, or 17%, depressed
individuals, we must adjust the constant. Since $a = -0.676, \hat{P} = 0.20, n_0 = 244$,

and $n_1 = 50$, we compute the adjusted constant as

$$
\begin{aligned}
a^* &= -0.676 + \ln[(0.2)(244)/(0.8)(50)] \\
&= -0.676 + \ln(1.22) \\
&= -0.676 + 0.199 \\
&= -0.477
\end{aligned}
$$

Therefore the equation used for estimating the probability for being depressed is

Prob(depressed)

$$
= \frac{\exp\{-0.477 - 0.021(\text{age}) - 0.037(\text{income}) + 0.929(\text{sex})\}}{1 + \exp\{-0.477 - 0.021(\text{age}) - 0.037(\text{income}) + 0.929(\text{sex})\}}
$$

Another method of sampling is to select **pairs** of individuals **matched** on certain characteristics. One member of the pair has the event and the other does not. Data from such matched studies may be analyzed by the logistic regression programs described in the next paragraph after making certain adjustments. Holford, White and Kelsey (1978) and Kleinbaum and Klein (2010) describe these adjustments for one-to-one matching; Breslow and Day (1993) give a theoretical discussion of the subject. Woolson and Lachenbruch (1982) describe adjustments that allow a least-squares approach to be used for the same estimation, and Hosmer and Lemeshow (2000) discuss many-to-one matching.

As an example of one-to-one matching, consider a study of melanoma among workers in a large company. Since this disease is not common, all the new cases are sampled over a five-year period. These workers are then matched by gender and age intervals to other workers from the same company who did not have melanoma. The result is a matched-paired sample. The researchers may be concerned about a particular set of exposure variables, such as exposure to some chemicals or working in a particular part of the plant. There may be other variables whose effect the researcher also wants to control for, although they were not considered matching variables. All these variables, except those that were used in the matching, can be considered as independent variables in a logistic regression analysis. Three adjustments have to be made to running the usual logistic regression program in order to accommodate matched-pairs data.

1. For each of the N pairs, compute the difference $X(\text{case}) - X(\text{control})$ for each X variable. These differences will form the new variables for a single sample of size N to be used in the logistic regression equation.

2. Create a new outcome variable Y with a value of 1 for each of the N pairs.

3. Set the constant term in the logistic regression equation equal to zero.

Schlesselman (1982) compares the output that would be obtained from a paired-matched logistic regression analysis to that obtained if the matching is ignored. He explains that if the matching is ignored and the matching variables are associated with the exposure variables, then the estimates of the odds ratio of the exposure variables will be biased toward no effect (or an odds ratio of 1). If the matching variables are unassociated with the exposure variables, then it is not necessary to use the form of the analysis that accounts for matching. Furthermore, in the latter case, using the analysis that accounts for matching would result in larger estimates of the standard error of the slope coefficients.

Use in imputation

In Section 9.2 we discussed the use of regression analysis to impute missing X variables. The methods given in Section 9.2 are appropriate to use when the X variables are interval data. But often the variables are categorical or nominal data, particularly in questionnaire data. The hot deck method given in that section can be used for categorical data, but the usual regression analysis appears not to be suitable since it assumes that the outcome variable is interval or continuous data. However, it has been used for binomial 0,1 data by simply using the 0,1 outcome variable as if it were continuous data and computing a regression equation using predictor variables for which data exist (see Allison, 2001). Then the predicted values are rounded to 0 or 1 depending on whether the predicted outcome variable is < 0.5 or ≥ 0.5.

Alternatively, logistic regression analysis can be used to predict categorical data, particularly if the variable that has missing data is a yes/no binary variable. The estimated probability that the missing value takes on a value of zero or one can be obtained from the logistic regression analysis with the observed variables as covariates. Then this probability can be used to make the assignment.

Logistic regression analysis is also used for another method of imputation called the **propensity scores** method (see Rubin, 1987 and Little and Rubin, 2002). Here for each variable, X, continuous or discrete, that has missing values we make up a new outcome variable. This new variable is assigned a score of 0 if the data are missing and a score of 1 if data are present. The next step is to find if there are other variables for which we have data that will predict the new 0, 1 variable using logistic regression analysis. This method can be used if there is a set of variables that can predict missingness. For example, if people who do not answer a particular question are similar to each other, this method may be useful. Using logistic regression analysis, we calculate the probability that a respondent will have a missing value for the particular variable. Propensity is the conditional probability of being missing given the results from the other variables used to predict it.

A sorted set of the propensity scores is made. Then, each missing value for variable X is imputed from a value randomly drawn from the subset of observed values of X with a computed probability close to the missing X value. For example, this can be done by dividing the sorted propensity scores into equal sized classes of size n (see Section 9.2). If a single imputation is desired, a single value could be imputed from a single random sample of the cases in the class that are nonmissing in the X variable. Alternatively, an X value could be chosen from within the class that is closest to the missing value in terms of its propensity score. For multiple imputation the sampling scheme given in Section 9.2 is suggested. Both SAS and SOLAS provide options for doing the propensity score method.

This method would take quite a bit of effort if considerable data are missing on several variables. For a discussion of when not to use this method see Allison (2000).

12.11 Poisson regression

In Chapters 6 through 9 on simple and multiple regression analysis, we assumed that the outcome variable was normally distributed. In this chapter we discussed a method of performing regression analysis where the outcome variable is a discrete variable that can only take on one of two (or more) values. If there are only two values, this is called a binary, dichotomous, or dummy variable. Examples would be gender (male, female) or being classified as depressed (yes, no).

Here we will describe how to perform regression analysis when the outcome variable is the number of times an event occurs. This is often called **counted** (or **count**) **data**. For example, we may want to analyze the number of cells counted within a square area by a laboratory technician, the number of telephone calls per second from a company, the number of times students at a college visit a dentist per year, or the number of defects per hour that a production worker makes.

This type of data is usually assumed to follow a **Poisson distribution**, especially if the events are rare (van Belle *et al.*, 2004). The Poisson distribution has several interesting properties. One such property is that the population mean is equal to the population variance and, consequently, the standard deviation is equal to the square root of the mean. Once we compute the mean we know the shape of the distribution as there are no other parameters that need to be estimated. Also, negative values are not possible. Another property is that the shape of the distribution varies greatly with the size of the mean. When the mean is large the distribution is slightly skewed, but is quite close to a normal distribution. When the mean (μ) is small, the most common count is zero. For example, if $\mu = 0.5$, then the probabilities of obtaining a count of zero is 0.61, a count of one is 0.33, a count of two is 0.12, a count of three is 0.03, and counts of four or more have a very low probability. Hence, the distribution is highly skewed with the lowest value, zero, the most apt to occur.

On the other hand, Figure 12.7 illustrates how a histogram from a sample drawn from a Poisson distribution with $\mu = 5$ appears. The bell-shaped normal curve is superimposed on it, showing reasonably good fit, except in the low end.

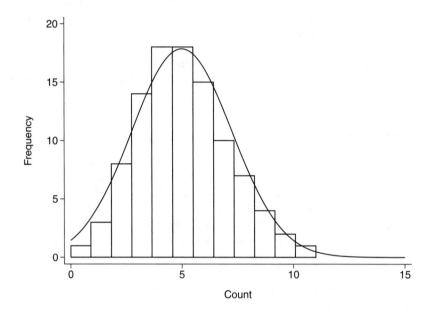

Figure 12.7: *Plot of Histogram of a Poisson Distribution with a Mean of 5 and a Normal Curve*

Figure 12.8 depicts a histogram with $\mu = 30$. Note that the histogram shown in Figure 12.8 looks close to a normal bell-shape curve. From a practical viewpoint, none of the area under the normal curve falls to the left of zero when $\mu = 30$ but some does when the $\mu = 5$.

Figure 12.8 shows that one can either use the square root transformation given in Section 4.3 or use the method given in this section when $\mu = 30$. Using the transformation tends to reduce the skewness and also assures that the size of the variance is not dependent on the mean. Performing the square root transformation enables us to use multiple linear regression programs with their numerous features.

If we were to use a typical regression program when the outcome variable is a Poisson distributed variable with a very small mean, say 0.5, then the square root transformation made on the counts would not achieve the bell-

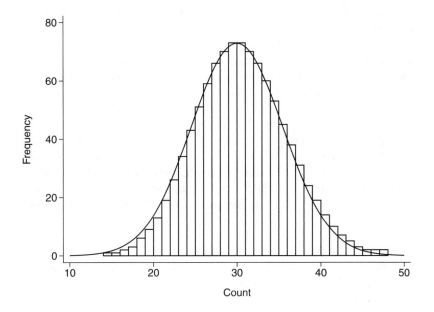

Figure 12.8: *Plot of Histogram of a Poisson Distribution with a Mean of 30 and a Normal Curve*

shaped curve that we associate with the normal distribution. Also, some of the predicted values of our outcome variable will be likely to have negative values and hence not make any physical sense. When the mean of the observations is small, say less than 25, a typical regression program should not be used.

On the other hand, if all the observations are quite large, say greater than 100, the square root transformation does not affect the relative size of the observations very much. This can be seen in Figure 4.2b where the curve depicting X and the $\sqrt{X}$ begins to look almost like a straight line. In that case, it is not crucial to make the square root transformation.

If the mean is small, i.e., the event is rare, then a **generalized linear model** (GLM) procedure should be used. This procedure has three components. The first is a random component consisting of the outcome variable Y. The second component is the set of predictor variables $X_1, \cdots, X_P$ that we wish to use as predictors of Y. For a given set of values $X_1, \cdots, X_P$, we assume that the distribution of Y is Poisson with mean μ. The GLM uses a linear predictor $\alpha + \beta_1 X_1 + \cdots + \beta_P X_P$ to indirectly predict Y. Specifically, the third component is a **link** function $g(\mu)$ that relates μ to the linear predictor, i.e., $g(\mu) = \alpha + \beta_1 X_1 + \cdots + \beta_P X_P$. When the outcome variable has a Poisson distribution, the link function is $\ln(\mu)$. Hence, a generalized linear model (or regression) program is used, specifying either the Poisson distribution or the link function.

The estimates of the slope coefficients require a more complicated numerical method than ordinary least squares regression. In the software packages, the regression coefficients are estimated using the maximum likelihood (ML) procedure. To understand the ML estimates recall that the probability (or likelihood) of obtaining the particular results obtained in our sample depends on the values of $\alpha, \beta_1, \beta_2, \cdots, \beta_P$. The ML procedure looks for estimates of these quantities that maximize this probability or likelihood, and hence the name. In some situations, such as estimating the mean of a normal distribution, it is possible to write the formula for the ML estimate explicitly. However, in many situations, including the Poisson regression, we must resort to approximate methods that rely on iterative procedures. That is, initial estimates are used, which are then improved in successive steps, or iterations, until no further improvements are useful. The procedure is then said to have converged.

To perform a Poisson regression, you can use a program written specifically for that purpose (such as Stata poisson) or a GLM program (such as SAS GENMOD). If a GLM program is used, you simply specify that the distribution is Poisson and the link function is the log. In R and S-PLUS the family argument is set to poisson. Note that LogXact 5 from Cytel Software Corporation allows the user to compute a Poisson regression using exact methods. This is useful when sample sizes are small or unbalanced.

Whichever program is used, the output typically includes the parameter estimates, say $A, B_1, B_2, \cdots, B_P$, and their standard errors, as well as possibly some tests of hypotheses or confidence intervals. For example, we can use an approximate normal test statistic $z = B_i/\mathrm{SE}(B_i)$ to test the hypothesis that $\beta_i = 0$. The 95% confidence interval for B_i is $B_i \pm 1.96\ \mathrm{SE}(B_i)$. The Wald and likelihood ratio tests presented for logistic regression can be used here as well.

Also, similar to logistic regression, the exponentiated coefficients can be given useful interpretations. If X is dichotomous, taking on values 0 or 1, then e raised to the power of its slope β is the **ratio** of μ_1, the mean of Y corresponding to $X = 1$, to μ_0, the mean of Y corresponding to $X = 0$. If X is continuous, then $\exp(\beta)$ is the ratio of μ at $(X + 1)$ to μ at X. As in logistic regression, if other independent variables are in the equation, then these means are adjusted to them.

In many Poisson regression situations, the period of observation is not the same for all individuals. For example, suppose that we are interested in the number of seizures occurring in an epileptic patient per day. If we observe a patient for t days, then the number of seizures Y has mean $\lambda = (\mu t)$, where μ is the mean number per day. To use Poisson regression, for patients observed at different times t, we use the model:

$$\ln(\lambda/t) = \alpha + \beta_1 X_1 + \cdots + \beta_P X_P$$

or, equivalently,

$$\ln(\lambda) = \ln(t) + \alpha + \beta_1 X_1 + \cdots + \beta_P X_P$$

The quantity $\ln(t)$ is called the **offset** and must be supplied as input to the program. In this case, the exponentiated coefficient is the ratio of the **rates** of occurrence per unit time. Thus, for Poisson regressions, exponentiated coefficients are **rate ratios** as contrasted with the **odds ratios** we discussed for logistic regression.

For more information on Poisson regression and examples of its use see Cameron and Trivedi (1998) or Greene (2008). Winkelmann (2008) provides a thorough theoretical background.

12.12 Discussion of computer programs

STATISTICA performs logistic regression as a part of general regression analysis. In STATISTICA, logistic regression can be selected as an option in the Nonlinear Estimation module. STATISTICA will include odds ratios. STATISTICA provides many diagnostic residuals and plots that are useful in finding cases that do not fit the model. If the General Linear-Nonlinear Models is used, ordinal logistic can be obtained.

SAS, SPSS, and Stata have specific logistic regression programs. SAS has LOGISTIC, which is a general purpose program with numerous features. In addition to the binary logistic regression model, it also performs ordinal logistic regression for the case where the outcome variable can take on two or more ordinal values. It can also be used to do probit regression analysis and extreme value analysis (not discussed in this book).

SAS also provides global measures for assessing the fit of the model, namely, Akaike's Information Criterion (AIC), Schwartz's Criterion, and $-2\log(\text{likelihood})$ where a small value of the statistic indicates a desirable model. It also provides several rank correlation statistics between observed and predicted probabilities. Stepwise options are available. SAS has an especially rich mix of diagnostic statistics that may be useful in detecting observations that are outliers or do not fit the model.

The SPSS LOGISTIC program also has a good choice of available options. It performs stepwise selection and can generate dummy variables using numerous methods (see the CONTRASTS subcommand). The PLUM command is used to obtain an ordinal logistic regression. It also provides a rich assortment of regression diagnostics for each case.

Stata provides an excellent set of options for logistic regression analysis. It has almost all the options available in any of the other packages and is very easy to use. It is one of the three packages that provide the odds ratios, their standard errors, z values, and 95% confidence limits, which are widely used in medical applications. It provides both Pearson and deviance residuals along with a range of other diagnostic statistics. It also prints a summary of the residuals that is useful in assessing individual cases, and computes an ROC curve.

In R and S-PLUS, a logistic regression model can be fit using the glm function and setting the family argument equal to binomial. The summary function will produce the results. The estimated regression coefficients, their standard errors, t test of their significance, correlation matrix of the coefficients, and four types of residual plots are available.

Finally, Table 12.6 shows the programs that can be used to perform ordinal and nominal logistic regression, as well as Poisson regression. Note that all six software packages will perform Poisson regression analysis.

12.13   What to watch out for

When researchers first read that they do not have to assume a multivariate normal distribution to use logistic regression analysis, they often leap to the conclusion that no assumptions at all are needed for this technique. Unfortunately, as with any statistical analysis, we still need some assumptions: a simple random sample, correct knowledge of group membership, and data that are free of miscellaneous errors and outliers. Additional points to watch out for include the following.

1. The model assumes that ln(odds) is linearly related to the independent variables. This should be checked using goodness-of-fit measures or other means provided by the program. It may require transformations of the independent variables to achieve this.

2. Inadequate sample size can be a problem in logistic regression analysis. One method of judging whether or not the sample size is adequate is given in Section 12.7 as suggested in Hosmer and Lemeshow (2000). If their criterion cannot be met, the use of an exact program such as LogXact 5 from CYTEL Software Corporation in Cambridge, Massachusetts is suggested. The LogXact program is also recommended when the data are all discrete and there are numerous empty cells (see Hirji *et al.* (1987) and Mehta and Patel (1995) for an explanation of the method of computation). King and Ryan (2002) compare the results from logistic regression analysis when the LogXact program and maximum likelihood method is used for estimation when there are rare outcome events and when a condition they describe as overlap occurs.

3. Logistic regression should not be used to evaluate risk factors in longitudinal studies in which the studies are of different durations (see Woodbury, Manton and Stallard (1981) for an extensive discussion of problems in interpreting the logistic function in this context). However, mixed and hierarchical linear models can be used to analyze such data (see Snijders and Bosker, 1999).

Table 12.6: *Software commands and output for logistic regression analysis*

	S-PLUS/R	SAS	SPSS	Stata	STATISTICA
Beta coefficients	glm	LOGISTIC	LOGISTIC	logit	Nonlin. Estimation
Standard error of coefficients	glm	LOGISTIC	LOGISTIC	logit	Nonlin. Estimation
Odds ratios		LOGISTIC	LOGISTIC	logistic	Nonlin. Estimation
Stepwise variable selection	step.glm (S-PLUS only) step & glm (R)	LOGISTIC	LOGISTIC	stepwise	Gen. Linear-
Model chi-square	glm	LOGISTIC	LOGISTIC	logistic	Nonlin. Estimation
GOF* chi-square	chisq.gof (S-PLUS only)	LOGISTIC	LOGISTIC	estat gof	Gen. Linear-
Hosmer–Lemeshow GOF*		LOGISTIC	LOGISTIC	estat gof	Nonlin. Estimation
Residuals	residuals	LOGISTIC	LOGISTIC	predict	Nonlin. Estimation
Classification table cutpoint		LOGISTIC	LOGISTIC	estat class	Nonlin. Estimation
Correlation matrix of coefficients	glm	LOGISTIC	LOGISTIC	estat vce	Nonlin. Estimation
ROC curve		LOGISTIC	ROC	lroc	Nonlin. Estimation
Predicted probability for cases	predict.glm	LOGISTIC	LOGISTIC	predict	Nonlin. Estimation
Nominal logistic regression	Design library	CATMOD	NOMREG	mlogit	Gen. Linear-Nonlinear Models
Ordinal logistic regression	Design library	LOGISTIC	PLUM	ologit	Gen. Linear-Nonlinear Models
Poisson regression	glm	GENMOD COUNTREG CATMOD	HILOGLINEAR GENLIN	poisson or glm	Gen. Linear-Nonlinear Models

* GOF: Goodness of fit

4. The coefficient for a variable in a logistic regression equation depends on the other variables included in the model. The coefficients for the same variable, when included with different sets of variables, could be quite different. The epidemiologic term for this phenomenon is confounding. For a detailed discussion of confounding in logistic regression analysis see, e.g., Hosmer and Lemeshow (2000).

5. If a matched analysis is performed, any variable used for matching cannot also be used as an independent variable.

6. There are circumstances where the maximum likelihood method of estimating coefficients will not produce estimates. If one predictor variable becomes a perfect predictor (say the variable seldom is positive but when it is, it has a one-to-one relationship with the outcome) then that variable must be excluded to get a maximum likelihood solution. This case can be handled by LogXact.

7. Some statistical software packages parameterize logistic regression models differently from the notation used in most books. For example, for binary logistic regression SAS models by default the probability of the dependent variable having a value of zero. The descending option in the PROC LOGISTIC statement needs to be specified to model the probability of the dependent variable having a value of one.

8. In the presence of an interaction, odds ratios for comparisons of interest should be obtained manually. We have not included how to calculate confidence intervals in the presence of an interaction, since we believe that the steps involved are beyond the scope of this book. We refer the reader to Hosmer and Lemeshow (2000) or to a biostatistician colleague.

9. Nominal and ordinal logistic regression are conceptually straightforward extensions of binary logistic regression. Nevertheless, model building, goodness-of-fit analyses, and interpretation of results can easily become very involved and complex. Thus, obtaining help from a biostatistician regarding the analyses is highly recommended.

10. If the assumptions of binary logistic regression or discriminant analysis cannot be justified, you can consider using classification and regression trees (CART), mentioned in Section 9.3. Instead of a regression tree, a program such as CART can produce a branching process (a tree) that partitions the space of the X variables into distinct cells. The algorithm then classifies the cases in each cell into one of two possible groups, without the need to assume any underlying model. Interested readers should consult Breiman *et al.* (1984) or Zhang and Singer (1999).

11. As discussed in Section 12.10, the interpretation of the results of logistic regression depends on the type of sample selected. The sample may result from a **population-based** or cross-sectional study, i.e., it is ideally a simple

random sample of the population under study. In that case, all of the estimates and other inference procedures discussed in this chapter are valid. In a **cohort study**, separate samples of population subgroups, such as those exposed or not exposed to a particular risk factor (e.g., smokers vs. non-smokers) are collected. In this case, inferences regarding odds ratios and risk ratios are valid. For case-control studies, inference for odds ratios is valid but for risk ratios additional information is needed to adjust the results (see Section 12.10). For additional discussion of these topics, see Hosmer and Lemeshow (2000), Jewell (2004), or Rothman *et al.* (2008).

12.14 Summary

In this chapter we presented a type of regression analysis that can be used when the outcome variable is a categorical variable. This analysis can also be used to classify individuals into one of two (or more) populations. It is based on the assumption that the logarithm of the odds of belonging to a specific population compared to a referent population are a linear function of the independent variables. The result is that the probability of belonging to the specific population is a multiple logistic function.

Logistic regression is appropriate when both categorical and continuous independent variables are used, while the linear discriminant function is preferable when the multivariate normal model can be assumed. We discussed calculation and interpretation of odds ratios in the presence of an interaction and described approaches to check the fit of a logistic model. For more than two outcome groups we introduced nominal logistic regression. For outcome groups which can be ordered ordinal logistic regression was presented. We also described situations in which logistic regression is a useful method for imputation and provided an introduction to Poisson regression.

Demaris (1992) discusses logistic regression as one of a variety of models used in situations where the dependent variable is dichotomous. Agresti (2002), Hosmer and Lemeshow (2000), and Kleinbaum and Klein (2010) provide an extensive treatment of the subject of logistic regression.

12.15 Problems

12.1 If the probability of an individual getting a hit in baseball is 0.20, then the odds of getting a hit are 0.25. Check to determine that the previous statement is true. Would you prefer to be told that your chances are one in five of a hit or that for every four hitless times at bat you can expect to get one hit?

12.2 Using the formula odds $= P/(1-P)$, fill in the accompanying table.

Odds	P
0.25	0.20
0.5	
1.0	0.5
1.5	
2.0	
2.5	
3.0	0.75
5.0	

12.3 The accompanying table presents the number of individuals by smoking and disease status. What are the odds that a smoker will get disease A? That a nonsmoker will get disease A? What is the odds ratio?

	Disease A		
Smoking	Yes	No	Total
Yes	80	120	200
No	20	280	300
Total	100	400	500

12.4 Perform a logistic regression analysis with the same variables and data used in the example in Problem 11.13.

12.5 Perform a logistic regression analysis on the data described in Problem 11.2.

12.6 Repeat Problem 11.5, using stepwise logistic regression.

12.7 Repeat Problem 11.7, using stepwise logistic regression.

12.8 Repeat Problem 11.10, using stepwise logistic regression.

12.9 (a) Using the depression data set, fill in the following table:

Regular drinker	Sex		
	Female	Male	Total
Yes			
No			
Total			

What are the odds that a woman is a regular drinker? That a man is a regular drinker? What is the odds ratio?

(b) Repeat the tabulation and calculations for part (a) separately for people who are depressed and those who are not. Compare the odds ratios for the two groups.

(c) Run a logistic regression analysis with DRINK as the dependent variable and CESD and SEX as the independent variables. Include an interaction term. Is it significant? How does this relate to part (b) above?

12.10 Using the depression data set and stepwise logistic regression, describe the probability of an acute illness as a function of age, education, income, depression, and regular drinking.

12.11 Repeat Problem 12.10, but for chronic rather than acute illness.

12.12 Repeat Problem 11.13(a), using stepwise logistic regression.

12.13 Repeat Problem 11.14, using stepwise logistic regression.

12.14 Define low FEV1 to be an FEV1 measurement below the median FEV1 of the fathers in the family lung function data set given in Appendix A. What are the odds that a father in this data set has low FEV1? What are the odds that a father from Glendora has low FEV1? What are the odds that a father from Long Beach does? What is the odds ratio of having low FEV1 for these two groups?

12.15 Using the definition of low FEV1 given in Problem 12.14, perform a logistic regression of low FEV1 on area for the fathers. Include all four areas and use a dummy variable for each area. What is the intercept term? Is it what you expected (or could have expected)?

12.16 For the family lung function data set, define a new variable VALLEY (residence in San Gabriel or San Fernando Valley) to be one if the family lives in Burbank or Glendora and zero otherwise. Perform a stepwise logistic regression of VALLEY on mother's age and FEV1, father's age and FEV1, and number of children (1, 2, or 3). Are these useful predictor variables?

12.17 Assume a logistic regression model includes a continuous variable like age and a categorical variable like gender and an interaction term for these. Is the P value for any of the main effects helpful in determining whether a) the interaction is significant, b) whether the main effect should be included in the model? c) Would your answers to a) or b) change if the categorical variable were RACE with three categories?

12.18 Perform a logistic regression analysis for the depression data which includes income and sex and models age as an a) quadratic or b) cubic function. Use likelihood ratio test statistics to determine whether these models are significantly better than a model which includes income and sex and models age as a linear function.

12.19 Generate a graph for income similar to Figure 12.2 to assess whether modeling income linearly seems appropriate.

12.20 For the depression data perform an ordinal logistic regression analysis using the same outcome categories as in Section 12.9 which reverses the order and hence compares more severe depression to less severe depression. Use age, income, and sex as the independent variables. Compare your results to those given in the example in Section 12.9.

12.21 For the family lung function data perform a nominal logistic regression where the outcome variable is place of residence and the predictors are those defined in Problem 12.16. Test the hypothesis that each of the variables can be dropped, while keeping all the others in the model. Compare your results to those in Problem 12.16.

12.22 Perform a nominal and ordinal logistic regression analysis using the health scale as the outcome variable and age and income as independent variables. Present and interpret the results for an increase in age of 10 years and an increase in income of $5,000. How do the results from the ordinal logistic regression compare to the results from the nominal logistic regression?

12.23 Perform a binary logistic regression analysis using the Parental HIV data to model the probability of having been absent from school without a reason (variable HOOKEY). Find the variables that best predict whether an adolescent had been absent without a reason or not. Assess goodness-of-fit for the final model (overall and influence of patterns).

12.24 For the model in 12.23 find an appropriate cutoff point to discriminate between adolescents who were absent without a reason and those who were not. Assess how well the model predicts the outcome using sensitivity, specificity, and the ROC curve.

12.25 For the Parental HIV data perform a Poisson regression on the number of days the adolescents were absent from school without a reason. For this analysis assume that the referent time period was one month for all adolescents. As independent variables consider gender, age, and to what degree the adolescents liked school. Present rate ratios, including confidence intervals, and interpret the results.

12.26 For the model in 12.25 use the referent time as given by the variable HMONTH to define the offset. Run the same model as in 12.25. Present rate ratios, including confidence intervals, and interpret the results. Describe any discrepancies compared to the results obtained in 12.25.

12.27 This problem and the following ones use the Northridge earthquake data set. We wish to answer the questions: Were homeowners more likely than renters to report emotional injuries as a result of the Northridge earthquake, controlling for age (RAGE), gender (RSEX), and ethnicity (NEWETHN)? Use V449, rent/own, as the outcome variable. To begin with, create dummy variables for each ethnic group. Calculate frequencies for each categorical variable, and histograms and descriptive statistics for each interval or ratio variable used. Comment on any unusual values.

12.28 (Problem 12.27 continued) Fit a logistic regression model using emotional injury (yes/no, W238) as an outcome and using home ownership status (rent/own, V449), age (RAGE), gender (RSEX), and ethnicity (NEWETHN) as independent variables. Give parameter estimates and their standard errors. What are the odds ratios and their 95% confidence intervals? Interpret each.

12.29 (Problem 12.28 continued) Based on your results, what is the estimated probability of reporting emotional injuries for a 30-year-old white female renter? For a 50-year-old Latino home owner?

12.30 (Problem 12.28 continued) Are the effects of ethnicity upon reporting emotional injuries statistically significant, controlling for home ownership status, age, and gender? Use a likelihood ratio test to answer this question.

12.31 (Problem 12.28 continued) Is there an interaction effect between gender and home ownership? That is, are the estimated effects of home ownership upon reporting emotional injuries different for men and women, controlling for age and ethnicity?

12.32 (Problem 12.28 continued) Is there an interaction effect between age and home ownership, controlling for gender and ethnicity?

12.33 (Problem 12.28 continued) Perform diagnostic procedures to identify influential observations. Remove the four (4) most influential observations using the delta chi-square method. Rerun the analysis and compare the results. What is your conclusion?

12.34 This problem and the following ones also use the Northridge earthquake data set. Perform an appropriate regression analysis using variable selection techniques for the following outcome: evacuate home (use V173 of the questionnaire). Choose from the following predictor variables: MMI (which is a localized measure of earthquake strength), home damage (V127), physical injuries (W220), emotional injuries (W238), age (RAGE), sex (RSEX), education (V461), ethnicity (NEWETHN), marital status (V455), home owner status (V449). You may or may not want to categorize some of the continuous measurements. Where appropriate, consider interaction terms as well. Feel free to look at additional predictors.

12.35 (Problem 12.34 continued) Fit a logistic regression model (again using as the outcome "evacuate home" (V173)) which includes as the only covariates home owner status (rent/own, V449) and status of home damage (V127) and an appropriate interaction term. Is there a statistically significant interaction between homeowner status (use either the Wald or likelihood ratio test statistic)? Whether the interaction is statistically significant or not, present and interpret the odds ratio for each of the following comparisons: (a) home owners whose home was not damaged versus renters whose home was not damaged, (b) home owners whose home was damaged versus renters whose home was not damaged, (c) home owners whose home was damaged versus renters whose home was not damaged. Obtain these odds ratios by hand and show your calculations.

12.36 (Problem 12.35 continued) Using an appropriate method in your software package, obtain confidence intervals for the odds ratios you computed in parts a, b and c of Problem 12.35.

Chapter 13

Regression analysis with survival data

13.1 Chapter outline

In Chapters 11 and 12, we presented methods for classifying individuals into one of two possible populations and for computing the probability of belonging to one of the populations. We examined the use of discriminant function analysis and logistic regression for identifying important variables and developing functions for describing the risk of occurrence of a given event. In this chapter, we present another method for further quantifying the probability of occurrence of an event, such as death or termination of employment, but here the emphasis will be on the length of time until the event occurs, not simply whether or not it occurs.

Section 13.2 gives examples of when survival analysis is used and explains how it is used. Section 13.3 presents a hypothetical example of survival data and defines censoring. In Section 13.4 we describe four types of distribution-related functions used in survival analysis and show how they relate to each other. The Kaplan-Meier method of estimating the survival function is illustrated. The normal distribution that is often assumed in statistical analysis is seldom used in survival analysis, so it is necessary to think in terms of other distributions. Section 13.5 presents the distributions commonly used in survival analysis and includes figures to show how the two most commonly used distributions look. Section 13.6 discusses tests used to determine if the survival distributions in two or more groups are significantly different.

Section 13.7 introduces a log-linear regression model to express an assumed relationship between the predictor variables and the log of the survival time. Section 13.8 describes an alternative model to fit the data, called Cox's proportional hazards model. The use of the Cox model when there are interactions between two explanatory variables and when explanatory variables change over time is explained. Methods for checking the fit of the model are given. In Section 13.9, we discuss the relationship between the log-linear model and Cox's model, and between Cox's model and logistic regression.

The output of the statistical packages is described in Section 13.10 and what to watch out for is given in Section 13.11.

13.2   When is survival analysis used?

Survival analysis can be used to analyze data on the length of time it takes for a specific event to occur. This technique takes on different names, depending on the particular application on hand. For example, if the event under consideration is the death of a person, animal, or plant, then the name **survival analysis** is used. If the event is the failure of a manufactured item, e.g., a light bulb, then one speaks of **failure time analysis** or reliability theory (Smith, 2002). The term **event history analysis** is used by social scientists to describe applications in their fields (Yamaguchi, 1991). For example, analysis of the length of time it takes an employee to retire or resign from a given job could be called event history analysis. In this chapter, we will use the term survival analysis to mean any of the analyses just mentioned.

Survival analysis is a way of describing the distribution of the length of time to a given event. Suppose the event is termination of employment. We could simply draw a histogram of the length of time individuals are employed. Alternatively, we could use log length of employment as a dependent variable and determine if it can be predicted by variables such as age, gender, educational level, or type of position (this will be discussed in Section 13.7). Another possibility would be to use the Cox regression model as described in Section 13.8.

Readers interested in a comprehensive coverage of the subject of survival analysis are advised to study one of the texts referenced in this chapter. In this chapter, our objective is to describe regression-type techniques that allow the user to examine the relationship between length of survival and a set of explanatory variables. The explanatory variables, often called covariates, can be either continuous, such as age or income, or they can be discrete, such as dummy variables that denote a treatment group. The material in Sections 13.3–13.5 is intended as a summary of the background necessary to understand the remainder of the chapter.

13.3   Data examples

In this section, we begin with a description of a hypothetical situation to illustrate the types of data typically collected. This will be followed by a real data example taken from a cancer study.

In a survival study we can begin at a particular point in time, say 1990, and start to enroll patients. For each patient we collect baseline data, that is, data at the time of enrollment, and begin follow-up. We then record other information about the patient during follow-up, especially the time at which the patient dies if this occurs before termination of the study, say 2000. If some patients are still alive, we simply record the fact that they survived up to ten years. However, we may enroll patients throughout a certain period, say the first eight years of the study. For each patient we similarly record the baseline and other data, the length of survival, or the length of time in the study (if the patient does not die during the period of observation).

Typical examples of five patients are shown in Figure 13.1 to illustrate the various possibilities. This type of graph has been called a **calendar event chart** (see Lee *et al.*, 2000). Patient 1 started in 1990 and died after two years. Patient 2 started in 1992 and died after six years. Patient 3 was observed for two years, but was lost to follow-up. All we know about this patient is that survival was more than two years. This is an example of what is known as a **censored** observation. Another censored observation is represented by patient 4, who enrolled for three years and was still alive at the end of the study. Patient 5 entered in 1998 and died one year later. Patients 1, 2, and 5 were observed until death.

The data shown in Figure 13.1 can be represented as if all patients enrolled at the same time (Figure 13.2). This makes it easier to compare the length of time the patients were followed. Note that this representation implicitly assumes that there have been no changes over time in the conditions of the study or in the types of patients enrolled.

We end this section by describing a real data set used in various illustrations in this chapter. The data were taken from a multicenter clinical trial of patients with lung cancer who, in addition to the standard treatment, received either the experimental treatment known as BCG or the control treatment consisting of a saline solution injection (Gail *et al.*, 1984). The total length of the study was approximately ten years. We selected a sample of 401 patients for analysis. The patients in the study were selected to be without distant metastasis or other life threatening conditions, and thus represented early stages of cancer. Data selected for presentation in this book are described in a codebook shown as Table 13.1. In the binary variables, we used 0 to denote good or favorable values and 1 to denote unfavorable values. For HIST, subtract 1 to obtain 0, 1 data. The data are described in Appendix A. The full data set can be obtained from the web site mentioned in Appendix A.

13.4 Survival functions

The concept of a frequency histogram should be familiar to the reader. If not, the reader should consult an elementary statistics book. When the variable under consideration is the length of time to death, a histogram of the data from a sample can be constructed either by hand or by using a packaged program. If we imagine that the data were available on a very large sample of the population, then we can construct the frequency histogram using a large number

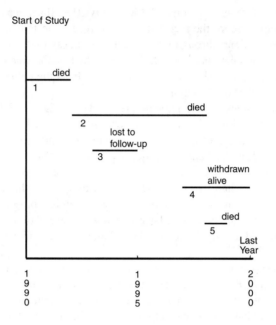

Figure 13.1: *Calendar Event Chart of Five Patients in a Survival Study*

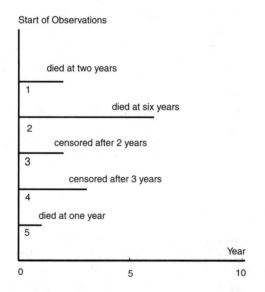

Figure 13.2: *Five Patients Represented by a Common Starting Time*

Table 13.1: *Codebook for lung cancer data*

Variable number	Variable name	Description
1	ID	Identification number from 1 to 401
2	Staget	Tumor size
		0 = small
		1 = large
3	Stagen	Stage of nodes
		0 = early
		1 = late
4	Hist	Histology
		1 = squamous cells
		2 = other types of cells
5	Treat	Treatment
		0 = saline (control)
		1 = BCG (experiment)
6	Perfbl	Performance status at baseline
		0 = good
		1 = poor
7	Poinf	Post-operative infection
		0 = no
		1 = yes
8	Smokfu	Smoking status at follow-up
		1 = smokers
		2 = ex-smokers
		3 = never smokers
9	Smokbl	Smoking status at baseline
		1 = smokers
		2 = ex-smokers
		3 = never smokers
10	Days	Length of observation time in days
11	Death	Status at end of observation time
		0 = alive (censored)
		1 = dead

of narrow frequency intervals. In constructing such a histogram, we use the relative frequency as the vertical axis so that the total area under the frequency curve is equal to one. This relative frequency curve is obtained by connecting the midpoints of the tops of the rectangles of the histogram (Figure 13.3). This curve represents an approximation of what is known as the **death density function**. The actual death density function can be imagined as the result of letting the sample get larger and larger – the number of intervals gets larger and larger while the width of each interval gets smaller and smaller. In the limit, the death density is therefore a smooth curve.

Figure 13.4 presents an example of a typical death density function. Note that this curve is not symmetric, since most observed death density functions are not. In particular, it is not the familiar bell-shaped normal density function.

The death density is usually denoted by $f(t)$. For any given time t, the area under the curve to the left of t is the proportion of individuals in the population who die up to time t. As a function of time t, this is known as the **cumulative**

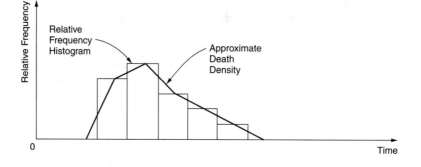

Figure 13.3: *Relative Frequency Histogram and Death Density*

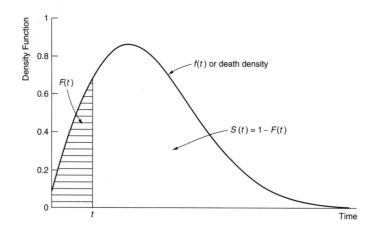

Figure 13.4: *Graph of Death Density, Cumulative Death Distribution Function $F(t)$, and Survival Function $S(t)$*

death distribution function and is denoted by $F(t)$. The area under the curve to the right of t is $1 - F(t)$, since the total area under the curve is one. This latter proportion of individuals, denoted by $S(t)$, is the proportion of those surviving at least to time t and is called the **survival function**. A graphic presentation of both $F(t)$ and $S(t)$ is given in Figure 13.5. In many statistical analyses, the cumulative distribution function $F(t)$ is presented as shown in Figure 13.5(a). In survival analysis, it is more customary to plot the survival function $S(t)$ as shown in Figure 13.5(b).

We will next define a quantity called the hazard function. Before doing so, we return to an interpretation of the death density function, $f(t)$. If time is measured in very small units, say seconds, then $f(t)$ is the likelihood, or probability, that a randomly selected individual from the population dies in the interval between t and $t+1$, where 1 represents one second. In this context, the population consists of all individuals regardless of their time of death. On the other hand, we may restrict our attention to only those members of the population who are known to be alive at time t. The rate at which members die at time t, given that they have lived up to time t, is described by the **hazard function** and is denoted by $h(t)$. Mathematically, we can write $h(t) = f(t)/S(t)$. Other names of the hazard function are the **force of mortality, instantaneous death rate**, or **failure rate**.

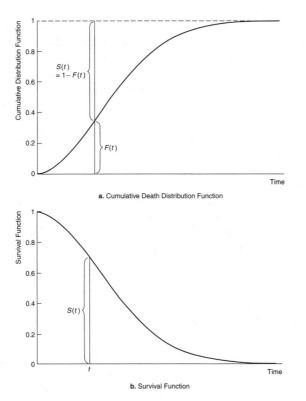

Figure 13.5: *Cumulative Death Distribution Function and Survival Function*

To summarize, we described four functions: $f(t), F(t), S(t)$, and $h(t)$. Mathematically, any three of the four can be obtained from the fourth (Lee and Wang, 2003; Miller, 1981). In survival studies it is important to compute an estimated survival function. A theoretical distribution can be assumed, for example, one of those discussed in the next section. In that case, we need to estimate the parameters of the distribution and substitute in the formula for $S(t)$ to obtain its estimate for any time t. This is called a **parametric** estimate.

If a theoretical distribution cannot be assumed based on our knowledge of the situation, we can obtain a **nonparametric** estimated survival function.

Most statistical software packages use the **Kaplan-Meier** method (also called **product-limit estimator**) to obtain a nonparametric survival function. The main idea of the Kaplan-Meier estimate of the survival function is to first estimate the probability of surviving beyond a specific time point, conditional on surviving up to that time point, and then to compute the products of the conditional probabilities over time. This approach allows for incorporating both censored and non-censored observations.

To illustrate how the Kaplan-Meier estimate is calculated, consider the example data presented in Figure 13.2. The basic steps in the calculation are the following (see Table 13.2). We begin by sorting the unique event times in ascending order. Here the event is death. We start the table with a row which describes the time interval from time zero up to the first observed unique event (death) time. In our example the first observed event time is one year for patient 5. Each following row describes the intervals between the next consecutive death times. In our example the second row covers the time from year one up to year two. The last row represents a single time point, the largest observed survival time, which in our example is six years. With the exception of this last "interval," which represents only a single time point, each of the intervals should include the starting time point and exclude the ending time point. In our example this means that, e.g., the second interval covers the time from year 1 (inclusive) up to just before year 2. As a result, all time intervals taken together should cover the complete observation period seamlessly.

Table 13.2: *Kaplan-Meier estimates for data from Figure 13.2*

Time interval (1)	n=No. at risk (2)	d=No. of deaths (3)	No. censored (4)	$(n-d)/n$ (5)	$\hat{S}$ (6)
0 up to 1	5	0	0	5/5=1	1
1 up to 2	5	1	0	(5-1)/5=0.8	(1)(0.8)=0.8
2 up to 6	4	1	2	(4-1)/4=0.75	(0.8)(0.75)=0.6
6 and up	1	1	0	(1-1)/1=0	(0.6)(0)=0

We then provide for each time interval the number of patients (n) who were known to be still alive just before the start of the interval, which is also called the number at risk (column 2), the number (d) of patients who died during the interval (column 3), and the number of patients who were censored during the interval (column 4). Note that if a censored observation and a death happen to occur after exactly the same length of time, it is standard to assume that the censored observation occurs after the death. For each of the rows, the number at risk can be obtained by subtracting the number who died (column 3) and the number who were censored (column 4) in the previous interval from the

number who were at risk (column 2) in the previous interval. After filling in
the first four columns for each row, we can proceed to estimating the conditional probability of surviving beyond the starting point of each interval, given
survival until just before the interval (column 5). This is done by subtracting
the number who died in the interval (column 3) from the number who are at
risk of dying in the interval (column 2) and dividing the result by the number who are at risk of dying in the interval. The Kaplan-Meier estimate of the
survival function is then calculated iteratively by multiplying the conditional
probability by the preceding Kaplan-Meier estimate (column 5) and noting that
the Kaplan-Meier estimate for the first row is one.

If we calculate Kaplan-Meier curves for two subgroups of the observations
and plot them on the same graph, we can obtain a graphical comparison of potential differences in survival experience between the subgroups. For the lung
cancer data presented in Section 13.3, a comparison between survival of patients with large or small tumor size might be of interest. Figure 13.6 shows
Kaplan-Meier estimates of the survival functions for the group of patients ex-
hibiting a small tumor size versus a large tumor size.

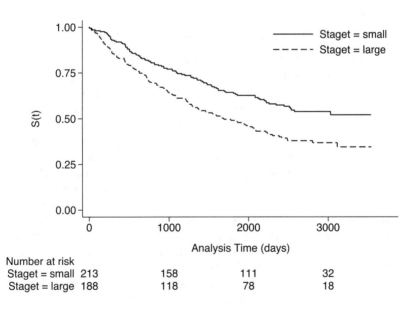

Figure 13.6: *Kaplan-Meier Estimates of the Survival Functions for Lung Cancer Data*

Note that the graph gives the impression of a significant difference in survival experience between patients with a small tumor size (staget 0) and a large

tumor size (staget 1). The unit of time is days. The graph appears to show that patients with small tumors have a better survival function than those with large tumors. In Sections 13.7 and 13.8, we will show how this can be tested. If the gap between the two survival curves widens appreciably over time then we could say that having a small tumor becomes more important over time.

If the two groups represent two treatments, the following might be a potential interpretation for crossing survival curves. Assume that the curves cross with one dropping substantially at first and then staying almost level over time while the second curve continues to drop. Then the first treatment may be hazardous initially but if one survives he/she may be cured later. An example of such treatments might be a surgical versus a nonsurgical treatment. Comparisons between two survival curves are sometimes difficult to interpret since both curves decrease over time.

Readers interested in further details of survival function estimation using the Kaplan-Meier method may consult Kleinbaum and Klein (2005) or other texts cited in this chapter. Many of these texts also discuss estimation of the hazard function and death density function. In addition, Hosmer, Lemeshow and May (2008) discuss how to obtain confidence limits for the estimated Kaplan-Meier survival function.

13.5   Common survival distributions

The simplest model for survival analysis assumes that the hazard function is constant over time, that is, $h(t) = \lambda$, where λ is any constant greater than zero. This results in the **exponential** death density $f(t) = \lambda \exp(-\lambda t)$ and the exponential survival function $S(t) = \exp(-\lambda t)$. Graphically, the hazard function and the death density function are displayed in part (a) of Figure 13.7 and Figure 13.8. This model assumes that having survived up to a given point in time has no effect on the probability of dying in the next instant. Although simple, this model has been successful in describing many phenomena observed in real life. For example, it has been demonstrated that the exponential distribution closely describes the length of time from the first to the second myocardial infarction in humans and the time from diagnosis to death for some cancers. The exponential distribution can be easily recognized from a flat (constant) hazard function plot. Such plots are available in the output of many software programs, as we will discuss in Section 13.10.

If the hazard function is not constant, the **Weibull distribution** should be considered. For this distribution, the hazard function may be expressed as $h(t) = \alpha\lambda(\lambda t)^{\alpha-1}$. The expressions for the density function, the cumulative distribution function, and the survival function can be found in specialized texts, e.g., Kalbfleisch and Prentice (2002). Figures 13.7(b) and 13.8(b) show plots of the hazard and density functions for $\lambda = 1$ and $\alpha = 0.5, 1.5$, and 2.5. The value of α determines the shape of the distribution and for that reason it is

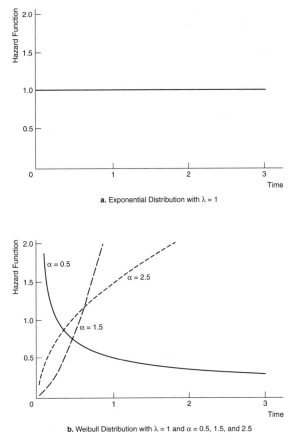

a. Exponential Distribution with $\lambda = 1$

b. Weibull Distribution with $\lambda = 1$ and $\alpha = 0.5, 1.5,$ and 2.5

Figure 13.7: *Hazard Functions for the Exponential and Weibull Distributions*

called the **shape parameter** or **index**. Furthermore, as may be seen in Figure 13.7(b), the value of α determines whether the hazard function increases or decreases over time. Namely, when $\alpha < 1$ the hazard function is decreasing and when $\alpha > 1$ it is increasing. When $\alpha = 1$ the hazard is constant, and the Weibull and exponential distributions are identical. In that case, the exponential distribution is used. The value of λ determines how much the distribution is stretched, and therefore it is called the **scale parameter**.

The Weibull distribution is used extensively in practice. Section 13.7 describes a model in which this distribution is assumed. However, the reader should note that other distributions are sometimes used. These include the lognormal, gamma, and others (e.g., Kalbfleisch and Prentice, 2002 or Andersen *et al.*, 1993).

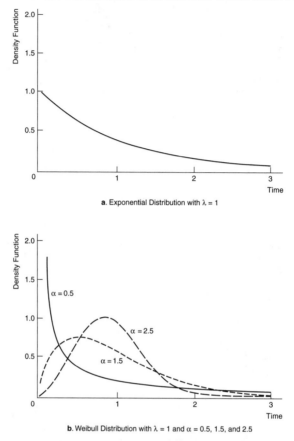

Figure 13.8: *Death Density Functions for the Exponential and Weibull Distributions*

A good way for deciding whether or not the Weibull distribution fits a set of data is to obtain a plot of $\log(-\log S(t))$ versus log time, and check whether the graph approximates a straight line (Section 13.10). If it does, then the Weibull distribution is appropriate and the methods described in Section 13.7 can be used. If not, either another distribution may be assumed or the method described in Section 13.8 can be used.

13.6   Comparing survival among groups

Most studies involving survival data aim not only at describing survival over time, but also at assessing differences in survival between subgroups in the study. For a graphical comparison of potential differences, Kaplan-Meier curves are often used (Section 13.4). To test the null hypothesis that the survival distribution is the same in two or more groups, a number of tests are available. The test statistic most commonly used is the **log-rank test** (also called **Mantel test**). This test can be used to compare survival between two or more groups. The main idea behind the log-rank test is to compare the number of observed events to the number of expected events (based on Kaplan-Meier curves) at all time points where events are observed. The numerical value of the test statistic is compared to a χ^2 distribution with degrees of freedom that are equal to the number of groups being compared minus one.

If we want to assess differences in survival across a continuous explanatory variable (such as age) via a log-rank test, we first need to generate a categorical variable from the continuous variable. This categorization can be done by dividing the range of values into contextually meaningful intervals or intervals of equal size (e.g., by using quartiles). Other test statistics are available to assess differences between survival in two or more groups (e.g., **Peto**, **Wilcoxon** or **Tarone-Ware**; see Klein and Moeschberger, 2003). The main difference between these tests and the log-rank test is that these tests use different weights for the individual observations, whereas the log-rank test uses the same weight (namely the weight of one) for each of the observations. The Peto test places more weight in the left tail of the survival curve where there are more observations and the log-rank test places more weight on the right tail.

Referring again to the lung cancer example, we might have suspected from Figure 13.6 (as well as from general knowledge) that the overall survival experience is different for patients presenting with a large tumor compared to patients presenting with a small tumor. Calculating the log-rank test statistic for this example yields a test statistic value of 12.7 which, compared to a chi-square distribution with one degree of freedom, results in a p value of less than 0.001. Therefore, in our example, we would reject the hypothesis that the survival experience is the same for the two groups.

We note that the log-rank test makes the assumption that the hazard for one group at any time point is proportional to the hazard for the other group. This is the same assumption made for the Cox proportional hazards model discussed in Section 13.8. In that section, we present a method for checking this assumption. The same procedure can be used to check for the appropriateness of using the log-rank test. If the proportional hazards assumption is not satisfied, we may use the Wilcoxon test instead.

13.7   The log-linear regression model

In this section, we describe the use of multiple linear regression to study the relationship between survival time and a set of explanatory variables. Suppose that t is survival time and $X_1, X_2, \ldots, X_P$ are the independent or explanatory variables. Let $Y = \log(t)$ be the dependent variable, where natural logarithms are used. Then the model assumes a linear relationship between $\log(t)$ and the X's. The model equation is

$$\log(t) = \alpha + \beta_1 X_1 + \beta_2 X_2 + \cdots + \beta_P X_P + e$$

where e is an error term. This model is known as the **log-linear regression model** since the log of survival time is a linear function of the X's. If the distribution of $\log(t)$ were normal and if no censored observations exist in the data set, it would be possible to use the regression methods described in Chapter 7 to analyze the data. However, in most practical situations some of the observations are censored, as was described in Section 13.3. Furthermore, $\log(t)$ is usually not normally distributed (t is often assumed to have a Weibull distribution). For those reasons, the method of maximum likelihood is used to obtain estimates of β_i's and their standard errors. When the Weibull distribution is assumed, the log-linear model is sometimes known as the **accelerated life** or **accelerated failure time model** (Kalbfleisch and Prentice, 2002).

For the data example presented in Section 13.3, it is of interest to study the relationship between length of survival time (since admission to the study) and the explanatory variables (variables 2–9) of Table 13.1. For the sake of illustration, we restrict our discussion to the variables Staget (tumor size: 0 = small, 1 = large), Perfbl (performance status at baseline: 0 = good, 1 = poor), Treat (treatment: 0 = control, 1 = experimental), and Poinf (post operative infection: 0 = no, 1 = yes).

A simple analysis that can shed some light on the effect of these variables is to determine the percentage of those who died in the two categories of each of the variables. Table 13.3 gives the percentage dying and the results of a χ^2 test of the hypothesis that the proportion of deaths is the same for each category. Based on this analysis, we may conclude that of the four variables, all except Treat may affect the likelihood of death.

Table 13.3: *Percentage of deaths versus explanatory variables*

Variable	Outcome	Deaths (%)	P value
Staget	Small	42.7	<0.01
	Large	60.1	
Perfbl	Good	48.4	0.02
	Poor	64.5	
Treat	Control	49.2	0.52
	Experiment	52.4	
Poinf	No	49.7	0.03
	Yes	75.0	

This simple analysis may be misleading, since it does not take into account the length of survival or the simultaneous effects of the explanatory variables. Previous analysis of these data (Gail *et al.*, 1984) has demonstrated that the data fit the Weibull model well enough to justify this assumption. We estimated $S(t)$ from the data set and plotted $\log(-\log S(t))$ versus $\log(t)$ for small and large tumor sizes separately. The plots are approximately linear, further justifying the Weibull distribution assumption.

Assuming a Weibull distribution, the LIFEREG procedure of SAS was used to analyze the data. Table 13.4 displays some of the resulting output. Shown are the maximum-likelihood estimates of the intercept (α) and regression (β_i) coefficients along with their estimated standard errors, and P values for testing whether the parameters are zero.

Since all of the regression coefficients are negative except Treat, a value of one for any of the three status variables is associated with shorter survival time than is a value of zero. For example, for the variable Staget, a large tumor is associated with a shorter survival time. Furthermore, the P values confirm the same results obtained from the previous simple analysis: three of the variables are significantly associated with survival (at the 5% significance level), whereas treatment is not. The information in Table 13.4 can also be used to obtain approximate confidence intervals for the parameters.

Table 13.4: *Log-linear model for lung cancer data: Results*

Variable	Estimate	Standard error	Two-sided P value
Intercept	8.64	0.21	<0.01
Staget	−0.59	0.16	<0.01
Perfbl	−0.60	0.20	<0.01
Poinf	−0.71	0.31	<0.02
Treat	0.08	0.15	0.59

Again, for the variable Staget, an approximate 95% confidence interval for β_1 is

$$-0.59 \pm (1.96)(0.16)$$

that is, $-0.90 < \beta_1 < -0.28$.

The log-linear regression equation can be used to estimate typical survival times for selected cases. For example, for the least-serious case when each explanatory variable equals zero, we have $\log(t) = 8.64$. Since this is a natural logarithm, $t = \exp(8.64) = 5653$ days $= 15.9$ years. This may be an unrealistic estimate of survival time caused by the extremely long tail of the distribution. On the other extreme, if every variable has a value of one, $t = \exp(6.82) = 916$ days, or somewhat greater than two and a half years.

13.8   The Cox regression model

The standard Cox model

In this section, we describe the use of another method of modelling the relationship between survival time and a set of explanatory variables. In Section 13.4, we defined the hazard function and used the symbol $h(t)$ to indicate that it is a function of time t. Suppose that we use X, with no subscripts, as shorthand for all the X_i variables. Since the hazard function may depend on t and X, we now need to use the notation $h(t,X)$. The idea behind the **Cox model** is to express $h(t,X)$ as the product of two parts: one that depends on t only and another that depends on the X_i variables only. In symbols, the basic model is

$$h(t,X) = h_0(t)\exp(\beta_1 X_1 + \beta_2 X_2 + \cdots + \beta_P X_P)$$

where $h_0(t)$ does not depend on the X_i variables and $\exp(\beta_1 X_1 + \beta_2 X_2 + \cdots + \beta_P X_P)$ does not depend on t. Implicitly it is assumed here that the values of the explanatory variables do not change over time. The case where we allow for a change in values over time is discussed below.

If all X_i's are zero, then the second part of the equation would be equal to 1 and $h(t,X)$ reduces to $h_0(t)$. For this reason, $h_0(t)$ is sometimes called the **baseline hazard function**. In order to further understand the model, suppose that we have a single explanatory variable X_1, such that $X_1 = 1$ if the subject is from group 1 and $X_1 = 0$ if the subject is from group 2. For group 1, the hazard function is

$$h(t,1) = h_0(t)\exp(\beta_1)$$

Similarly, for group 2, the hazard function is

$$h(t,0) = h_0(t)\exp(0) = h_0(t)$$

The ratio of these two hazard functions is

$$h(t,1)/h(t,0) = \exp(\beta_1)$$

which is a constant that does not depend on time. In other words, the hazard function for group 1 is **proportional** to the hazard function for group 2. This property motivated D.R. Cox, the inventor of this model, to call it the **proportional hazards regression model**.

The hazard ratio (HR), i.e., the ratio of two hazard functions, is typically reported as the estimated effect that group 1 has relative to group 2. If the event under study is death or a different adverse event, then a hazard ratio between zero and one is interpreted as protective, whereas a hazard ratio larger than one is interpreted as harmful. If the event under study is considered beneficial, the interpretation is reversed.

Approximate confidence intervals for the hazard ratio for a binary variable like the above group indicator variable can be computed using the coefficient b and its standard error listed in the output. For example, 95% approximate confidence limits for the hazard ratio can be computed as $\exp(b \pm 1.96 \times \text{standard error of } b)$.

Another way to understand the model is to think in terms of two individuals in the study, each with a given set of values of the explanatory variables. The Cox model assumes that the ratio of the hazard functions for the two individuals, say $h_1(t)/h_2(t)$, is a constant that does not depend on time. We will discuss different approaches to checking this assumption later in this section.

Statistical software programs use the maximum likelihood method to obtain estimates of the parameters and their standard errors. Some programs also allow variable selection by stepwise procedures similar to those we described for other models. One such program is Stata. In this analysis, we did not use the stepwise feature but rather obtained the output for the same variables as in Section 13.6. The results are given in Table 13.5.

Table 13.5: *Cox's model for lung cancer data: Results*

Variable	Estimate	Standard error	HR	Two-sided P value
Staget	0.54	0.14	1.72	<0.01
Perfbl	0.53	0.19	1.70	<0.01
Poinf	0.67	0.28	1.95	<0.025
Treat	0.07	0.14	1.07	>0.50

Note that all except one of the signs of the coefficients in Table 13.5 are the opposites of those in Table 13.4. This is due to the fact that the Cox model describes the hazard function, whereas the log-linear model describes survival time. The reversal in sign indicates that a long survival time corresponds to low hazard and *vice versa*. The hazard ratio for all the variables except Treat is appreciably larger than one.

In Section 13.9, we compare the two models with each other as well as with the logistic regression model presented in Chapter 12.

The Cox model and interactions

The interpretation of the coefficients of a Cox model as described in the preceding section needs to be revised if there is evidence that two (or more) variables are interacting. The inclusion of an interaction is necessary if the effect of an independent variable depends on the level of another independent variable. Like the logistic regression model, the Cox model is multiplicative. Thus, combined effects of independent variables cannot be determined by multiplying the individual effects if an interaction is present. We will illustrate this with a simulated data set.

Suppose we are interested in studying time to divorce for first marriages and, particularly, the effect of age at marriage given in years and whether the level of education was similar at the time of marriage. For the purpose of this example we generated a data set with 1801 individuals, each representing one of the two partners of a first marriage. We could think of these individuals as part of a cross-sectional study. Only individuals who were ever married are included in the analysis. Individuals are asked at what age they were first married, and whether they are still married to the same partner or whether they were divorced, widowed, or separated. The time to divorce is measured in months. Time to divorce is censored for individuals who are not divorced; thus time to censoring is recorded for these individuals. For widowed individuals, the time of death of the partner is when the observation is considered censored. For individuals who are still married or who are separated, the censoring time is the interview date. Thus, they are withdrawn from the study (censored) at their interview date. Also included in the data set is a variable indicating whether the educational level of the partners was similar (value one) at the time of marriage or not (value zero).

If we estimate the effects of age at marriage and similar education with a proportional hazards model without including an interaction between the two factors, we obtain the following results.

Term	Coefficient	Standard error	HR
Age at marriage	−0.0281	0.0087	0.972
Similar education	−0.1536	0.1081	0.858

The hazard ratios (HR) for an increase in age of one year and for comparing similar to different educational level are $\exp(-0.0281) = 0.972$ and $\exp(-0.1536) = 0.858$, respectively. We might prefer to present a hazard ratio for an increase in age of 10 years; the resulting hazard ratio is $\exp(-0.0281 \times 10) = 0.755$. Thus, the rate of divorce of a first marriage is estimated to decrease by approximately 24% for an increase in age at marriage of 10 years. Also, the rate of divorce is estimated to decrease by approximately 14% if the partners have attained a similar educational level at the time of marriage.

Ninety-five percent confidence intervals for the estimated hazard ratio for an increase in age of k years can be computed as $\exp[kb \pm k(1.96)\text{SE}(b)]$, where "SE" stands for standard error. For example, for an increase in age of 10 years

$$\exp(10 \times -0.0281 \pm 10 \times 1.96 \times 0.0087) \quad \text{or}$$
$$\exp(-0.4515 \text{ to} - 0.1105) \quad \text{or}$$
$$0.637 < \text{hazard ratio} < 0.895$$

Note that we assume for this model that age and educational level do not interact. In other words, the effect of age on the divorce rate is assumed to be independent of educational level and *vice versa*. Thus, the hazard ratio of 0.858 comparing similar to different educational level is assumed to be the same whether the compared individuals are 20 years old at the time of first marriage or whether they are 50 years old or any other age, as long as they are of the same age.

Under the assumption of no interaction, an estimate of the combined effect of age and educational level would be calculated as follows. When comparing the rate of divorce for an increase in age of 10 years and comparing similar to different educational level, we obtain a hazard ratio:

HR($age = (a + 10)$ and educational level $= 1$ (similar)
versus $age = a$ and educational level $= 0$ (different))

$$= \frac{\exp(-0.0281 \times (a + 10) - 0.1563 \times 1)}{\exp(-0.0281 \times a - 0.1563 \times 0)}$$

$$= \frac{\exp(-0.0281 \times a - 0.0281 \times 10) - 0.1563)}{\exp(-0.0281 \times a)}$$

$$= \exp(-0.281 - 0.1563)$$

$$= \exp(-0.281) \times \exp(-0.1563)$$

$$= 0.755 \times 0.858$$

$$= 0.648$$

Note that a is considered a placeholder for any age within the range observed in the data set. Assuming that there is no interaction, the combined effect of an increase in age of 10 years and a similar educational level can be calculated by multiplying the individual effects. With no interaction, the combined HR $= 0.648$ or an effect that is lower than either age or similar educational level alone.

An important next question is: how do we assess whether the assumption of "no interaction" is adequate? The answer is similar to what we described for multiple linear regression (see Section 9.3). To formally test whether an interaction term is necessary we can add such an interaction term to the model and assess whether the coefficient for the interaction term is significantly different from zero. The interaction term in our example represents a variable which is created by multiplying the variable age by the dummy variable for educational level. Estimates from a model including the interaction term are as follows:

Term	Coefficient	Standard error	p value
Age	−0.0139	0.0101	0.169
Educational level	1.1804	0.5352	0.027
Age × Educational level	−0.0508	0.0202	0.012

Obviously, the coefficient for educational level when the interaction term is included is very different from the corresponding coefficient if it is not included. Similar to what was discussed for interactions for logistic regression, this is not a contradiction. Even though the labelling of the included variables stays the same, the interpretation of the coefficients is different in the presence of an interaction term. We will return to the interpretation shortly, but first we describe two methods of determining whether the interaction term is necessary, the Wald and the likelihood ratio test, which were discussed in Section 12.7 of the logistic regression chapter. The next three paragraphs review these tests.

For each variable in the model, most statistical software packages report a so-called Wald statistic. This statistic is calculated as

$$\frac{b^2}{[\text{SE}(b)]^2}$$

where b represents the estimated coefficient β and $\text{SE}(b)$ is its standard error.

Under the null hypothesis of a zero slope (i.e., no significant association between the corresponding variable and the outcome) and based on asymptotic theory, this quantity follows a chi-square distribution with one degree of freedom when the sample size is large. Roughly speaking, if the estimated value of the slope is small and its estimated variability is large we do not have enough evidence to conclude that the slope is significantly different from zero and vice versa. Some statistical software packages report a "z" score, which is the square root of the Wald test statistic. In this case the test statistic value would be compared to a standard normal distribution and would yield exactly the same p value. We have reported p values from the Wald test with the estimated coefficients in the preceding table.

The Wald test is an approximate test in the sense that it might not be "right on target," but hopefully it is close. Unfortunately, there is no way to measure exactly how close it is. One other approximate test to assess whether a coefficient is zero is the **likelihood ratio test**. The general idea of this test is to compare a global measure of fit of the data without the interaction term to a global measure of fit of the data with the interaction term. If the inclusion of the interaction term improves the fit of the model significantly, then we would conclude that there is sufficient evidence that the interaction term is necessary. The test statistic is calculated as two times the difference of the logarithms of the likelihood ratio test statistic between the model excluding, and the model including, the interaction term. The test statistic is assumed to follow a chi-square distribution with one degree of freedom. Most statistical software packages provide this test statistic as part of the default output. Note that some software packages (like Stata) provide the test statistic value as the logarithm of the likelihood ratio test, whereas others provide minus two times the logarithm of the likelihood ratio test statistic (e.g., SAS). Some statistical software packages provide automated commands to calculate the test statistic value and the corresponding p value (e.g., Stata) whereas others (e.g., SAS) only provide the minus two log-likelihood for each model, and the user has to calculate the value of the test statistic and the corresponding p value.

In our example, the value of the Wald test statistic for the difference in slopes is $z = -2.51$, which corresponds to a p value of 0.012, whereas the value of the likelihood ratio χ^2 test statistic is 6.66 with a corresponding p value of 0.01. Thus, based on either test, we would conclude that it is necessary to include the interaction term. These two tests rarely lead to different conclusions, but if they do, the likelihood ratio test is preferable.

Now, how do we interpret the results in the presence of the interaction? In our example, the effect of age depends on whether the educational level is similar or not. Thus we calculate two hazard ratios for age, one for an increase in age of 10 years given that the educational level is similar and one for an increase in age of 10 years given that the educational level is different.

$$\text{HR}(age = (a + 10) \text{ versus } age = a$$
$$\text{given educational level} = 1 \text{ (similar)})$$

$$
\begin{aligned}
&= \frac{\exp(-0.0139 \times (a + 10) + 1.1804 \times 1 - 0.0508 \times (a + 10) \times 1)}{\exp(-0.0139 \times a + 1.1804 \times 1 - 0.0508 \times a \times 1)} \\
&= \exp(-0.0139 \times 10 - 0.0508 \times 10) \\
&= \exp(-0.139 - 0.508) \\
&= \exp(-0.647) \\
&= 0.524
\end{aligned}
$$

$$\text{HR}(age = (a + 10) \text{ versus } age = a$$
$$\text{given educational level} = 0 \text{ (different)})$$

$$
\begin{aligned}
&= \frac{\exp(-0.0139 \times (a + 10) + 1.1804 \times 0 - 0.0508 \times (a + 10) \times 0)}{\exp(-0.0139 \times a + 1.1804 \times 0 - 0.0508 \times a \times 0)} \\
&= \exp(-0.0139 \times 10) \\
&= 0.870
\end{aligned}
$$

When educational levels are similar the hazard ratio for being 10 years older is lower than when educational levels are different.

The effect of educational level also depends on the age at first marriage. In the presence of interactions, we need to consider a specific age at which we want to estimate the effect of educational level. Since age is a continuous variable, we could, e.g., present estimates of the effect of educational level at the 25th, 50th (median), and 75th percentiles of age. In our case these would be at ages 23, 26, and 31. When dealing with age, it might be more desirable to present estimates of an effect at round numbers like, e.g., 20, 30, 40, and 50. We will proceed according to the latter and calculate hazard ratios for similar versus different educational level at ages 20, 30, 40, and 50.

HR(educational level = similar (1) vs different (0) for $age = 20$)

$$
\begin{aligned}
\text{HR} &= \frac{\exp\left(-0.0139 \times 20 + 1.1804 \times 1 - 0.0508 \times 20 \times 1\right)}{\exp\left(-0.0139 \times 20 + 1.1804 \times 0 - 0.0508 \times 20 \times 0\right)} \\
&= \exp\left(1.1804 - 0.0508 \times 20\right) \\
&= 1.179
\end{aligned}
$$

Similar calculations for $age = 30$, $age = 40$, and $age = 50$ result in

HR(educational level = similar (1)
versus different (0) for $age = 30$) $= 0.709$

HR(educational level = similar (1)
versus different (0) for $age = 40$) $= 0.427$

HR(educational level = similar (1)
versus different (0) for $age = 50$) $= 0.257$

In our example data set a similar educational level is associated with a slightly harmful effect if an individual married at a young age, whereas an increasingly protective effect is observed if the individual first married at an older age. In presenting these effects one should keep in mind that we are assuming that age has a linear effect (on the log hazard). This assumption needs to be verified. Furthermore, we calculated an effect for educational level at age 50, but there are only nine (out of 1801) subjects in the data set who were 50 years or older when they first got married. Thus, estimates relating to this age should be viewed with caution.

A common error is to interpret the coefficients for the variable's age and educational level as if an interaction term was not included or as if they would represent an overall estimate. Another common error is to interpret the exponentiated coefficient for the interaction term as a hazard ratio. These errors are easily made since statistical software packages often report hazard ratios as the exponentiated coefficients whether or not they represent a true hazard ratio. For this reason, we have not included the exponentiated coefficients (HRs) with the estimated coefficients in the above table. If we had exponentiated the coefficient for age we would have obtained the estimated hazard ratio for an increase in age of one year given that educational level is different (value zero). If we had exponentiated the coefficient for educational level we would have obtained the estimated hazard ratio for educational level given that age is zero. Since an age of zero is not meaningful in our example, this latter estimate is not meaningful. In the presence of an interaction it is best to determine which hazard ratios are of interest and then calculate them manually as shown above.

Calculation of appropriate confidence intervals for hazard ratios which involve an interaction is complicated and involves estimating the covariance between estimated coefficients. These covariances are typically not available as part of the default output accompanying survival regression analysis results, but can be requested through separate options or commands. A description of calculating confidence intervals for hazard ratios in the presence of interactions is beyond the scope of this book. The interested reader is referred to Hosmer, Lemeshow and May (2008).

The Cox model and time-dependent explanatory variables

When discussing the Cox model, we noted that the first part of $h(t)$ depends only on t and that the second part depends only on X_i's. However, the Cox model can also accommodate the situation where the X_i values may change over time. In this case the X_i's are called **time dependent** and we use the notation $X_i(t)$. Time-dependent explanatory variables can be divided into two major types. One type is assumed to vary continuously over time, like blood pressure or blood cholesterol level. The other type represents a one time change in status.

A classical example of the latter type occurred in the Stanford heart transplant study (Crowley and Hu, 1977). For this study, a patient's survival time started with enrollment into the Stanford heart transplant program. Initially each patient was considered to be in the control group. The change in status to the treatment group occurred at the time when the patient received a heart transplant. In this case, it would have been incorrect to simply compare survival of the group of patients who received a transplant to the group of patients who did not receive a transplant. To assess the effect of the transplant on survival correctly, it is necessary to use a variable indicating the change in status over time. Such a variable would begin, e.g., with a value of zero and would change to a value of one from the time of the transplant onward.

A crucial aspect of time-dependent explanatory variables is that estimation for the associated parameters becomes very involved. This is mainly due to the fact that the values for all explanatory variables need to be known or calculated for all individuals in the study at each event time. This requirement is not only essential for estimation, but is also important from a conceptual perspective. For a given study design, this requirement might or might not hold. For the Stanford heart transplant data it holds and means that the status "transplant received" or "transplant not yet received" is known for all patients who are still alive at each of the time points of death of *another* patient. An example of where the requirement might not hold is the following. Assume that level of blood cholesterol is one of the time-dependent explanatory variables that we assume to be important for survival. In this case, we would need to know the cholesterol level for each patient still alive at each death time of another patient. This is hardly ever possible. In such a case we would need to make some assumptions about how to impute (estimate) the value of the explanatory variable between available measurements. One possibility would be to assume a linear change from one measurement to the next. Another possibility is to assume that the value stays constant until the next measurement.

Interrelationships between the event under study and any time-dependent variables can lead to biased or distorted conclusions. Consider a study where we are interested in time to death after a specific surgery. If we were to include a time-dependent variable representing the heart rate of each patient we would expect to see a large effect for this variable. Since the heart rate of a patient will be closely related to a patient's death, the estimation of the effect of heart rate over time will not be meaningful. In addition, the inclusion of a variable that is so closely related to the outcome will prevent us from obtaining meaningful results for other variables under consideration. For a detailed discussion of issues relating to the use of time-dependent variables see Fisher and Lin (1999).

If time-dependent explanatory variables are an essential part of the design and analysis, either special formats for entering the data or special commands or options need to be used for an appropriate analysis. The level of complexity for the implementation of time-dependent variables varies considerably among the different statistical software packages. For a complete description of such commands and options consult the manual of the specific statistical package. For an overview of the available options see Section 13.10. Another area where time-dependent variables are used is in evaluating the fit of a model. We will discuss this and other approaches for assessing the appropriateness of the model in the next section.

Evaluating appropriateness of the model

An important assumption underlying the Cox model is the assumption of proportional hazards. An indirect method to check this assumption is to obtain a plot of $\log(-\log S(t))$ versus t for different values of a given explanatory variable (note that this is the natural logarithm). This plot should exhibit an approximately constant difference between the curves corresponding to the levels of the explanatory variable. To check the assumption of proportionality for the lung cancer data, we obtained a plot of $\log(-\log S(t))$ versus t for small and large tumors (variable Staget). The results, given in Figure 13.9, show that an approximately constant difference exists between the two curves, making the proportionality assumption a reasonable one.

To statistically verify the visual impression, we use the concept of time-

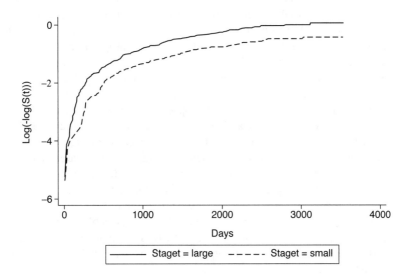

Figure 13.9: *Graph of log(−logS(t)) versus t for Lung Cancer Data*

dependent variables. The underlying idea is to allow the hazard function to be non-proportional and to assess whether the non-proportionality is strong enough so as to reject the hypothesis of proportional hazards. This is accomplished by multiplying the explanatory variable by a function of time. Typical functions that are used are a linear function of time t or the logarithm of time $\log(t)$. As mentioned in the preceding section, inclusion of time-dependent variables requires specific commands or options to be used. For SAS, such time-dependent variables need to be generated within the PROC PHREG statement, while for Stata the options *tvc* and *texp* of the *stcox* statement need to be used.

To check the proportionality assumption for the variable Staget in the lung cancer data we fit a model which includes Staget as well as an interaction term for Staget and the logarithm of time in addition to the variables Perfbl, Poinf, and Treat. In this model we are only interested in the coefficient for the interaction term, which in our example is estimated to have a value of −0.07 with an associated p value of 0.50; therefore the proportionality assumption for the effect of a large versus a small tumor is acceptable.

What alternatives would be available if we were to reject the assumption of proportional hazards for an explanatory variable? One often-used approach is to stratify the model on the explanatory variables which exhibit non-proportional hazards. Stratification for the Cox model allows the baseline hazard function to be different for each stratum. The estimation and interpretation of coefficients for variables which are not stratum variables would remain the same. We are not able to estimate the effect of the stratum variable though. Note that, within each stratum, we would still assume proportional hazards for the explanatory variables included in the model.

Residuals play an important role in assessing goodness-of-fit for linear regression and logistic regression. The same is true for survival analysis. Procedures based on residuals have been proposed to evaluate the overall fit of a model, to assess the functional form of a continuous variable, and to identify observations with extreme influence on estimated parameters. Nevertheless, few procedures have been implemented in statistical software and different statistical software packages have implemented different procedures. We will not discuss these topics further, but refer the interested reader to Hosmer, Lemeshow, and May (2008) and Kleinbaum and Klein (2005).

A test statistic to evaluate the overall fit of a model has been developed by Grønnesby and Borgan (1996). This test statistic is similar to the Hosmer—Lemeshow test for logistic regression. May and Hosmer (1998) show that this test statistic can be calculated using any statistical software package. The basic idea of the test statistic is to group observations by their estimated risk score $(\beta_1 X_1 + \beta_2 X_2 + \cdots + \beta_P X_P)$ and to compare the number of observed and the number of model-based expected events across the risk score groups. A large discrepancy between the observed and expected number of events would indicate a poor fit. Grønnesby and Borgan (1996) use four risk score groups, whereas May and Hosmer (1998) suggest using ten risk score groups. Parzen and Lipsitz (1999) also develop this test for assessing goodness-of-fit for the Cox proportional hazards model and suggest using ten risk score groups as well. The optimal number of risk score groups is still under investigation.

The test statistic value is calculated in the following way. Once a model is obtained that is considered final, the risk score $(\beta_1 X_1 + \beta_2 X_2 + \cdots + \beta_P X_P)$ is estimated for each individual. The observations are then ordered by their estimated risk score and groups are formed of approximately equal size. Then, indicator or dummy variables (see Section 9.3) are generated for each of the risk score groups. A model is then estimated, which includes all variables from the final model and in addition all except one of the group indicator variables. A likelihood ratio (or Wald) test is performed to assess the combined significance of the added group indicator variables. If the resulting test statistic is significant, we would interpret this as evidence that the model does not fit adequately. Note that the second model, which includes the group indicator variables, is only performed to obtain the goodness-of-fit test statistic value. The estimated coefficients for variables that were also included in the final model are not meaningful.

This overall test has some disadvantages. The test cannot be used if there are only a few categorical variables included in the model. The test also shares with other overall goodness-of-fit tests that the power to detect model violations is expected to be lower than for more specialized tests. In addition, if we conclude that there is no evidence, that the model fit is inappropriate, this does not guarantee that there is no model violation. If we conclude, on the other hand, that the fit is inappropriate, this test does not give any indication why the fit is inappropriate. Why would such a test be useful? In complicated multivariate models even gross violations of the assumptions are typically not obvious. An overall test statistic would be expected to detect gross model violations.

In the divorce data example using the model that includes age, educational level, and an interaction term, the likelihood ratio test statistic for this overall goodness-of-fit test has a value of 11.03, which, based on a chi-square distribution with 9 degrees of freedom, is associated with a p value of 0.27. Thus, we would conclude that there is not sufficient evidence of gross model violations.

For further details regarding the above goodness-of-fit tests, and additional approaches to assessing model adequacy, see Hosmer, Lemeshow and May (2008). Finally, we note that, for the lung cancer data set, it is reasonable to interpret the information in terms of the results of either the Cox or the log-linear model. The results from both models are similar.

13.9 Comparing regression models

Log-linear versus Cox

In Tables 13.4 and 13.5, the estimates of the coefficients from the log-linear and Cox models are reported for the lung cancer data. As noted in the previous section, the coefficients in the two tables have opposite signs except for the variable Treat, which was nonsignificant. Also, in this particular example the magnitudes of the coefficients are not too different. For example, for the variable Staget, the **magnitude** of the coefficient was 0.59 for the log-linear model and 0.54 for the Cox model. One could ask, is there some general relationship between the coefficients obtained by using these two different techniques? An answer to this question can be found when the Weibull distribution is assumed in fitting the log-linear model. For this distribution, the **population** coefficients β have the following relationship

$$\beta(\text{Cox}) = -(\text{Shape})\beta(\text{log-linear})$$

where (Shape) is the value of the shape parameter α. For a mathematical proof of this relationship, see Kalbfleisch and Prentice (2002, page 45). The sample coefficients will give an approximation to this relationship. For example, the sample estimate of shape is 0.918 for the lung cancer data and $-(0.918)(-0.59) = +0.54$, satisfying the relationship exactly for the Staget variable. But for Perfbl, $(-0.918)(-0.60) = +0.55$, it is close but not exactly equal to $+0.53$ given in Table 13.4.

If one is trying to decide which of these two models to use, then there are several points to consider. A first step would be to determine if either model fits the data, by obtaining the plots and statistics suggested in Sections 13.5 and 13.8:

1. Plot $\log(-\log S(t))$ versus t for the Cox model separately for each category of the major dichotomous independent variables to see if the proportionality assumption holds. Assess the assumption statistically by adding an interaction term for each independent variable with a function of time and testing whether the coefficient for the interaction term is zero.

2. Plot $\log(-\log S(t))$ versus $\log (t)$ to see if a Weibull fits, or plot $-\log S(t)$ versus t to see if a straight line results, implying an exponential distribution fits. (The Weibull and exponential distributions are commonly used for the log-linear model.)

In practice, the plots are sometimes helpful in discriminating between the two models, but often both models appear to fit equally well. This is the case with the lung cancer data. Neither plot looks perfect and neither fit is hopelessly poor. The statistical test, on the other hand, can shed more light on the appropriateness of specifically the proportionality assumption.

Cox versus logistic regression

In Chapter 12, logistic regression analysis was presented as a method of classifying individuals into one of two categories. In analyzing data on survival, one possibility is to classify individuals as dead or alive after a **fixed** duration of follow-up. Thus we create two groups: those who survived and those who did not survive beyond this fixed duration. Logistic regression can be used to analyze the data from those two groups with the objective of estimating the probability of surviving the given length of time. In the lung cancer data set, for example, the patients could be separated into two groups, those who died in less than one year and those who survived one year or more. In using this method, only patients whose results are known, or could have been known, for at least one year should be considered for analysis. These are the same patients that could be used to compute a one-year survival rate as discussed in Section 13.3. Thus, excluded would be all patients who enrolled less than one year from the end of the study (whether they lived or died) and patients who had a censored survival time of less than one year. Logistic regression coefficients can be computed to describe the relationship between the independent variables and the log odds of death.

Several statisticians (Ingram and Kleinman, 1989; Green and Symons, 1983) have examined the relationships between the results obtained from logistic regression analysis and those from the Cox model. Ingram and Kleinman (1989) assumed that the true population distribution followed a Weibull distribution with one independent variable that took on two values, and that the distribution could be thought of as group membership. Using a technique known as Monte Carlo simulation, the authors obtained the following results.

1. **Length of follow-up**. The estimated regression coefficients were similar (same sign and magnitude) for logistic regression and the Cox model when the patients were followed for a short time. They classified the cases as alive if they survived the follow-up period. Note that for a short time period, relatively few patients die. The range of survival times for those who die would be less than for a longer period. As the length of follow-up increased, the logistic regression coefficients increased in magnitude but those for the Cox model stayed the same. The magnitude of the standard error of the coefficients was somewhat larger for the logistic regression model. The estimates of the standard error for the Cox model decreased as the follow-up time increased.

2. **Censoring**. The logistic regression coefficients became very biased when there was greater censoring in one group than in the other, but the regression coefficients from the Cox model remained unbiased (50% censoring was used).

3. When the proportion dying is small (10% and 19% in the two groups), the estimated regression coefficients were similar for the two methods. As the proportion dying increased, the regression coefficients for the Cox model stayed the same and their standard errors decreased. The logistic regression coefficients increased as the proportion dying increased.

4. Minor violations of the proportional hazards assumption had only a small effect on the estimates of the coefficients for the Cox model. More serious violations resulted in bias when the proportion dying is large or the effect of the independent variable is large.

Green and Symons (1983) have concluded that the regression coefficients for the logistic regression and the Cox model will be similar when the outcome event is uncommon, the effect of the independent variables is weak, and the follow-up period is short.

The use of the Cox model is preferable to logistic regression if one wishes to compare results with other researchers who chose different follow-up periods. Logistic regression coefficients may vary with follow-up time or proportion dying, and therefore use of logistic regression requires care in describing the results and in making comparisons with other studies.

For a short follow-up period of fixed duration (when there is an obvious choice of the duration), some researchers may find logistic regression simpler to interpret and the options available in logistic regression programs useful. If it is thought that different independent variables best predict survival at different time periods (for example, less than two months versus greater than two months after an operation), then a separate logistic regression analysis on each of the time periods may be sensible.

13.10   Discussion of computer programs

Table 13.6 summarizes the options available in the six statistical packages that perform the regression analyses presented in this chapter. Some statistical packages have a variety of commands or programs for survival analysis techniques (R, S-PLUS). Others group them under a few major themes (LIFEREG, LIFETEST, and PHREG for SAS; LIFE TABLE, KM, and COXREG for SPSS). Stata groups them under *st* commands and STATISTICA under *Survival Analysis*.

Before entering the data for regression analysis of survival data, the user should check what the options are for entering dates in the program. Most programs have been designed so that the user can simply type the date of the starting and end points and the program will compute the length of time until the event such as death occurs or until the case is censored. But the user has to follow the precise form of the date that the program allows. Some have several options and others are quite restricted. The manual should be checked to avoid later problems.

LIFEREG from SAS was used to obtain the log-linear coefficients in the cancer example. The program uses the Weibull distribution as the default option. In addition, a wide range of distributions can be assumed, namely, the exponential, log-normal, log-logistic, gamma, normal, and logistic. The INFORMAT statement should be checked to find the options for dates. This program can use data that are either censored on the left or right. The *survfit* function or kaplanMeier in R and S-PLUS produces the Kaplan-Meier survival curve. The Cox proportional hazard model may be fit using the *stcox* routine and the *streg* routine can be used if the user wishes to use a log-linear model. S-PLUS, SAS, Stata, and SPSS can compute residuals and/or perform goodness-of-fit tests.

Stata also has a wide range of features. The software can handle dates. It has both Cox's regression analysis and the log-linear model (Weibull, exponential, and normal (see cnreg) distributions). For a specified model, it computes estimated hazard ratios for a one-unit change in the independent variable. Stepwise regression is also available for these models and time-varying covariates can be included for Cox's model.

STATISTICA also can compute either the Cox or log-linear (exponential, log-normal, or normal distribution) model. The user first chooses Survival Analysis, then Regression Models, and then scrolls to the model of choice. Survival curves are available in the Kaplan–Meier or Life Table options listed under Survival Analysis.

Table 13.6: *Software commands and output for survival regression analysis*

	S-PLUS/R	SAS	SPSS	Stata	STATISTICA
Data entry with dates	chron	INFORMAT	FORMATS	dates	Survival Analysis
Cox's proportional hazards Model	coxph	PHREG	COXREG	stcox	Survival Analysis
HR	coxph	PHREG	COXREG	stcox	Survival Analysis
Log-linear model	survReg	LIFEREG	HILOGLINEAR	streg	Survival Analysis
Number of censored cases	censorReg	LIFEREG	COXREG	tabulate	Survival Analysis
Graphical goodness-of-fit	resid	PLOT		predict	Survival Analysis
Covariance matrix of coeff.	survReg	LIFEREG,PHREG	COXREG	correlate	Survival Analysis
Coefficients and std errors	survReg,coxph	LIFEREG,PHREG	COXREG	stcox or streg	Survival Analysis
Test of coefficients	survReg,coxph	LIFEREG,PHREG	COXREG	stcox or streg	Survival Analysis
Stepwise available		PHREG	COXREG	sw:stcox	
Weights of cases	survReg,coxph	LIFEREG,PHREG	WEIGHT	stcox	Survival Analysis
Quantiles	predict	LIFEREG	KM	predict	Survival Analysis
Diagnostic plots	plot	LIFEREG,PHREG	COXREG	stphplot,stcoxkm	Survival Analysis
Time dependent covariates	coxph	PHREG	COXREG	stcox	Survival Analysis
Kaplan-Meier estimator	kaplanMeier*, survfit	LIFETEST	KM	sts list,sts graph	Survival Analysis
Log-rank test	survdiff	LIFETEST	KM	sts test	Survival Analysis
Peto test	survdiff	LIFETEST	LIFETABLE	sts test	Survival Analysis
Wilcoxon test		LIFETEST	KM	sts test	Survival Analysis
Tarone-Ware test		LIFETEST		sts test	

*kaplanMeier is in S-PLUS only.

13.11   What to watch out for

In addition to the remarks made in the regression analysis chapters, several potential problems exist for survival regression analysis.

1. If subjects are entered into the study over a long time period, the researcher needs to check that those entering early are like those entering later. If this is not the case, the sample may be a mixture of persons from different populations. For example, if the sample was taken from employment records 1980–2000 and the employees hired in the 1990s were substantially different from those hired in the 1980s, putting everyone back to a common starting point does not take into account the change in the type of employees over time. It may be necessary to stratify by year and perform separate analyses, or to use a dummy independent variable to indicate the two time periods.

2. The analyses described in this chapter assume that censoring occurs independently of any possible treatment effect or subject characteristic (including survival time). If persons with a given characteristic that relates to the independent variables, including treatment, are censored early, then this assumption is violated and the above analyses are not valid. Whenever considerable censoring exists, it is recommended to check this assumption by comparing the patterns of censoring among subgroups of subjects with different characteristics. A researcher can reverse the roles of censored and noncensored values by temporarily switching the labels and calling the censored values "deaths." The researcher can then use the methods described in this chapter to study patterns in time to "death" in subgroups of subjects. If such an examination reveals that the censoring pattern is not random, then other methods need to be used for analysis. Unfortunately, the little that is known about analyzing nonrandomly censored observations is beyond the level of this book (e.g., Kalbfleisch and Prentice (2002)). Also, to date such analyses have not been incorporated in general statistical packages.

3. In comparisons among studies it is critical to have the starting and end points defined in precisely the same way in each study. For example, death is a common end point for medical studies of life-threatening diseases such as cancer. But there is often variation in how the starting point is defined. It may be time of first diagnosis, time of entry into the study, just prior to an operation, post operation, etc. In medical studies, it is suggested that a starting point be used that is similar in the course of the disease for all patients.

4. If the subjects are followed until the end point occurs, it is important that careful and comprehensive follow-up procedures be used. One type of subject should not be followed more carefully than another, as differential censoring can occur. Note that although the survival analysis procedures allow for censoring, more reliable results are obtained if there is less censoring.

5. The investigator should be cautious concerning stepwise results with small sample sizes or in the presence of extensive censoring.

13.12 Summary

In this chapter, we presented two methods for performing regression analysis where the dependent variable was time until the occurrence of an event, such as death or termination of employment. These methods are called the log-linear or accelerated failure time model, and the Cox or proportional hazards model. One special feature of measuring time to an event is that some of the subjects may not be followed long enough for the event to occur, so they are classified as censored. The regression analyses allow for such censoring as long as it is independent of the characteristics of the subjects or the treatment they receive. We introduced time-dependent variables in the context of the Cox proportional hazards model and discussed interactions and how to check the assumptions underlying the model. In addition, we discussed the relationship among the results obtained when the log-linear, Cox, or logistic regression models are used.

Further information on survival analysis at a modest mathematical level can be found in Allison (2010), Hosmer, Lemeshow and May (2008), Kleinbaum and Klein (2005), or Lee and Wang (2003). Applied books also include Harris and Albert (1991) and Yamaguchi (1991).

13.13 Problems

The following problems all refer to the lung cancer data described in Table 13.1 and Appendix A.

13.1 (a) Find the effect of Stagen and Hist upon survival by fitting a log-linear model. Check any assumptions and evaluate the fit using the graphical methods described in this chapter.

(b) What happens in part (a) if you include Staget in your model along with Stagen and Hist as predictors of survival?

13.2 Repeat Problem 13.1, using a Cox proportional hazards model instead of a log-linear. Compare the results.

13.3 Do the patterns of censoring appear to be the same for smokers at baseline, ex-smokers at baseline, and nonsmokers at baseline? What about for those who are smokers, ex-smokers, and nonsmokers at follow-up?

13.4 Assuming a log-linear model for survival, does smoking status (i.e., the variables Smokbl and Smokfu) significantly affect survival?

13.5 Repeat Problem 13.4 assuming a proportional hazards model.

13.6 Assuming a log-linear model, do the effects of smoking status upon survival change depending on the tumor size at diagnosis?

13.7 Repeat Problem 13.6 assuming a proportional hazards model.

13.8 Define a variable Smokchng that measures change in smoking status between baseline and follow-up, so that Smokchng equals 1 if a person changes from being a smoker to being an ex-smoker and equals 0 otherwise.
(a) Is there an association between Staget and Smokchng?
(b) What effects does Smokchng have upon survival? Is it significant? Is the effect the same if the smoking status variables are also included in the model as independent variables?

13.9 Evaluate graphically and statistically the proportional hazards assumption for the variables Perfbl, Poinf, and Treat in the model presented in Table 13.4 using the methods described in Section 13.8.

13.10 Perform a proportional hazards regression using the variables Staget, Treat, and Perfbl and stratify the model by Poinf. Compare the estimated hazard ratios to the ones obtained in Problem 13.9.

Chapter 14

Principal components analysis

14.1 Chapter outline

Principal components analysis is used when a simpler representation is desired for a set of intercorrelated variables. Section 14.2 explains the advantages of using principal components analysis and Section 14.3 gives a hypothetical data example used in Section 14.4 to explain the basic concepts. Section 14.5 contains advice on how to determine the number of components to be retained, how to transform coefficients to compute correlations, and how to use standardized variables. It also gives an example from the depression data set. Section 14.6 discusses some uses of principal components analysis including how to handle multicollinearity in a regression model. Section 14.7 summarizes the output of the statistical computer packages and Section 14.8 lists what to watch out for.

14.2 When is principal components analysis used?

Principal components analysis is performed in order to simplify the description of a set of interrelated variables. In principal components analysis the variables are treated equally, i.e., they are not divided into dependent and independent variables, as in regression analysis.

The technique can be summarized as a method of transforming the original variables into new, uncorrelated variables. The new variables are called the **principal components**. Each principal component is a linear combination of the original variables. One measure of the amount of information conveyed by each principal component is its variance. For this reason the principal components are arranged in order of decreasing variance. Thus the most informative principal component is the first, and the least informative is the last (a variable with zero variance does not distinguish between the members of the population).

An investigator may wish to reduce the dimensionality of the problem, i.e., reduce the number of variables without losing much of the information. This objective can be achieved by choosing to analyze only the first few principal components. The principal components not analyzed convey only a small amount of information since their variances are small. This technique is attractive for another reason, namely, that the principal components are not intercorrelated. Thus instead of analyzing a large number of original variables with complex interrelationships, the investigator can analyze a small number of uncorrelated principal components.

The selected principal components may also be used to test for their normality. If the principal components are not normally distributed, then neither are the original variables. Another use of the principal components is to search for outliers. A histogram of each of the principal components can identify those individuals with very large or very small values; these values are candidates for outliers or blunders.

In regression analysis it is sometimes useful to obtain the first few principal components corresponding to the X variables and then perform the regression on the selected components. This tactic is useful for overcoming the problem of multicollinearity since the principal components are uncorrelated (Chatterjee and Hadi, 2006). Principal components analysis can also be viewed as a step toward factor analysis (Chapter 15).

Principal components analysis is considered to be an exploratory technique that may be useful in gaining a better understanding of the interrelationships among the variables. This idea will be discussed further in Section 14.5.

The original application of principal components analysis was in the field of educational testing. Hotelling (1933) developed this technique and showed that there are two major components to responses on entry-examination tests: verbal and quantitative ability. Principal components and factor analysis are also used extensively in psychological applications in an attempt to discover underlying structure. In addition, principal components analysis is frequently used in biological and medical applications.

14.3   Data example

The depression data set will be used later in the chapter to illustrate the technique. However, to simplify the exposition of the basic concepts, we generated a hypothetical data set. These hypothetical data consist of 100 random pairs of observations, X_1 and X_2. The population distribution of X_1 is normal with mean 100 and variance 100. For X_2 the distribution is normal with mean 50 and variance 50. The population correlation between X_1 and X_2 is $1/2^{1/2} = 0.707$. Figure 14.1 shows a scatter diagram of the 100 random pairs of points. The two variables are denoted in this graph by NORM1 and NORM2. The sample statistics are shown in Table 14.1. The sample correlation is $r = 0.675$.

The advantage of this data set is that it consists of only two variables, so it is easily plotted. In addition, it satisfies the usual normality assumption made in statistical theory.

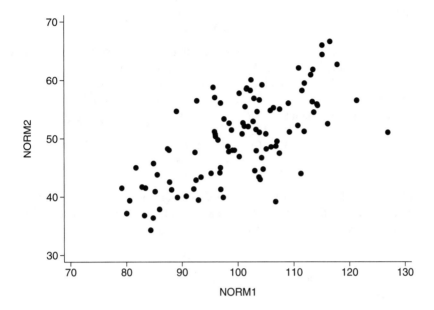

Figure 14.1: *Scatter Diagram of NORM1 versus NORM2*

Table 14.1: *Sample statistics for hypothetical data set*

	Variable	
Statistic	NORM1	NORM2
N	100	100
Mean	100.24	49.99
Standard deviation	10.20	7.31
Variance	103.98	53.51

14.4 Basic concepts

Again, to simplify the exposition of the basic concepts, we present first the case of two variables X_1 and X_2. Later we discuss the general case of P variables.

Suppose that we have a random sample of N observations on X_1 and X_2. For ease of interpretation we subtract the sample mean from each observation, thus obtaining

$$x_1 = X_1 - \bar{X}_1$$

and

$$x_2 = X_2 - \bar{X}_2$$

Note that this technique makes the means of x_1 and x_2 equal to zero but does not alter the sample variances S_1^2 and S_2^2 or the correlation r.

The basic idea is to create two new variables, C_1 and C_2, called the **principal components**. These new variables are linear functions of x_1 and x_2 and can therefore be written as

$$C_1 = a_{11}x_1 + a_{12}x_2$$

$$C_2 = a_{21}x_1 + a_{22}x_2$$

We note that for any set of values of the coefficients $a_{11}, a_{12}, a_{21}, a_{22}$, we can introduce the N observed x_1 and x_2 and obtain N values of C_1 and C_2. The means and variances of the N values of C_1 and C_2 are

$$
\begin{aligned}
\text{mean } C_1 &= \text{mean } C_2 = 0 \\
\text{Var } C_1 &= a_{11}^2 S_1^2 + a_{12}^2 S_2^2 + 2a_{11}a_{12}rS_1S_2 \\
\text{Var } C_2 &= a_{21}^2 S_1^2 + a_{22}^2 S_2^2 + 2a_{21}a_{22}rS_1S_2
\end{aligned}
$$

where $S_i^2 = \text{Var } X_i$.

The **coefficients** are chosen to satisfy three requirements.

1. Var C_1 is as large as possible.

2. The N values of C_1 and C_2 are uncorrelated.

3. $a_{11}^2 + a_{12}^2 = a_{21}^2 + a_{22}^2 = 1$.

The mathematical solution for the coefficients was originally derived by Hotelling (1933). The solution is illustrated graphically in Figure 14.2. Principal components analysis amounts to rotating the original x_1 and x_2 axes to new C_1 and C_2 axes. The angle of rotation is determined uniquely by the requirements just stated. For a given point x_1, x_2 (Figure 14.2) the values of C_1 and C_2 are found by drawing perpendicular lines to the new C_1 and C_2 axes. The N values of C_1 thus obtained will have the largest variance according to requirement 1. The N values of C_1 and C_2 will have zero correlation.

In our hypothetical data example, the two principal components are

$$
\begin{aligned}
C_1 &= 0.851x_1 + 0.525x_2 \\
C_2 &= -0.525x_1 + 0.851x_2
\end{aligned}
$$

where

$$x_1 = \text{NORM1} - \text{mean}(\text{NORM1})$$

and

$$x_2 = \text{NORM2} - \text{mean}(\text{NORM2})$$

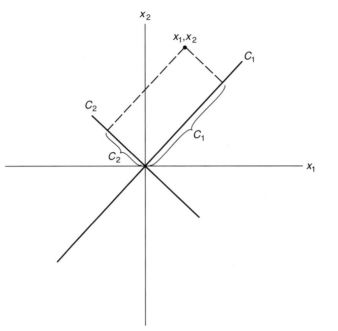

Figure 14.2: *Illustration of Two Principal Components C_1 and C_2*

Note that
$$0.851^2 + 0.525^2 = 1$$
and
$$(-0.525)^2 + 0.851^2 = 1$$
as required by requirement 3 above. Also note that, for the **two**-variable case only, $a_{11} = a_{22}$ and $a_{12} = -a_{21}$.

Figure 14.3 illustrates the x_1 and x_2 axes (after subtracting the means) and the rotated C_1 and C_2 principal component axes. Also drawn in the graph is an ellipse of concentration of the original bivariate normal distribution.

The variance of C_1 is 135.04, and the variance of C_2 is 22.45. These two variances are commonly known as the first and second **eigenvalues** of the co-variance matrix of X_1 and X_2, respectively. Synonymous names used for eigenvalue are characteristic root, latent root, and proper value. Note that the sum of these two variances is 157.49. This quantity is equal to the sum of the original two variances ($103.98 + 53.51 = 157.49$). This result will always be the case, i.e., the total variance is preserved under rotation of the principal components. Note also that the lengths of the axes of the ellipse of concentration (Figure 14.3) are proportional to the sample standard deviations of C_1 and C_2, respec-

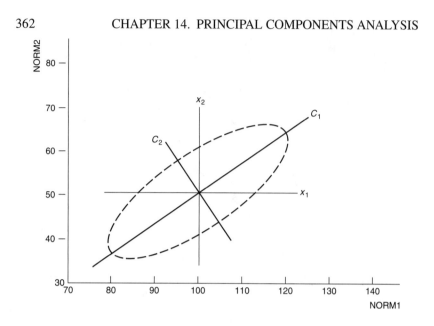

Figure 14.3: *Plot of Principal Components for Bivariate Hypothetical Data*

tively. These standard deviations are the square roots of the eigenvalues. It is therefore easily seen from Figure 14.3 that C_1 has a larger variance than C_2. In fact, C_1 has a larger variance than either of the original variables x_1 and x_2.

These ideas are extended to the case of P variables $x_1, x_2, \ldots, x_P$. Each principal component is a linear combination of the x variables. Coefficients of these linear combinations are chosen to satisfy the following three requirements:

1. $\text{Var}C_1 \geq \text{Var}C_2 \geq \cdots \geq \text{Var}C_P$.

2. The values of any two principal components are uncorrelated.

3. For any principal component the sum of the squares of the coefficients is one.

In other words, C_1 is the linear combination with the largest variance. Subject to the condition that it is uncorrelated with C_1, C_2 is the linear combination with the largest variance. Similarly, C_3 has the largest variance subject to the condition that it is uncorrelated with C_1 and C_2, etc. The $\text{Var}C_i$ are the **eigenvalues** of the covariance matrix of $X_1, X_2, \cdots X_P$. These P variances add up to the original total variance. In some packaged programs the set of coefficients of the linear combination for the ith principal component is called the ith **eigenvector** (also known as the characteristic or latent vector).

14.5   Interpretation

In this section we discuss how many components should be retained for further analysis and we present the analysis for standardized x variables. Application to the depression data set is given as an example.

Number of components retained

As mentioned earlier, one of the objectives of principal components analysis is **reduction of dimensionality**. Since the principal components are arranged in decreasing order of variance, we may select the first few as representatives of the original set of variables. The number of components selected may be determined by examining the proportion of total variance explained by each component. The cumulative proportion of total variance indicates to the investigator just how much information is retained by selecting a specified number of components.

In the hypothetical example the total variance is 157.49. The variance of the first component is 135.04, which is $135.04/157.49 = 0.857$, or 85.7%, of the total variance. It can be argued that this amount is a sufficient percentage of the total variation, and therefore the first principal component is a reasonable representative of the two original variables NORM1 and NORM2 (see Eastment and Krzanowski (1982) for further discussion).

Various rules have been proposed for deciding how many components to retain but none of them appear to work well in all circumstances. Nevertheless, they do provide some guidance on this topic. The discussion given above relates to one rule, that of keeping a sufficient number of principal components to explain a certain percentage of the total variance. One common cutoff point is 80%. If we used that cutoff point, we would only retain one principal component in the above example. But the use of that rule may not be satisfactory. You may have to retain too many principal components, each of which explains only a small percentage of the variation, to reach the 80% limit. Or there may be a principal component you wish to retain that puts you over the cutoff point.

Other rules have approached the subject from the opposite end, that is, to discard principal components with the smallest variances. Dunteman (1989) discusses several of these rules. One is to discard all components that have a variance less than $70/P$ percent of the total variance. In our example, that would be 35% of the total variance of 55.1, which is larger than the variance of the second component or 22.45, so we would not retain the second component. Other statisticians use $100/P$ instead of $70/P$ percent of the total variance. Other variations of this type of rule are not to keep any components that explain small proportions of the variance since they may represent simply "random" variation in the data. For example, some users do not retain any principal component that has a variance of less than 5% of the total variance.

Others advocate plotting the principal component number on the horizontal axis versus the individual eigenvalues on the vertical axis (see Figure 14.5b for an example using the depression data). The idea is to use a cutoff point where lines joining consecutive points are relatively steep left of the cutoff point and relatively flat right of the cutoff point. Such plots are called *scree plots* and this rule is called the *elbow rule* since the point where the two slopes meet resembles an elbow. In practice this sometimes works and at other times there is no clear change in the slope.

Since principal components analysis is often an exploratory method, many investigators argue for not taking any of the above rules seriously, and instead they will retain as many components as they can either interpret or are useful in future analyses.

Transforming coefficients to correlations

To interpret the meaning of the first principal component, we recall that it was expressed as

$$C_1 = 0.851x_1 + 0.525x_2$$

The coefficient 0.851 can be transformed into a correlation between x_1 and C_1. In general, the correlation between the ith principal component and the jth x variable is

$$r_{ij} = \frac{a_{ij}(\text{Var } C_i)^{1/2}}{(\text{Var } x_j)^{1/2}}$$

where a_{ij} is the coefficient of x_j for the ith principal component. For example, the correlation between C_1 and x_1 is

$$r_{11} = \frac{0.85(135.04)^{1/2}}{(103.98)^{1/2}} = 0.969$$

and the correlation between C_1 and x_2 is

$$r_{12} = \frac{0.525(135.04)^{1/2}}{(53.51)^{1/2}} = 0.834$$

Note that both of these correlations are fairly high and positive. As can be seen from Figure 14.3, when either x_1 or x_2 increases, so will C_1. This result occurs often in principal components analysis whereby the first component is positively correlated with all of the original variables.

Using standardized variables

Investigators frequently prefer to **standardize** the x variables prior to performing the principal components analysis. Standardization is achieved by dividing each variable by its sample standard deviation. This analysis is then equivalent to analyzing the correlation matrix instead of the covariance matrix. When we derive the principal components from the correlation matrix, the interpretation becomes easier in two ways.

1. The total variance is simply the number of variables P, and the proportion explained by each principal component is the corresponding eigenvalue divided by P.

2. The correlation between the ith principal component C_i and the jth variable x_j is

$$r_{ij} = a_{ij}(\text{Var } C_i)^{1/2}$$

Therefore for a given C_i we can compare the a_{ij} to quantify the relative degree of dependence of C_i on each of the standardized variables. This correlation is called the **factor loading** in some computer programs.

In our hypothetical example the correlation matrix is as follows:

x_1/S_1	x_2/S_2
1.000	0.675
0.675	1.000

Here S_1 and S_2 are the standard deviations of the first and second variables. Analyzing this matrix results in the following two principal components:

$$C_1 = 0.707 \frac{\text{NORM1}}{S_1} + 0.707 \frac{\text{NORM2}}{S_2}$$

which explains $1.675/2 \times 100 = 83.8\%$ of the total variance, and

$$C_2 = 0.707 \frac{\text{NORM1}}{S_1} - 0.707 \frac{\text{NORM2}}{S_2}$$

explaining $0.325/2 \times 100 = 16.3\%$ of the total variance. So the first principal component is equally correlated with the two standardized variables. The second principal component is also equally correlated with the two standardized variables, but in the opposite direction. The case of two standardized variables is illustrated in Figure 14.4.

Forcing both standard deviations to be one causes the vast majority of the data to be contained in the square shown in the figure (e.g., 99.7% of normal data must fall within plus or minus three standard deviations of the mean). Because of the symmetry of this square, the first principal component will be in the direction of the 45° line for the case of two variates.

Using the original variables NORM1 and NORM2, the principal components based on the correlation matrix are

$$C_1 = 0.707 \frac{\text{NORM1}}{S_1} + 0.707 \frac{\text{NORM2}}{S_2}$$

$$= \quad 0.707\,\frac{\text{NORM1}}{10.20} + 0.707\,\frac{\text{NORM2}}{7.31}$$

$$= \quad 0.0693\,\text{NORM1} + 0.0967\,\text{NORM2}$$

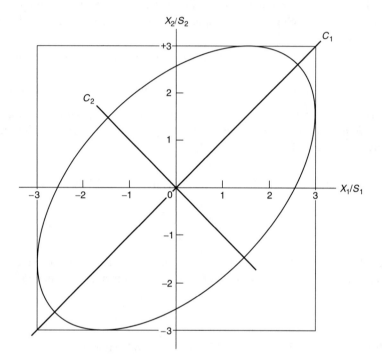

Figure 14.4: *Principal Components for Two Standardized Variables*

Similarly,

$$C_2 = +0.0693\,\text{NORM1} - 0.0967\,\text{NORM2}$$

Note that these results are very different from the principal components obtained from the covariance matrix (Section 14.4). This is the case in general. In fact, there is no easy way to convert the results based on the covariance matrix into those based on the correlation matrix, or *vice versa*. Note that if we use the covariance matrix and change the scale of one of the variables, say from inches to centimeters, this will change the results of the principal components (see van Belle *et al.*, 2004). The majority of researchers prefer to use the correlation matrix because it compensates for the units of measurement of the different variables. But if it is used, then all interpretations must be made in terms of the standardized variables.

Analysis of depression data set

Next, we present a principal components analysis of a real data set, the depression data. We select for this example the 20 items that make up the CESD scale. Each item is a statement to which the response categories are ordinal. The answer "rarely or none of the time" (less than 1 day) is coded as 0, "some or a little of the time" (1–2 days) as 1, "occasionally or a moderate amount of the time" (3–4 days) as 2, and "most or all of the time" (5–7 days) as 3. The values of the response categories are reversed for the positive affect items (see Table 14.2 for a listing of the items) so that a high score indicates likelihood of depression. The CESD score is simply a sum of the scores for these 20 items.

We emphasize that these variables do not satisfy the assumptions often made in statistics of a multivariate normal distribution. In fact, they cannot even be considered to be continuous variables. However, they are typical of what is found in real-life applications.

In this example we used the correlation matrix in order to be consistent with the factor analysis we present in Chapter 15. The eigenvalues (variances of the principal components) are plotted in Figure 14.5 (plot b). Since the correlation matrix is used, the total variance is the number of variables, 20. By dividing each eigenvalue by 20 and multiplying by 100, we obtain the percentage of total variance explained by each principal component. Adding these percentages successively produces the cumulative percentages plotted in Figure 14.5 (plot a).

These eigenvalues and cumulative percentages are found in the output of standard packaged programs. They enable the user to determine whether and how many principal components should be used. If the variables were uncorrelated to begin with, then each principal component, based on the correlation matrix, would explain the **same** percentage of the total variance, namely, $100/P$. If this were the case, a principal components analysis would be unnecessary. Typically, the first principal component explains a much larger percentage than the remaining components, as shown in Figure 14.5.

Ideally, we wish to obtain a small number of principal components, say two or three, which explain a large percentage of the total variance, say 80% or more. In this example, as is the case in many applications, this ideal is not achieved. We therefore must compromise by choosing as few principal components as possible to explain a reasonable percentage of the total variance. A rule of thumb adopted by many investigators is to select only the principal components explaining at least $100/P$ percent of the total variance (at least 5% in our example). This rule applies whether the covariance or the correlation matrix is used. In our example we would select the first five principal components if we followed this rule. These five components explain 59% of the total variance, as seen in Figure 14.5. Note, however, that the next two components

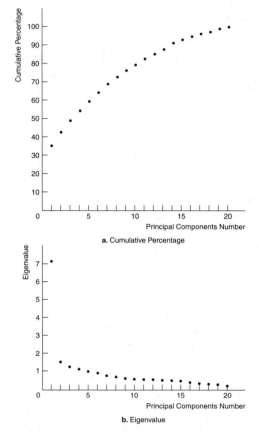

Figure 14.5: *Eigenvalues and Cumulative Percentages of Total Variance for Depression Data*

each explains nearly 5%. Some investigators would therefore select the first seven components, which explain 69% of the total variance.

It should be explained that the eigenvalues are estimated variances of the principal components and are therefore subject to large sample variations. Arbitrary cutoff points should thus not be taken too seriously. Ideally, the investigator should make the selection of the number of principal components on the basis of some underlying theory.

Once the number of principal components is selected, the investigator should examine the coefficients defining each of them in order to assign an interpretation to the components. As was discussed earlier, a high coefficient of a principal component on a given variable is an indication of high correlation between that variable and the principal component (see the formulas given

Table 14.2: *Principal components analysis for standardized CESD scale items (depression data set)*

Item	Principal component				
	1	2	3	4	5
Negative affect					
I felt that I could not shake off the blues even with the help of my family or friends.	0.2774	0.1450	0.0577	−0.0027	0.0883
I felt depressed.	0.3132	−0.0271	0.0316	0.2478	0.0244
I felt lonely.	0.2678	0.1547	0.0346	0.2472	−0.2183
I had crying spells.	0.2436	0.3194	0.1769	−0.0716	−0.1729
I felt sad.	0.2868	0.0497	0.1384	0.2794	−0.0411
I felt fearful.	0.2206	−0.0534	0.2242	0.1823	−0.3399
I thought my life had been a failure.	0.2844	0.1644	−0.0190	−0.0761	−0.0870
Positive affect					
I felt that I was as good as other people.	0.1081	0.3045	0.1103	−0.5567	−0.0976
I felt hopeful about the future.	0.1758	0.1690	−0.3962	−0.0146	0.5355
I was happy.	0.2766	0.0454	−0.0835	0.0084	0.3651
I enjoyed life.	0.2433	0.1048	−0.1314	0.0414	0.2419
Somatic, retarded activity					
I was bothered by things that usually don't bother me.	0.1790	−0.2300	0.1634	0.1451	0.0368
I did not feel like eating; my appetite was poor.	0.1259	−0.2126	0.2645	−0.5400	0.0953
I felt that everything was an effort.	0.1803	−0.4015	−0.1014	−0.2461	−0.0847
My sleep was restless.	0.2004	−0.2098	0.2703	0.0312	0.0834
I could not "get going."	0.1924	−0.4174	−0.1850	−0.0467	−0.0399
I had trouble keeping my mind on what I was doing.	0.2097	−0.3905	−0.0860	−0.0684	−0.0499
I talked less than usual.	0.1717	−0.0153	0.2019	−0.0629	0.2752
Interpersonal					
People were unfriendly.	0.1315	−0.0569	−0.6326	−0.0232	−0.3349
I felt that people disliked me.	0.2357	0.2283	−0.1932	−0.2404	−0.2909
Eigenvalues or Var C_i	7.055	1.486	1.231	1.066	1.013
Cumulative proportion	0.353	0.427	0.489	0.542	0.593
$0.5/(\text{VAR } C_i)^{1/2}$	0.188	0.410	0.451	0.484	0.497

earlier). Principal components are interpreted in the context of the variables with high coefficients.

For the depression data example, Table 14.2 shows the coefficients for the first five components. For each principal component the variables with a correlation greater than 0.5 with that component are underlined. For the sake of illustration this value was taken as a cutoff point. Recall that the correlation r_{ij} is $a_{ij} (\text{Var } C_i)^{1/2}$ and therefore a coefficient a_{ij} is underlined if it exceeds $0.5/(\text{Var } C_i)^{1/2}$ (Table 14.2).

As the table shows, many variables are highly correlated (greater than 0.5) with the first component. Note also that the correlations of all the variables with the first component are positive (recall that the scaling of the response for items

8–11 was reversed so that a high score indicates likelihood of depression). The first principal component can therefore be viewed as a weighted average of most of the items. A high value of C_1 is an indication that the respondent had many of the symptoms of depression.

On the other hand, the only item with more than 0.5 absolute correlation with C_2 is item 16, although items 17 and 14 have absolute correlations close to 0.5. The second principal component, therefore, can be interpreted as a measure of lethargy or energy. A low value of C_2 is an indication of a lethargic state and a high value of C_2 is an indication of a high level of energy. By construction, C_1 and C_2 are uncorrelated.

Similarly, C_3 measures the respondent's feeling toward how others perceive him or her; a low value is an indication that the respondent believes that people are unfriendly. Similar interpretations can be made for the other two principal components in terms of the items corresponding to the underlined coefficients.

The numerical values of C_i for $i = 1$ to 5, or 1 to 7, for each individual can be used in subsequent analyses. For example, we could use the value of C_1 instead of the CESD score as a dependent variable in a regression analysis such as that given in Problem 8.1. Had the first component explained a higher proportion of the variance, this procedure might have been better than simply using a sum of the scores.

The depression data example illustrates a situation in which the results are **not** clear-cut. The conclusion may be reached from observing Figure 14.5, where we saw that it is difficult to decide how many components to use. It is not possible to explain a very high proportion of the total variance with a small number of principal components. Also, the interpretation of the components in Table 14.2 is not straightforward. This is frequently the case in real-life situations. Occasionally, situations do come up in which the results are clear-cut. For examples of such situations, see Jolliffe (2010).

In biological examples where distance measurements are made on particular parts of the body, the first component is often called the size component. For example, in studying the effects of operations in young children who have deformed skulls, various points on the skull are chosen and the distances between these points are measured. Often, 15 or more distances are computed. When principal components analysis is done on these distances, the first component would likely have only positive coefficients. It is called the size component, since children who have overall large skulls would score high on this component. The second component would likely be a mixture of negative and positive coefficients and would contrast one set of distances with another. The second component would then be called shape since it contrasts, say, children who have prominent foreheads with those who don't. In some examples, a high proportion of the variance can be explained by a size and shape component, thus not only reducing the number of variables to analyze but also providing insight into the objects being measured.

Tests of hypotheses regarding principal components assuming multivariate normality are discussed in Jackson and Hearne (1973), Jackson (2003), and Jolliffe (2010). These tests should be interpreted with caution since the data are often not multivariate normal, and most users view principal components analysis as an exploratory technique.

14.6   Other uses

As mentioned in Section 9.5, when multicollinearity is present, principal components analysis may be used to alleviate the problem. The advantage of using principal components analysis is that it both helps in understanding what variables are causing the multicollinearity and provides a method for obtaining stable (though biased) estimates of the slope coefficients. If serious multicollinearity exists, the use of these coefficients will result in larger residuals in the data from which they were obtained and smaller multiple correlation than when least squares is used (the same holds for ridge regression). But the estimated standard error of the slopes could be smaller and the resulting equation could predict better in a fresh sample. In Chapter 9 we noted that STATISTICA regression performs ridge regression directly. SAS REG does print the eigenvalues, thus alerting the user if principal components regression is sensible, and provides information useful in deciding how many principal components to retain.

So far in this chapter we have concentrated our attention on the first few components that explain the highest proportion of the variance. To understand the use of principal components in regression analysis, it is useful to consider what information is available in, say, the last or Pth component. The eigenvalue of that component will tell us how much of the total variance in the X's it explains. Suppose we consider standardized data to simplify the magnitudes. If the eigenvalue of the last principal component (which must be smallest) is almost one, then the simple correlations among the X variables must be close to zero. With low or zero simple correlations, the lengths of the principal component axes within the ellipse of concentration (Figure 14.3) are all nearly the same and multicollinearity is not a problem.

At the other extreme, if the numerical value of the last eigenvalue is close to zero, then the length of its principal axis within the ellipse of concentration is very small. Since this eigenvalue can also be considered as the variance of the Pth component, C_P, the variance of C_P is close to zero. For standardized data with zero mean, the mean of C_P is zero and its variance is close to zero. Approximately, we have

$$0 = a_{P1}x_1 + a_{P2}x_2 + \cdots + a_{PP}x_P$$

a linear relationship among the variables. The values of the coefficients of the principal components can give information on this interrelationship. For example, if two variables were almost equal to each other, then the value of C_P might be $0 = 0.707x_1 - 0.707x_2$. For an example taken from actual data, see Chatterjee and Hadi (2006). In other words, when multicollinearity exists, the examination of the last few principal components will provide information on which variables are causing the problem and on the nature of the interrelationship among them. It may be that two variables are almost a linear function of each other, or that two variables can almost be added to obtain a third variable.

In principal component regression analysis programs, first the principal components of the X or independent variables are found. Then the regression equation is computed using the principal components instead of the original X variables. From the regression equation using all the principal components, it is possible to obtain the original regression equation in the X's, since there is a linear relationship between the X's and the C's (Section 14.4). But what the special programs mentioned above allow the researcher to do is to obtain a regression equation using the X variables where one or more of the last principal components are not used in the computation. Based on the size of the eigenvalue (how small it is) and the possible relationship among the variables displayed in the later principal components, the user can decide how many components to discard. A regression equation in the X's can be obtained even if only one component is kept, but usually only one or two are discarded. The standard error of the slope coefficients when the components with very small eigenvalues are discarded tends to be smaller than when all are kept.

The use of principal component regression analysis is especially useful when the investigator understands the interrelationships of the X variables well enough to interpret these last few components. Knowledge of the variables can help in deciding what to discard.

To perform principal components regression analysis, you can first perform a principal components analysis, examine the eigenvalues, and discard the components with eigenvalues less than, say, 0.01. Suppose the last eigenvalue whose value is greater than 0.01 is the kth one. Then, compute and save the values of $C_1, C_2, \ldots, C_k$ for each case. These can be called **principal component scores**. Note that programs such as SAS PRINCOMP can do this for you. Using the desired dependent variable, you can then perform a regression analysis with the k C_i's as independent variables. After you obtain the regression analysis, replace the C's by X's using the k equations expressing C's as a function of X's (see Section 14.4 for examples of the equations). You will now have a regression equation in terms of the X's. If you begin with standardized variables, you must use standardized variables throughout. Likewise, if you begin with unstandardized variables you must use them throughout.

Principal components analysis can be used to assist in a wide range of statistical applications (Dunteman, 1989). One of these applications is finding outliers among a set of variables. It is easier to see outliers in two-dimensional space in the common scatter diagram than in other graphical output. For multivariate data, if the first two components explain a sizable proportion of the variance among the P variables then a display of the first two principal component scores in a scatter diagram should give a good graphical indication of the data set. Values of C_1 for each case are plotted on the horizontal axis and values of C_2 on the vertical axis. Points can be considered outliers if they do not lie close to the bulk of the others.

Scatter diagrams of the first two components have also been used to spot separate clusters of cases (see Figure 16.8 (Section 16.5) for an example of what clusters look like).

In general, when you want to analyze a sizable number of correlated variables, principal components analysis can be considered. For example, if you have six correlated variables and you consider an analysis of variance on each one, the results of these six analyses would not be independent since the variables are correlated. But if you first do a principal components analysis and come up with fewer uncorrelated components, interpreting the results may be more valid and there would be fewer analyses to do.

The same would hold true for discriminant function or logistic regression analysis using highly correlated covariates. If you use the principal component scores derived from the total data set instead of the original X's, you will have uncorrelated components and their contribution to the analysis can be evaluated using analysis of variance. Note that the first component may not be the one that is most useful in discriminating among the two or more groups or outcomes, since group membership is a separate variable from the others.

Similarly, principal components analysis can be performed on the dependent and independent variables separately prior to canonical correlation analysis. If each set of variables is highly intercorrelated, then the two principal component analyses can reduce the number of variables in each set and could possibly result in a canonical correlation analysis that is easier to interpret.

But if the correlations among a set of variables are near zero, then there is no advantage in performing principal components analysis since you already have variables with low intercorrelations and you would need to keep a lot of components to explain a high proportion of the variance.

14.7   Discussion of computer programs

The principal components output for the six computer packages is summarized in Table 14.3. The most straightforward program is the PRINCOMP procedure in SAS. In the depression example, we stored a SAS data file called "depress" in the computer and it was only necessary to type

```
proc princomp data = in.depress;
Var c1–c20;
```

to tell the computer what to run. As can be seen from Table 14.3, all software packages can perform a principal components analysis. In STATISTICA,
PCCA stands for Principal Components and Classification Analysis.

The FACTOR program in SPSS and pca in Stata can be used to perform
principal components analysis. The only option available in SPSS is correlation; so the results are standardized. The user should specify "EXTRACTION
= PC" and also "CRITERIA = FACTORS(P)," where P is the total number of
variables. Adjusting the scores is necessary. This can be done by multiplying
the scores by the square root of the corresponding eigenvalue. Finally, we note
that S-PLUS/R can produce all of the outputs discussed in this chapter.

14.8   What to watch out for

Principal components analysis is mainly used as an exploratory technique with
little use of statistical tests or confidence intervals. Because of this, formal distributional assumptions are not necessary. Nevertheless, it is obviously simpler
to interpret if certain conditions hold.

1. As with all statistical analyses, interpreting data from a random sample of a
   well-defined population is simpler than interpreting data from a sample that
   is taken in a haphazard fashion.

2. If the observations arise from or are transformed to a symmetric distribution,
   the results are easier to understand than if highly skewed distributions are
   analyzed. Obvious outliers should be searched for and removed. If the data
   follow a multivariate normal distribution, then ellipses such as that pictured
   in Figure 14.4 are obtained and the principal components can be interpreted
   as axes of the ellipses. Statistical tests that can be found in some packaged
   programs usually require the assumption of multivariate normality. Note
   that the principal components offer a method for checking (in an informal
   sense) whether or not some data follow a multivariate normal distribution.
   For example, suppose that by using C_1 and C_2, 75% of the variance in the
   original variables can be explained. Their numerical values are obtained for
   each case. Then the N values of C_1 are plotted using a cumulative normal
   probability plot, and checked to see if a straight line is obtained. A similar
   check is done for C_2. If both principal components are normally distributed,
   this lends some credence to the data set having a multivariate normal distribution. A scatter plot of C_1 versus C_2 can also be used to spot outliers.
   PRINCOMP allows the user to store the principal component scores of each
   case for future analysis.

3. If principal components analysis is to be used to check for redundancy in
   the data set (as discussed in Section 14.6), then it is important that the ob-

Table 14.3: *Software commands and output for principal components analysis*

	S-PLUS/R	SAS	SPSS	Stata	STATISTICA
Covariance matrix	var	PRINCOMP	FACTOR	correlate	PCCA
Correlation matrix	cor	PRINCOMP	FACTOR	correlate	PCCA
Coefficients from raw data	princomp	PRINCOMP	FACTOR	pca	PCCA
Coefficients from standardized data	princomp	PRINCOMP	FACTOR	pca	PCCA
Correlation between xs and Cs	factanal	FACTOR	FACTOR	predict;correlate	PCCA
Eigenvalues	princomp	PRINCOMP	FACTOR	pca	PCCA
Cum. proportion of variance explained	princomp	PRINCOMP	FACTOR	pca	PCCA
Principal component scores saved	princomp	PRINCOMP	FACTOR	predict	PCCA
Plot of principal components	biplot	GPLOT	FACTOR	loadingplot	PCCA
Scree plots	screeplot	FACTOR	FACTOR	screeplot	PCCA

servations be measured accurately. Otherwise, it may be difficult to detect the interrelationships among the X variables.

14.9 Summary

In the two previous chapters we presented methods of selecting variables for regression and discriminant function analyses. These methods include stepwise and subset procedures. In those analyses a dependent variable is present, implicitly or explicitly. In this chapter we presented another method for summarizing the data. It differs from variable selection procedures in two ways.

1. No dependent variable exists.

2. Variables are not eliminated but rather summary variables, i.e., principal components, are computed from all of the original variables.

The major ideas underlying the method of principal components analysis were presented in this chapter. We also discussed how to decide on the number of principal components retained and how to use them in subsequent analyses. Further, methods for attaching interpretations or "names" to the selected principal components were given.

A detailed discussion (at an introductory level) and further examples of the use of principal components analysis can be found in Dunteman (1989) and Hatcher (1994). At a higher mathematical level, Jolliffe (2010) describes recent developments.

14.10 Problems

14.1 For the depression data set described in Appendix A, perform a principal components analysis on the last seven variables DRINK–CHRONILL (Table 3.3). Interpret the results.

14.2 (Continuation of Problem 14.1.) Perform a regression analysis of CASES on the last seven variables as well as on the principal components. What does the regression represent? Interpret the results.

14.3 For the data generated in Problem 7.7, perform a principal components analysis on $X_1, X_2, \ldots, X_9$. Compare the results with what is known about the population.

14.4 (Continuation of Problem 14.3.) Perform the regression of Y on the principal components. Compare the results with the multiple regression of Y on $X1$ to $X9$.

14.5 Perform a principal components analysis on the data in Table 8.1 (not including the variable P/E). Interpret the components. Then perform a regression analysis with P/E as the dependent variable, using the relevant principal components. Compare the results with those in Chapter 8.

14.6 Using the family lung function data described in Appendix A define a new variable RATIO = FEV1/FVC for the fathers. What is the correlation between RATIO and FEV1? Between RATIO and FVC? Perform a principal components analysis on FEV1 and FVC, plotting the results. Perform a principal components analysis on FEV1, FVC, and RATIO. Discuss the results.

14.7 Using the family lung function data, perform a principal components analysis on age, height, and weight for the oldest child.

14.8 (Continuation of Problem 14.7.) Perform a regression of FEV1 for the oldest child on the principal components found in Problem 14.7. Compare the results to those from Problem 7.15.

14.9 Using the family lung function data, perform a principal components analysis on mother's height, weight, age, FEV1, and FVC. Use the covariance matrix, then repeat using the correlation matrix. Compare the results and comment.

14.10 Perform a principal components analysis on AGE and INCOME using the depression data set. Include all the additional data points listed in Problem 6.9(b). Plot the original variables and the principal components. Indicate the added points on the graph, and discuss the results.

14.11 Perform a principal components analysis on all the items of the Parental Bonding scale for the Parental HIV data (see Appendix A and the codebook). How many principal components would you expect to retain for this scale? How many components should be retained based on the rules of thumb mentioned in Section 14.5 and based on the scree plot?

Chapter 15

Factor analysis

15.1 Chapter outline

From this chapter you will learn a useful extension of principal components analysis called factor analysis that will enable you to obtain more distinct new summary variables. The methods given in this chapter for describing the interrelationships among the variables and obtaining new variables have been deliberately limited to a small subset of what is available in the literature in order to make the explanation more understandable.

Section 15.2 covers when factor analysis is used and Section 15.3 presents a hypothetical data example that was generated to display the features of factor analysis. The basic model assumed when performing factor analysis is given in Section 15.4. Section 15.5 discusses initial factor extraction using principal components analysis and Section 15.6 presents initial factor extraction using iterated principal components (principal factor analysis). Section 15.7 presents a commonly used method of orthogonal rotation and a method of oblique rotation. Section 15.8 discusses the process of assigning factor scores to individuals. Section 15.9 contains an additional example of factor analysis using the CESD scale items introduced in Section 14.5. A summary of the output of the statistical packages is given in Section 15.10 and what to watch out for is given in Section 15.11.

15.2 When is factor analysis used?

Factor analysis is similar to principal components analysis in that it is a technique for examining the interrelationships among a set of variables. Both of these techniques differ from regression analysis in that we do not have a **dependent** variable to be explained by a set of **independent** variables. However, principal components analysis and factor analysis also differ from each other. In principal components analysis the major objective is to select a number of components that explain as much of the total variance as possible. The values of the principal components for a given individual are relatively simple to compute and interpret. On the other hand, the **factors** obtained in factor analysis are selected mainly to explain the interrelationships among the original variables. Ideally, the number of factors expected is known in advance. The major emphasis is placed on obtaining easily understandable factors that convey the essential information contained in the original set of variables.

Areas of application of factor analysis are similar to those mentioned in Section 14.2 for principal components analysis. Chiefly, applications have come from the social sciences, particularly psychometrics. It has been used mainly to explore the underlying structure of a set of variables. It has also been used in assessing what items to include in scales and to explore the interrelationships among the scores on different items. A certain degree of resistance to using factor analysis in other disciplines has been prevalent, perhaps because of the heuristic nature of the technique and the special jargon employed. Also, the multiplicity of methods available to perform factor analysis leaves some investigators uncertain of their results.

In this chapter no attempt will be made to present a comprehensive treatment of the subject. Rather, we adopt a simple geometric approach to explain only the most important concepts and options available in standard software packaged programs. Interested readers should refer to texts referenced in this chapter that give more comprehensive treatments of factor analysis. In particular, Bartholomew and Knott (1999) and Long (1983) include confirmatory factor analysis and linear structural models, subjects not discussed in this book. A discussion of the use of factor analysis in the natural sciences is given in Reyment and Jöreskog (1996). Some uses of factor analysis in health research are given in Pett, Lackey, and Sullivan (2003).

15.3   Data example

As in Chapter 14, we generated a hypothetical data set to help us present the fundamental concepts. In addition, the same depression data set used in Chapter 14 will be subjected to a factor analysis here. This example should serve to illustrate the differences between principal components and factor analysis. It will also provide a real-life application.

The hypothetical data set consists of 100 data points on five variables, $X_1, X_2, \ldots, X_5$. The data were generated from a multivariate normal distribution with zero means. Note that most statistical packages provide the option of generating normal data. The sample means and standard deviations are shown in Table 15.1.

The correlation matrix is presented in Table 15.2. We note, on examining the correlation matrix, that there exists a high correlation between X_4 and X_5, a moderately high correlation between X_1 and X_2, and a moderate correlation between X_3 and each of X_4 and X_5. The remaining correlations are fairly low.

Table 15.1: *Means and standard deviations of 100 hypothetical data points*

Variable	Mean	Standard deviation
X_1	0.163	1.047
X_2	0.142	1.489
X_3	0.098	0.966
X_4	-0.039	2.185
X_5	-0.013	2.319

Table 15.2: *Correlation matrix of 100 hypothetical data points*

	X_1	X_2	X_3	X_4	X_5
X_1	1.000				
X_2	0.757	1.000			
X_3	0.047	0.054	1.000		
X_4	0.115	0.176	0.531	1.000	
X_5	0.279	0.322	0.521	0.942	1.000

15.4 Basic concepts

In factor analysis we begin with a set of variables $X_1, X_2, \ldots, X_P$. These variables are usually standardized by the computer program so that their variances are each equal to one and their covariances are correlation coefficients. In the remainder of this chapter we therefore assume that each x_i is a **standardized variable**, i.e., $x_i = (X_i - \overline{X}_i)/S_i$. In the jargon of factor analysis the x_i's are called the original or **response variables.**

The objective of factor analysis is to represent each of these variables as a linear combination of a smaller set of **common factors** plus a factor unique to each of the response variables. We express this representation as

$$
\begin{aligned}
x_1 &= l_{11}F_1 + l_{12}F_2 + \cdots + l_{1m}F_m + e_1 \\
x_2 &= l_{21}F_1 + l_{22}F_2 + \cdots + l_{2m}F_m + e_2 \\
&\vdots \\
x_P &= l_{P1}F_1 + l_{P2}F_2 + \cdots + l_{Pm}F_m + e_P
\end{aligned}
$$

where the following assumptions are made:

1. m is the number of common factors (typically this number is much smaller than P).

2. $F_1, F_2, \ldots, F_m$ are the **common factors**. These factors are assumed to have zero means and unit variances.

3. l_{ij} is the coefficient of F_j in the linear combination describing x_i. This term is called the **loading** of the ith variable on the jth common factor.

4. $e_1, e_2, \ldots, e_P$ are **unique factors**, each relating to one of the original variables.

The above equations and assumptions constitute the **factor model**. Thus, each of the response variables is composed of a part due to the common factors and a part due to its own unique factor. The part due to the common factors is assumed to be a linear combination of these factors.

As an example, suppose that x_1, x_2, x_3, x_4, x_5 are the standardized scores of an individual on five tests. If $m = 2$, we assume the following model:

$$
\begin{aligned}
x_1 &= l_{11}F_1 + l_{12}F_2 + e_1 \\
x_2 &= l_{21}F_1 + l_{22}F_2 + e_2 \\
&\vdots \\
x_5 &= l_{51}F_1 + l_{52}F_2 + e_5
\end{aligned}
$$

Each of the five scores consists of two parts: a part due to the common factors F_1 and F_2 and a part due to the unique factor for that test. The common factors F_1 and F_2 might be considered the individual's verbal and quantitative abilities, respectively. The unique factors express the individual variation on each test score. The unique factor includes all other effects that keep the common factors from completely defining a particular x_i.

In a sample of N individuals we can express the equations by adding a subscript to each x_i, F_j and e_i to represent the individual. For the sake of simplicity this subscript was omitted in the model presented here.

The factor model is, in a sense, the mirror image of the principal components model, where each principal component is expressed as a linear combination of the variables. Also, the number of principal components is equal to the number of original variables (although we may not use all of the principal components). On the other hand, in factor analysis we choose the number of factors to be smaller than the number of response variables. Ideally, the number of factors should be known in advance, although this is often not the case. However, as we will discuss later, it is possible to allow the data themselves to determine this number.

The factor model, by breaking each response variable x_i into two parts, also breaks the variance of x_i into two parts. Since x_i is standardized, its variance is equal to one and is composed of the following two parts:

1. the **communality**, i.e., the part of the variance that is due to the common factors;

2. the **specificity**, i.e., the part of the variance that is due to the unique factor e_i.

Denoting the **communality** of x_i by h_i^2 and the **specificity** by u_i^2, we can write the variance of x_i as $\text{Var } x_i = 1 = h_i^2 + u_i^2$. In words, the variance of x_i equals the communality plus the specificity.

The numerical aspects of factor analysis are concerned with finding estimates of the **factor loadings** (l_{ij}) and the **communalities** (h_i^2). There are many ways available to numerically solve for these quantities. The solution process is called **initial factor extraction**. The next two sections discuss two such extraction methods. Once a set of initial factors is obtained, the next major step in the analysis is to obtain new factors, called the **rotated factors**, in order to improve the interpretation. Methods of rotation are discussed in Section 15.7.

In any factor analysis the number m of common factors is required. As mentioned earlier, this number is, ideally, known prior to the analysis. If it is not known, most investigators use a default option available in standard computer programs whereby the number of factors is the number of eigenvalues greater than one (see Chapter 14 for the definition and discussion of eigenvalues). Also, since the numerical results are highly dependent on the chosen number m, many investigators run the analysis with several values in an effort to get further insights into their data.

15.5 Initial extraction: principal components

In this section and the next we discuss two methods for the initial extraction of common factors. We begin with the **principal components analysis method**, which can be found in most of the standard factor analysis programs. The basic idea is to choose the first m principal components and modify them to fit the factor model defined in the previous section. The reason for choosing the first m principal components, rather than any others, is that they explain the greatest proportion of the variance and are therefore the most important. Note that the principal components are also uncorrelated and thus present an attractive choice as factors.

To satisfy the assumption of unit variances of the factors, we divide each principal component by its standard deviation. That is, we define the jth common factor F_j as $F_j = C_j/(\text{Var } C_j)^{1/2}$, where C_j is the jth principal component.

To express each variable x_i in terms of the F_j's, we first recall the relationship between the variables x_i and the principal components C_j. Specifically,

$$
\begin{aligned}
C_1 &= a_{11}x_1 + a_{12}x_2 + \cdots + a_{1P}x_P \\
C_2 &= a_{21}x_1 + a_{22}x_2 + \cdots + a_{2P}x_P \\
&\vdots
\end{aligned}
$$

$$C_P = a_{P1}x_1 + a_{P2}x_2 + \cdots + a_{PP}x_P$$

It may be shown mathematically that this set of equations can be inverted to express the x_i's as functions of the C_j's. The result is:

$$x_1 = a_{11}C_1 + a_{21}C_2 + \cdots + a_{P1}C_P$$
$$x_2 = a_{12}C_1 + a_{22}C_2 + \cdots + a_{P2}C_P$$
$$\vdots$$
$$x_P = a_{1P}C_1 + a_{2P}C_2 + \cdots + a_{PP}C_P$$

Note that the rows of the first set of equations become the columns of the second set of equations.

Now since $F_j = C_j/(\text{Var } C_j)^{1/2}$, it follows that $C_j = F_j(\text{Var } C_j)^{1/2}$, and we can then express the ith equation as

$$x_1 = a_{1i}F_1(\text{Var } C_1)^{1/2} + a_{2i}F_2(\text{Var } C_2)^{1/2} + \cdots + a_{Pi}F_P(\text{Var } C_P)^{1/2}$$

This last equation is now modified in two ways.

1. We use the notation $l_{ij} = a_{ji}(\text{Var } C_j)^{1/2}$ for the first m components.
2. We combine the last $P - m$ terms and denote the result by e_i. That is,

$$e_i = a_{m+1,i}F_{m+1}(\text{Var } C_{m+1})^{1/2} + \cdots + a_{Pi}F_P(\text{Var } C_P)^{1/2}$$

With these manipulations we have now expressed each variable x_i as

$$x_i = l_{i1}F_1 + l_{i2}F_2 + \cdots + l_{im}F_m + e_i$$

for $i = 1, 2, \ldots, P$. In other words, we have transformed the principal components model to produce the factor model. For later use, note also that when the original variables are standardized, the factor loading l_{ij} turns out to be the **correlation** between x_i and F_j (Section 14.5). The matrix of factor loadings is sometimes called the **pattern matrix**. When the factor loadings are correlations between the x_i's and F_j's as they are here, it is also called the **factor structure matrix**. Furthermore, it can be shown mathematically that the communality of x_i is $h_i^2 = l_{i1}^2 + l_{i2}^2 + \cdots + l_{im}^2$.

For example, in our hypothetical data set the eigenvalues of the correlation matrix (or Var C_j) are

$$
\begin{array}{rcl}
\text{Var } C_1 & = & 2.578 \\
\text{Var } C_2 & = & 1.567 \\
\text{Var } C_3 & = & 0.571 \\
\text{Var } C_4 & = & 0.241 \\
\text{Var } C_5 & = & \underline{0.043} \\
\text{Total} & = & 5.000
\end{array}
$$

Note that the sum 5.0 is equal to P, the total number of variables. Based on the rule of thumb of selecting only those principal components corresponding to eigenvalues of one or more, we select $m = 2$. We obtain the principal components analysis factor loadings, the l_{ij}'s, shown in Table 15.3. For example, the loading of x_1 on F_1 is $l_{11} = 0.511$ and on F_2 is $l_{12} = 0.782$. Thus, the first equation of the factor model is

$$x_1 = 0.511F_1 + 0.782F_2 + e_1$$

Table 15.3 also shows the variance explained by each factor. For example, the variance explained by F_1 is 2.578, or 51.6%, of the total variance of 5. The communality column, h_i^2, in Table 15.3 shows the part of the variance of each variable explained by the common factors. For example,

$$h_1^2 = 0.511^2 + 0.782^2 = 0.873$$

Finally, the specificity u_i^2 is the part of the variance not explained by the common factors. In this example we use standardized variables x_i whose variances are each equal to one. Therefore for each x_i, $u_i^2 = 1 - h_i^2$.

It should be noted that the sum of the communalities is equal to the cumulative part of the total variance explained by the common factors. In this example it is seen that

$$0.873 + 0.875 + 0.586 + 0.898 + 0.913 = 4.145$$

and

$$2.578 + 1.567 = 4.145$$

thus verifying the above statement. Similarly, the sum of the specificities is equal to the total variance minus the sum of the communalities.

A valuable graphical aid to interpreting the factors is the **factor diagram** shown in Figure 15.1. Unlike the usual scatter diagrams where each point represents an individual case and each axis a variable, in this graph each point represents a response variable and each axis a common factor. For example, the point labelled 1 represents the response variable x_1. The coordinates of that point are the loadings of x_1 on F_1 and F_2, respectively. The other four points represent the remaining four variables.

Since the factor loadings are the correlations between the standardized variables and the factors, the range of values shown on the axes of the factor diagram is -1 to $+1$. It can be seen from Figure 15.1 that x_1 and x_2 load more on F_2 than they do on F_1. Conversely, x_3, x_4 and x_5 load more on F_1 than on F_2. However, these distinctions are not very clear-cut, and the technique of factor rotation will produce clearer results (Section 15.7).

An examination of the correlation matrix shown in Table 15.2 confirms that x_1 and x_2 form a block of correlated variables. Similarly, x_4 and x_5 form

Table 15.3: *Initial factor analysis summary for hypothetical data set from principal components extraction method*

Variable	Factor loadings F_1	F_2	Communality h_i^2	Specificity u_i^2
x_1	0.511	0.782	0.873	0.127
x_2	0.553	0.754	0.875	0.125
x_3	0.631	−0.433	0.586	0.414
x_4	0.866	−0.386	0.898	0.102
x_5	0.929	−0.225	0.913	0.087
Variance explained	2.578	1.567	$\sum h_i^2 = 4.145$	$\sum u_i^2 = 0.855$
Percentage	51.6	31.3	82.9	17.1

another block, with x_3 also correlated to them. These two major blocks are only weakly correlated with each other. (Note that the correlation matrix for the standardized x's is the same as that for the unstandardized X's.)

15.6   Initial extraction: iterated components

The second method of extracting initial factors is a modification of the principal components analysis method. It has different names in different packages. It is called principal factor analysis or principal axis factoring approach in many texts.

To understand this method, you should recall that the communality is the part of the variance of each variable associated with the common factors. The principle underlying the **iterated solution** states that we should perform the factor analysis by using the communalities in place of the original variance. This principle entails substituting communality estimates for the 1's representing the variances of the standardized variables along the diagonal of the correlation matrix. With 1's in the diagonal we are factoring the total variance of the variables; with communalities in the diagonal we are factoring the variance associated with the common factors. Thus with communalities along the diagonal we select those common factors that maximize the total communality.

Many factor analysts consider maximizing the total communality a more attractive objective than maximizing the total proportion of the explained variance, as is done in the principal components method. The problem is that communalities are not known before the factor analysis is performed. Some initial estimates of the communalities must be obtained prior to the analysis. Various procedures exist, and we recommend, in the absence of *a priori* estimates, that the investigator use the default option in the particular program since the

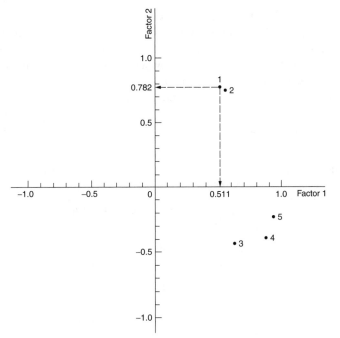

Figure 15.1: *Factor Diagram for Principal Components Extraction Method*

resulting factor solution is usually little affected by the initial communality estimates.

The steps performed by a packaged program in carrying out the **iterated factor extraction** are summarized as follows:

1. Find the initial communality estimates.

2. Substitute the communalities for the diagonal elements (1's) in the correlation matrix.

3. Extract m principal components from the modified matrix.

4. Multiply the principal components coefficients by the standard deviation of the respective principal components to obtain factor loadings.

5. Compute new communalities from the computed factor loadings.

6. Replace the communalities in step 2 with these new communalities and repeat steps 3, 4, and 5. This step constitutes an **iteration**.

7. Continue iterating, stopping when the communalities stay essentially the same in the last two iterations.

For our hypothetical data example the results of using this method are shown in Table 15.4. In comparing this table with Table 15.3, we note that the total communality is higher for the principal components method (82.9% versus 74.6%). This result is generally the case since in the iterative method we are factoring the total communality, which is by necessity smaller than the total variance. The individual loadings do not seem to be very different in the two methods; compare Figures 15.1 and 15.2. The only apparent difference in loadings is in the case of the third variable. Note that Figure 15.2, the factor diagram for the iterated method, is constructed in the same way as Figure 15.1.

Table 15.4: *Initial factor analysis summary for hypothetical data set from iterated principal factor extraction method*

Variable	Factor loadings F_1	F_2	Communality h_i^2	Specificity u_i^2
x_1	0.470	0.734	0.759	0.241
x_2	0.510	0.704	0.756	0.244
x_3	0.481	−0.258	0.598	0.702
x_4	0.888	−0.402	0.949	0.051
x_5	0.956	−0.233	0.968	0.032
Variance explained	2.413	1.317	$\sum h_i^2 = 3.730$	$\sum u_i^2 = 1.270$
Percentage	48.3	26.3	74.6	25.4

We point out that the factor loadings extracted by the iterated method depend on the number of factors extracted. For example, the loadings for the first factor would depend on whether we extract two or three common factors. This condition does not exist for the uniterated principal components method. It should also be pointed out that it is possible to obtain negative variances and eigenvalues. This occurs because the 1's in the diagonal of the correlation matrix have been replaced by the communality estimates which can be considerably less than one. This results in a matrix that does not necessarily have positive eigenvalues. These negative eigenvalues and the factors associated with them should not be used in the analysis.

Regardless of the choice of the extraction method, if the investigator does not have a preconceived number of factors (m) from knowledge of the subject matter, then several methods are available for choosing one numerically. Reviews of these methods are given in Gorsuch (1983). In Section 15.4 the commonly used technique of including any factor whose eigenvalue is greater than or equal to one was used. This criterion is based on theoretical rationales developed using true population correlation coefficients. It is commonly thought to yield about one factor to every three to five variables. It appears to correctly estimate the number of factors when the communalities are high and

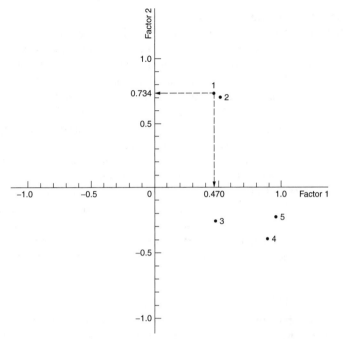

Figure 15.2: *Factor Diagram for Iterated Principal Factor Extraction Method*

the number of variables is not too large. If the communalities are low, then the number of factors obtained with a cutoff equal to the average communality tends to be similar to that obtained by the principal components method with a cutoff equal to one. As an alternative method, called the **scree method**, some investigators will simply plot the eigenvalues on the vertical axis versus the number of factors on the horizontal axis and look for the place where a change in the slope of the curve connecting successive points occurs. Examining the eigenvalues listed in Section 15.5, we see that compared with the first two eigenvalues (2.578 and 1.567), the values of the remaining eigenvalues are low (0.571, 0.241, and 0.043). This result indicates that $m = 2$ is a reasonable choice.

In terms of initial factor extraction methods the iterated principal factor solution is the method employed most frequently by social scientists, the main users of factor analysis. For theoretical reasons many mathematical statisticians are more comfortable with the principal components method. Many other methods are available that may be preferred in particular situations (Table 15.8 in Section 15.10). In particular, if the investigator is convinced that the factor model is valid and that the variables have a multivariate normal distribution, then the **maximum likelihood** (ML) method should be used. If these assumptions hold, then the ML procedure enables the investigator to perform certain tests of hypotheses or compute confidence intervals (Gorsuch, 1983).

In addition, the ML estimates of the factor loadings are invariant (do not change) with changes in scale of the original variables. Note that in Section 14.5, it was mentioned that for principal components analysis there was no easy way to go back and forth between the coefficients obtained from standardized and unstandardized data. The same is true for the principal components method of factor extraction. This invariance is therefore a major advantage of the ML extraction method. Finally, the ML procedure is the one used in confirmatory factor analysis (Long, 1983; Bartholomew and Knott, 1999).

In confirmatory factor analysis, the investigator specifies constraints on the outcome of the factor analysis. For example, zero loadings may be hypothesized for some variables on a specified factor or some of the factors may be assumed to be uncorrelated with other factors. Statistical tests can be performed to determine if the data "confirm" the assumed constraints (Long, 1983).

15.7   Factor rotations

Recall that the main purpose of factor analysis is to derive from the data easily interpretable common factors. The initial factors, however, are often difficult to interpret. For the hypothetical data example we noted that the first factor is essentially an average of all variables. We also noted earlier that the factor interpretations in that example are not clear-cut. This situation is often the case in practice, regardless of the method used to extract the initial factors.

Fortunately, it is possible to find new factors whose loadings are easier to interpret. These new factors, called the **rotated factors**, are selected so that (ideally) some of the loadings are very large (near ± 1) and the remaining loadings are very small (near zero). Conversely, we would ideally wish, for any given variable, that it have a high loading on only one factor. If this is the case, it is easy to give each factor an interpretation arising from the variables with which it is highly correlated (high loadings).

Theoretically, factor rotations can be done in an infinite number of ways. If you are interested in detailed descriptions of these methods, refer to the books referenced in this chapter. In this section we highlight only two typical factor rotation techniques. The most commonly used technique is the **varimax rotation**. This method is the default option in most packaged programs and we present it first. The other technique, discussed later in this section, is called **oblique rotation**; we describe one commonly used oblique rotation, the **direct quartimin rotation**.

Varimax rotation

We have reproduced in Figure 15.3 the principal component factors that were shown in Figure 15.1 so that you can get an intuitive feeling for what factor rotation does. As shown in Figure 15.3, factor rotation consists of finding new axes to represent the factors. These new axes are selected so that they go through clusters or subgroups of the points representing the response variables. The **varimax procedure** further restricts the new axes to being orthogonal (perpendicular) to each other. Figure 15.3 shows the results of the rotation. Note that the new axis representing rotated factor 1 goes through the cluster of variables x_3, x_4, and x_5. The axis for rotated factor 2 is close to the cluster of variables x_1 and x_2 but cannot go through them since it is restricted to being orthogonal to factor 1.

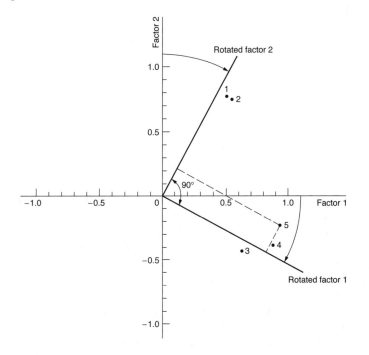

Figure 15.3: *Varimax Orthogonal Rotation for Factor Diagram of Figure 15.1; Rotated Axes Go Through Clusters of Variables and Are Perpendicular to Each Other*

The result of this varimax rotation is that variables x_1 and x_2 have high loadings on rotated factor 2 and nearly zero loadings on rotated factor 1. Similarly, x_3, x_4, and x_5 load heavily on rotated factor 1 but not on rotated factor 2. Figure 15.3 clearly shows that the rotated factors are orthogonal since the angle between the axes representing the rotated factors is $90°$. In statistical terms

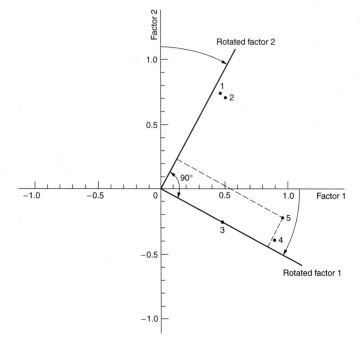

Figure 15.4: *Orthogonal Rotation for Factor Diagram of Figure 15.2*

the orthogonality of the rotated factors is equivalent to the fact that they are uncorrelated with each other.

Computationally, the **varimax rotation** is achieved by maximizing the sum of the variances of the squared factor loadings within each factor. Further, these factor loadings are adjusted by dividing each of them by the communality of the corresponding variable. This adjustment is known as the **Kaiser normalization** (Harman, 1976). It tends to equalize the impact of variables with varying communalities. If it were not used, the variables with higher communalities would highly influence the final solution.

Any method of rotation may be applied to any initially extracted factors. For example, in Figure 15.4 we show the varimax rotation of the factors initially extracted by the iterated principal factor method (Figure 15.2). Note the similarity of the graphs shown in Figures 15.3 and 15.4. This similarity is further reinforced by examination of the two sets of rotated factor loadings shown in Tables 15.5 and 15.6. In this example both methods of initial extraction produced a rotated factor 1 associated mainly with x_3, x_4 and x_5, and a rotated factor 2 associated with x_1 and x_2. The larger factor loadings are bold in Tables 15.5 and 15.6. The points in Figures 15.3 and 15.4 are plots of the rotated loadings (Tables 15.5 and 15.6) with respect to the rotated axes.

In comparing Tables 15.5 and 15.6 with Tables 15.3 and 15.4, we note that the **communalities are unchanged** after the varimax rotation. This result is always the case for any orthogonal rotation. Note also that the percentage of the variance explained by the rotated factor 1 is less than that explained by unrotated factor 1. However, the cumulative percentage of the variance explained by all common factors remains the same after orthogonal rotation. Furthermore, the loadings of any rotated factor depend on how many other factors are selected, regardless of the method of initial extraction.

Table 15.5: *Varimax rotated factors: Principal components extraction*

Variable	Factor loadings F_1	F_2	Communality h_i^2
x_1	0.055	**0.933**	0.873
x_2	0.105	**0.929**	0.875
x_3	**0.763**	−0.062	0.586
x_4	**0.943**	0.095	0.898
x_5	**0.918**	0.266	0.913
Variance explained	2.328	1.817	4.145
Percentage	46.6	36.3	82.9

Table 15.6: *Varimax rotated factors: Iterated principal factors extraction*

Variable	Factor loadings F_1	F_2	Communality h_i^2
x_1	0.063	**0.869**	0.759
x_2	0.112	**0.862**	0.756
x_3	**0.546**	0.003	0.298
x_4	**0.972**	0.070	0.949
x_5	**0.951**	0.251	0.968
Variance explained	2.164	1.566	3.730
Percentage	43.3	31.3	74.6

Oblique rotation

Some factor analysts are willing to relax the restriction of orthogonality of the rotated factors, thus permitting a further degree of flexibility. Nonorthogonal rotations are called **oblique rotations**. The origin of the term **oblique** lies in geometry, whereby two crossing lines are called oblique if they are not perpendicular to each other. Oblique rotated factors are correlated with each other, and in some applications it may be harmless or even desirable to have correlated common factors.

A commonly used oblique rotation is called the **direct quartimin procedure**. The results of applying this method to our hypothetical data example are shown in Figures 15.5 and 15.6. Note that the rotated factors go neatly through the centers of the two variable clusters. The angle between the rotated factors is not $90°$. In fact, the cosine of the angle between the two factor axes is the sample correlation between them. For example, for the rotated principal components the correlation between rotated factor 1 and rotated factor 2 is 0.165, which is the cosine of 80.5.

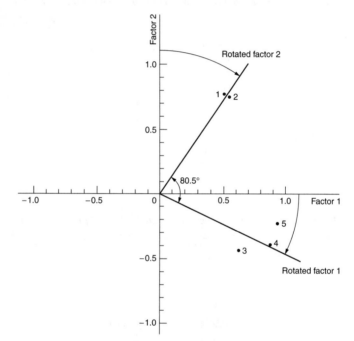

Figure 15.5: *Direct Quartimin Oblique Rotation for Factor Diagram of Figure 15.1; Rotated Axes Go Through Clusters of Variables But Are Not Required to Be Perpendicular to Each Other*

For oblique rotations, the reported pattern and structure matrices are somewhat different. The structure matrix gives the correlations between x_i and F_j. Also, the statements made concerning the proportion of the total variance explained by the communalities being unchanged by rotation do not hold for oblique rotations.

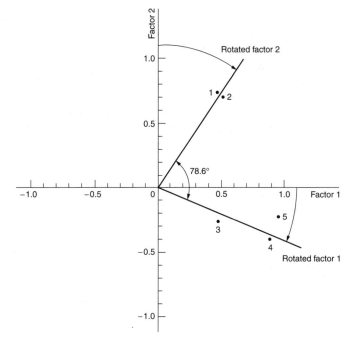

Figure 15.6: *Direct Quartimin Oblique Rotation for Factor Diagram of Figure 15.2*

In general, a factor analysis computer program will print the correlations between rotated factors if an oblique rotation is performed. The rotated oblique factor loadings also depend on how many factors are selected. Finally, we note that packaged programs generally offer a choice of orthogonal and oblique rotations, as discussed in Section 15.10.

15.8 Assigning factor scores

Once the initial extraction of factors and the factor rotations are performed, it may be of interest to obtain the score an individual has for each factor. For example, if two factors, verbal and quantitative abilities, were derived from a set of test scores, it would be desirable to determine equations for computing an individual's scores on these two factors from a set of test scores. Such equations are linear functions of the original variables.

Theoretically, it is conceivable to construct factor score equations in an infinite number of ways (Gorsuch, 1983). But perhaps the simplest way is to add the values of the variables loading heavily on a given factor. For example, for the hypothetical data discussed earlier we would obtain the score of rotated factor 1 as $x_3 + x_4 + x_5$ for a given individual. Similarly, the score of rotated factor 2 in that example would be $x_1 + x_2$. In effect, the factor analysis identifies x_3, x_4, and x_5 as a subgroup of intercorrelated variables. One way of combining the information conveyed by the three variables is simply to add them up. A similar approach is used for x_1 and x_2. In some applications such a simple approach may be sufficient.

More sophisticated approaches do exist for computing factor scores. The so-called **regression procedure** is commonly used to compute factor scores. This method combines the intercorrelations among the x_i variables and the factor loadings to produce quantities called **factor score coefficients**. These coefficients are used in a linear fashion to combine the values of the **standardized** x_i's into factor scores. For example, in the hypothetical data example using principal component factors rotated by the varimax method (Table 15.5), the scores for rotated factor 1 are obtained as

$$\text{factor score } 1 = -0.076x_1 - 0.053x_2 + 0.350x_3 + 0.414x_4 + 0.384x_5$$

The large factor score coefficients for x_3, x_4, and x_5 correspond to the large factor **loadings** shown in Table 15.5. Note also that the coefficients of x_1 and x_2 are close to zero and those of x_3, x_4, and x_5 are each approximately 0.4. Factor score 1 can therefore be approximated by $0.4x_3 + 0.4x_4 + 0.4x_5 = 0.4(x_3 + x_4 + x_5)$, which is proportional to the simple additive factor score given in the previous paragraph.

The factor scores can themselves be used as data for additional analyses. Many statistical packages facilitate this by offering the user the option of storing the factor scores in a file to be used for subsequent analyses. Before using the factor scores in other analyses, the user should check for possible outliers either by univariate methods such as box plots or displaying the factor scores two factors at a time in scatter diagrams. The discussion given in earlier chapters on detection of outliers and checks for normality and independence apply to the factor scores if they are to be used in later analyses.

15.9   Application of factor analysis

In Chapter 14 we presented the results of a principal components analysis for the 20 CESD items in the depression data set. Table 14.2 listed the coefficients for the first five principal components. However, to be consistent with published literature on this subject, we now choose $m = 4$ factors (not 5) and proceed to perform an orthogonal varimax rotation using the principal components as the method of factor extraction. The results for the rotated factors are given in Table 15.7.

In comparing these results with those in Table 14.1, we note that, as expected, rotated factor 1 explains less of the total variance than does the first principal component. The four rotated factors together explain the same proportion of the total variance as do the first four principal components (54.2%). This result occurs because the unrotated factors were the principal components.

In interpreting the factors, we see that factor 1 loads heavily on variables 1–7. These items are known as negative-affect items. Factor 2 loads mainly on items 12–18, which measure somatic and retarded activity. Factor 3 loads on the two interpersonal items, 19 and 20, as well as some positive-affect items (9 and 11). Factor 4 does not represent a clear pattern. Its highest loading is associated with the positive-affect item 8. The factors can thus be loosely identified as negative affect, somatic and retarded activity, interpersonal relations, and positive affect, respectively.

Overall, these rotated factors are much easier to interpret than the original principal components. This example is an illustration of the usefulness of factor analysis in identifying important interrelationships among measured variables. Further results and discussion regarding the application of factor analysis to the CESD scale may be found in Radloff (1977) and Clark *et al.* (1981).

15.10 Discussion of computer programs

Table 15.8 summarizes the computer output that relates to topics discussed in this chapter from the six statistical programs. R and S-PLUS offer a useful choice of options.

SAS FACTOR has a wide selection of methods of extraction and rotation as well as a variety of other options. One choice that is attractive if an investigator is uncertain whether to use an orthogonal or oblique rotation is to specify the promax rotation method. With this option a varimax rotation is obtained first, followed by an oblique rotation, so the two methods can be compared in a single run. SAS also prints residual correlations that are useful in assessing the goodness-of-fit of the factor model. These residual correlations are the differences between the actual correlations among the variables and correlations among the variables estimated from the factor model.

SPSS also has a wide selection of options especially for initial extraction methods. They include a test of the null hypothesis that all the correlations among the variables are zero. If this test is not rejected, then it does not make sense to perform factor analysis as there is no significant correlation to explain.

Stata has somewhat fewer options than the other four programs but it has an additional method of producing factor scores called Bartlett's method that is intended to produce unbiased factors.

In STATISTICA, the rotations labelled normalized will yield the results found in most programs. It also includes the centroid method, which is a multiple-group method (Gorsuch, 1983). STATISTICA also produces residual correlations.

In addition to the statistical packages used for exploratory factor analysis

Table 15.7: *Varimax rotation, principal component factors for standardized CESD: Scale items*

Item		F_1	F_2	F_3	F_4	h_i^2
	Negative affect					
1.	I felt that I could not shake off the blues even with the help of my family or friends.	0.638	0.146	0.268	0.280	0.5784
2.	I felt depressed.	0.773	0.296	0.272	−0.003	0.7598
3.	I felt lonely.	0.726	0.054	0.275	0.052	0.6082
4.	I had crying spells.	0.630	−0.061	0.168	0.430	0.6141
5.	I felt sad.	0.797	0.172	0.160	0.016	0.6907
6.	I felt fearful.	0.624	0.234	−0.018	0.031	0.4448
7.	I thought my life had been a failure.	0.592	0.157	0.359	0.337	0.6173
	Positive affect					
8.	I felt that I was as good as other people.	0.093	−0.051	0.109	0.737	0.5655
9.	I felt hopeful about the future.	0.238	0.033	0.621	0.105	0.4540
10.	I was happy.	0.557	0.253	0.378	0.184	0.5516
11.	I enjoyed life.	0.498	0.147	.407	0.146	0.4569
	Somatic and retarded activity					
12.	I was bothered by things that usually don't bother me.	0.449	0.389	−0.049	−0.065	0.3600
13.	I did not feel like eating; my appetite was poor.	0.070	0.504	−0.173	0.535	0.5760
14.	I felt that everything was an effort.	0.117	0.695	0.180	0.127	0.5459
15.	My sleep was restless.	0.491	0.419	−0.123	−0.089	0.4396
16.	I could not "get going."	0.196	0.672	0.263	−0.070	0.5646
17.	I had trouble keeping my mind on what I was doing.	0.270	0.664	0.192	0.000	0.5508
18.	I talked less than usual.	0.409	0.212	−0.026	0.223	0.2628
	Interpersonal					
19.	People were unfriendly.	−0.015	0.237	0.746	−0.088	0.6202
20.	I felt that people disliked me.	0.358	0.091	0.506	0.429	0.5770
	Variance explained	4.795	2.381	2.111	1.551	10.838
	Percentage	24.0	11.9	10.6	7.8	54.2

Table 15.8: *Software commands and output for factor analysis*

	S-PLUS/R	SAS	SPSS	Stata	STATISTICA
Matrix analyzed					
Correlation	factanal	FACTOR	FACTOR	factor	Factor Analysis
Covariance	factanal	FACTOR	FACTOR	factormat	Factor Analysis
Method of initial extraction					
Principal components	factanal	FACTOR	FACTOR	factor	Factor Analysis
Iterative principal factor	factanal	FACTOR	FACTOR	factor	Factor Analysis
Maximum likelihood	factanal	FACTOR	FACTOR	factor	Factor Analysis
Communality estimates					
Unaltered diagonal	factanal	FACTOR	FACTOR	factor	Factor Analysis
Multiple correlation squared		FACTOR	FACTOR	factor	Factor Analysis
Specify number of factors	factanal	FACTOR	FACTOR	factor	Factor Analysis
Orthogonal rotations					
Varimax	factanal	FACTOR	FACTOR	rotate	Factor Analysis
Equamax	factanal	FACTOR	FACTOR	rotate	Factor Analysis
Quartimax	factanal	FACTOR	FACTOR	rotate	Factor Analysis
Promax	factanal	FACTOR	FACTOR	rotate	
Oblique rotations					
Direct quartimin	factanal	FACTOR		rotate	
Promax	factanal	FACTOR	FACTOR	rotate	
Direct oblimin	factanal	FACTOR	FACTOR	rotate	
Factor scores					
Factor scores for individuals	factanal	FACTOR	FACTOR	predict	Factor Analysis
Factor score coefficients		FACTOR	FACTOR	predict	Factor Analysis
Other output					
Inverse correlation matrix		FACTOR	FACTOR		
Factor structure matrix	factanal	FACTOR	FACTOR	estat structure	Factor Analysis
Plots of factor loadings	biplot	FACTOR	FACTOR	loadingplot	Factor Analysis
Scree plots	princomp	FACTOR	FACTOR	screeplot	Factor Analysis

mentioned above, there are several packages that perform confirmatory factor analysis. The best known are:

1. LISREL, developed by K. Jőreskeg and D. Sőrbom. It is available from: `http://www.ssicentral.com/lisrel/index.html`

2. Mplus, developed by B. Muthén and L. Muthén. It is available from: `http://www.statmodel.com/`

Although more recent versions of the software have since appeared, the review by Waller (1993) presents useful comparisons of confirmatory factor analysis programs.

15.11   What to watch out for

In performing factor analysis, a linear model is assumed where each variable is seen as a linear function of common factors plus a unique factor. The method tends to yield more understandable results when at least moderate correlations exist among the variables. Also, since a simple correlation coefficient measures the full correlation between two variables when a linear relationship exists between them, factor analysis will fit the data better if only linear relationships exist among the variables. Specific points to look out for include:

1. The original sample should reflect the composition of the target population. Outliers should be screened, and linearity among the variables checked. It may be necessary to use transformations to linearize the relationships among the variables.

2. Do not accept the number of factors produced by the default options without checking that it makes sense. The results can change drastically depending on the number of factors used. Note that the choice among different numbers of factors depends mainly on what is most interpretable to the investigator. The cutoff points for the choice of the number of factors do not take into account the variability of the results and can be considered to be rather arbitrary. It is advised that several different numbers of factors be tried, especially if the first run yields results that are unexpected.

3. Ideally, factor analyses should have at least two variables with non-zero weights per factor. If each factor has only a single variable, it can be considered a unique factor and one might as well be analyzing the original variables instead of the factors.

4. If the investigator expects the factors to be correlated (as might be the case in a factor analysis of items in a scale), then oblique factor analysis should be tried.

5. The ML method assumes multivariate normality. It is the method to use if statistical tests are desired.

6. Usually, the results of a factor analysis are evaluated by how much sense they make to the investigator rather than by use of formal statistical tests of hypotheses. Gorsuch (1983) presents comparisons among the various methods of extraction and rotation that can help the reader in evaluating the results.

7. As mentioned earlier, the factor analysis discussed in this chapter is an exploratory technique that can be an aid to analytical thinking. Statisticians have criticized factor analysis because it has sometimes been used by some investigators in place of sound theoretical arguments. However, when carefully used, for example to motivate or corroborate a theory rather than replace it, factor analysis can be a useful tool to be employed in conjunction with other statistical methods.

15.12 Summary

In this chapter we presented the most essential features of exploratory factor analysis, a technique heavily used in social science research. The major steps in factor analysis are factor extraction, factor rotation, and factor score computation. We gave an example using the depression data set that demonstrated this process.

The main value of factor analysis is that it gives the investigator insight into the interrelationships present in the variables and it offers a method for summarizing them. Factor analysis often suggests certain hypotheses to be further examined in future research. On the one hand, the investigator should not hesitate to replace the results of any factor analysis by scientifically based theories derived at a later time. On the other hand, in the absence of such theory, factor analysis is a convenient and useful tool for searching for relationships among variables.

When factor analysis is used as an exploratory technique, there does not exist an optimal way of performing it. This is particularly true of methods of rotation. We advise the investigator to try various combinations of extraction and rotation of main factors (including oblique methods). Also, different choices of the number of factors should be tried. Subjective judgment is then needed to select those results that seem most appealing, on the basis of the investigator's own knowledge of the underlying subject matter.

Since either factor analysis or principal components analysis can be performed on the same data set, one obvious question is which should be used? The answer to that question depends mainly on the purpose of the analysis. Principal components analysis is performed to obtain a small set of uncorrelated linear combinations of the observed variables that account for as much of the total variance as possible. Since the principal components are linear combinations of observed variables, they are sometimes called **manifest variables**. They can be thought of as being obtained by a rotation of the axes.

In factor analysis, each observation is assumed to be a linear combination of the unobservable common factors and the unique factor. Here the user is assuming that underlying factors, such as intelligence or attitude, result in what is observed. The common factors are not observable and are sometimes called **latent factors**. When the iterative principal components method of extraction is performed, the variance of the observations is ignored and off-diagonal covariances (correlations) are used. (This is contrasted with principal components which are computed using the variances and the covariances.) Factor analysis attempts to explain the correlations among the variables with as few factors as possible. In general, the loadings for the factor analysis will be proportional to the principal component coefficients if the communalities are approximately equal. This often occurs when all the communalities are large.

In this chapter and the previous one, we discussed several situations in which factor analysis and principal components analysis can be useful. In some cases, both techniques can be used and are recommended. In other situations, only one technique is applicable. The above discussion, and a periodic scan of the literature, should help guide the reader in this part of statistics which may be considered more "art" than "science."

15.13   Problems

15.1 The CESD scale items (C1–C20) from the depression data set in Chapter 3 were used to obtain the factor loadings listed in Table 15.7. The initial factor solution was obtained from the principal components method, and a varimax rotation was performed. Analyze this same data set by using an oblique rotation such as the direct quartimin procedure. Compare the results.

15.2 Repeat the analysis of Problem 15.1 and Table 15.7, but use an iterated principal factor solution instead of the principal components method. Compare the results.

15.3 Another method of factor extraction, maximum likelihood, was mentioned in Section 15.6 but not discussed in detail. Use one of the packages which offers this option to analyze the data along with an oblique and orthogonal rotation. Compare the results with those in Problems 15.1 and 15.2, and comment on the tests of hypotheses produced by the program, if any.

15.4 Perform a factor analysis on the data in Table 8.1 and explain any insights this factor analysis gives you.

15.5 For the data generated in Problem 7.7, perform four factor analyses, using two different initial extraction methods and both orthogonal and oblique rotations. Interpret the results.

15.6 Separate the depression data set into two subgroups, men and women. Using four factors, repeat the factor analysis in Table 15.7. Compare the results of your two factor analyses to each other and to the results in Table 15.7.

15.7 For the depression data set, perform four factor analyses on the last seven variables DRINK–CHRONILL (Table 3.3 or 3.4). Use two different initial extraction methods and both orthogonal and oblique rotations. Interpret the results. Compare the results to those from Problem 14.1.

15.8 Perform a factor analysis on all of the items of the Parental Bonding scale for the Parental HIV data (see Appendix A and the codebook). Retain two factors. Rotate the factors using an orthogonal rotation. Do the items with the highest loadings for each of the factors correspond to the items of the overprotection and care scale? Interpret the findings.

15.9 Repeat Problem 15.8 using an oblique rotation. Do the substantive conclusions change?

15.10 This problem uses the Northridge earthquake data set. Perform a factor analysis on the following items in the Brief Symptom Inventory (BSI): V346, V357, V364, V383, V390, V394, V351, V358, V385, V386, V391. Note that the first six items relate to the anxiety subscale of BSI and the next five items relate to the hostility subscale of the BSI. Use (at least) these three methods: (a) principal components, (b) iterated principal components and (c) maximum likelihood. Spend a good amount of time interpreting your results. Do the substantive conclusions differ by method? In this analysis, delete missing data and recode all responses = 5 to be = 4 (this recoding makes possible response values 0, 1, 2, 3 or 4 rather than 0, 1, 2, 3 or 5).

Chapter 16

Cluster analysis

16.1   Chapter outline

Cluster analysis is usually done in an attempt to combine cases into groups when the group membership is not known prior to the analysis. Section 16.2 presents examples of the use of cluster analysis. Section 16.3 introduces two data examples to be used in the remainder of the chapter to illustrate cluster analysis. Section 16.4 describes two exploratory graphical methods for clustering and defines several distance measures that can be used to measure how close two cases are to each other. Section 16.5 discusses two widely used methods of cluster analysis: hierarchical (or join) clustering and K-means clustering. In Section 16.6, these two methods are applied to the second data set introduced in Section 16.3. A discussion and listing of the options in six statistical packages are given in Section 16.7. Section 16.8 presents what to watch out for in performing cluster analysis.

16.2   When is cluster analysis used?

Cluster analysis is a technique for grouping individuals or objects into *unknown* groups. It differs from other methods of classification, such as discriminant analysis, in that in cluster analysis the number and characteristics of the groups are to be derived from the data and are not usually known prior to the analysis.

In biology, cluster analysis has been used for decades in the area of taxonomy. In taxonomy, living things are classified into arbitrary groups on the basis of their characteristics. The classification proceeds from the most general to the most specific, in steps. For example, classifications for domestic roses and for modern humans are illustrated in Figure 16.1 (Wilson *et al.*, 1973). The most general classification is kingdom, followed by phylum, subphylum, etc.

Cluster analysis has been used in medicine to assign patients to specific diagnostic categories on the basis of their presenting symptoms and signs.

In particular, cluster analysis has been used in classifying types of depression (e.g., Andreasen and Grove, 1982 or Das-Munshi *et al.*, 2008). It has also

A.   Modern Humans

KINGDOM: Animalia (animals)
PHYLUM: Chordata (chordates)
SUBPHYLUM: Vertebrata (vertebrates)
CLASS: Mammalia (mammals)
ORDER: Primates (primates)
FAMILY: Hominidae (humans and close relatives)
GENUS: Homo (modern humans and precursors)
SPECIES: sapiens (modern humans)

B.   Domestic Rose

KINGDOM: Plantae (plants)
PHYLUM: Tracheophyta (vascular plants)
SUBPHYLUM: Pteropsida (ferns and seed plants)
CLASS: Dicotyledoneae (dicots)
ORDER: Rosales (saxifrages, psittosporums, sweet gum,
                                     plane trees, roses, and relatives)
FAMILY: Rosaceae (cherry, plum, hawthorn, roses,
                                     and relatives)
GENUS: Rosa (roses)
SPECIES: galliea (domestic roses)

Figure 16.1: *Example of Taxonomic Classification*

been used in anthropology to classify stone tools, shards, or fossil remains by the civilization that produced them. Cluster analysis is an important tool for investigators interested in data mining. For example, consumers can be clustered on the basis of their choice of purchases in marketing research. Here the emphasis may be on methods that can be used for large data sets. In short, it is possible to find applications of cluster analysis in virtually any field of research.

We point out that cluster analysis is highly empirical. Different methods can lead to different groupings, both in number and in content. Furthermore, since the groups are not known *a priori,* it is usually difficult to judge whether the results make sense in the context of the problem being studied.

It is also possible to cluster the variables rather than the cases. Clustering of variables is sometimes used in analyzing the items in a scale to determine which items tend to be close together in terms of the individual's response to them. Clustering variables can be considered an alternative to factor analysis, although the output is quite different. Some statistical packages allow the user to either cluster cases or variables in the same program while others have separate programs for clustering cases and variables (Section 16.7). In this chapter, we discuss in detail only clustering of cases.

16.3 Data example

A hypothetical data set was created to illustrate several of the concepts discussed in this chapter. Figure 16.2 shows a plot of five observations for the two variables X_1 and X_2. This small data set will simplify the presentation since the analysis can be performed by hand.

Another data set we will use includes financial performance data from the January 1981 issue of *Forbes*. The variables used are those defined in Section 8.3. Table 16.1 shows the data for 25 companies from three industries: chemical companies (the first 14 of the 31 discussed in Section 8.3), health care companies, and supermarket companies. The column labelled "Type" in Table 16.1 lists the abbreviations Chem, Heal, and Groc for these three industries. In Section 16.6 we will use two clustering techniques to group these companies and then check the agreement with their industrial type. These three industries were selected because they represent different stages of growth, different product lines, different management philosophies, different labor and capital requirements, etc. Among the chemical companies all of the large diversified firms were selected. From the major supermarket chains, the top six rated for return on equity were included. In the health care industry four of the five companies included were those connected with hospital management; the remaining company involves hospital supplies and equipment.

16.4 Basic concepts: initial analysis

In this section we present some preliminary graphical techniques for clustering. Then we discuss distance measures that will be useful in later sections.

Scatter diagrams

Prior to using any of the analytical clustering procedures (Section 16.5), most investigators begin with simple graphical displays of their data. In the case of two variables a scatter diagram can be very helpful in displaying some of the main characteristics of the underlying clusters. In the hypothetical data example shown in Figure 16.2, the points closest to each other are points 1 and 2. This observation may lead us to consider these two points as one cluster. Another cluster might contain points 3 and 4, with point 5 perhaps constituting a third cluster. On the other hand, some investigators may consider points 3, 4, and 5 as the second cluster. This example illustrates the indeterminacy of cluster analysis, since even the number of clusters is usually unknown. Note that the concept of closeness was implicitly used in defining the clusters. Later in this section we expand on this concept by presenting several definitions of distance.

If the number of variables is small, it is possible to examine pairwise scatter diagrams using each variable and search for possible clusters. But this tech-

Table 16.1: *Financial performance data for chemical, health, and supermarket companies (Source: Forbes, vol. 127, no. 1 (January 5, 1981))*

Type	Symbol	Num	ROR5	D/E	SALESGR5	EPS5	NPM1	P/E	PAYOUTR1
Chem	dia	1	13.0	0.7	20.2	15.5	7.2	9	0.426398
Chem	dow	2	13.0	0.7	17.2	12.7	7.3	8	0.380693
Chem	stf	3	13.0	0.4	14.5	15.1	7.9	8	0.406780
Chem	dd	4	12.2	0.2	12.9	11.1	5.4	9	0.568182
Chem	uk	5	10.0	0.4	13.6	8.0	6.7	5	0.324544
Chem	psm	6	9.8	0.5	12.1	14.5	3.8	6	0.508083
Chem	gra	7	9.9	0.5	10.2	7.0	4.8	10	0.378913
Chem	hpc	8	10.3	0.3	11.4	8.7	4.5	9	0.481928
Chem	mtc	9	9.5	0.4	13.5	5.9	3.5	11	0.573248
Chem	acy	10	9.9	0.4	12.1	4.2	4.6	9	0.490798
Chem	cz	11	7.9	0.4	10.8	16.0	3.4	7	0.489130
Chem	ald	12	7.3	0.6	15.4	4.9	5.1	7	0.272277
Chem	rom	13	7.8	0.4	11.0	3.0	5.6	7	0.315646
Chem	rei	14	6.5	0.4	18.7	-3.1	1.3	10	0.384000
Heal	hum	15	9.2	2.7	39.8	34.4	5.8	21	0.390879
Heal	hca	16	8.9	0.9	27.8	23.5	6.7	22	0.161290
Heal	nme	17	8.4	1.2	38.7	24.6	4.9	19	0.303030
Heal	ami	18	9.0	1.1	22.1	21.9	6.0	19	0.303318
Heal	ahs	19	12.9	0.3	16.0	16.2	5.7	14	0.287500
Groc	lks	20	15.2	0.7	15.3	11.6	1.5	8	0.598930
Groc	win	21	18.4	0.2	15.0	11.6	1.6	9	0.578313
Groc	sgl	22	9.9	1.6	9.6	24.3	1.0	6	0.194946
Groc	slc	23	9.9	1.1	17.9	15.3	1.6	8	0.321070
Groc	kr	24	10.2	0.5	12.6	18.0	0.9	6	0.453731
Groc	sa	25	9.2	1.0	11.6	4.5	0.8	7	0.594966
Means			10.4	0.7	16.8	13.2	4.3	10	0.408
SD			2.6	0.5	7.9	8.4	2.2	5	0.124

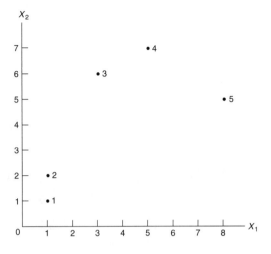

Figure 16.2: *Plot of Hypothetical Cluster Data Points*

nique may become unwieldy if the number of variables exceeds four and the number of points is large.

Profile diagram

A helpful technique for a moderate number of variables is a **profile diagram**. To plot a profile of an individual case in the sample, the investigator customarily first standardizes the data by subtracting the mean and dividing by the standard deviation for each variable. However, this step is omitted by some researchers, especially if the units of measurement of the variables are comparable. In the financial data example the units are not the same, so standardization seems helpful. The standardized financial data for the 25 companies are shown in Table 16.2.

A profile diagram, as shown in Figure 16.3, lists the variables along the horizontal axis and the standardized value scale along the vertical axis. Each point on the graph indicates the value of the corresponding variable. The profile for the first company in the sample has been plotted in the figure. The points are connected in order to facilitate the visual interpretation. We see that this company hovers around the mean, being at most 1.3 standard deviations away on any variable.

A preliminary clustering procedure is to graph the profiles of all cases on the same diagram. To illustrate this procedure, we plotted the profiles of seven companies (15–21) in Figure 16.4 using Stata. To avoid unnecessary clutter, we reversed the sign of the values of ROR5 and PAYOUTR1. Using the original data on these two variables would have caused the lines connecting the

Table 16.2: Standardized financial performance data for diversified chemical, health, and supermarket companies (Source: *Forbes, vol. 127, no. 1 (January 5, 1981)*)

Type	Symbol	Num	ROR5	D/E	SALESGR5	EPS5	NPM1	P/E	PAYOUTR1
Chem	dia	1	0.963	-0.007	0.431	0.277	1.289	-0.237	0.151
Chem	dow	2	0.963	-0.007	0.052	-0.057	1.334	-0.442	-0.193
Chem	stf	3	0.963	-0.559	-0.290	0.230	1.601	-0.442	-0.007
Chem	dd	4	0.661	-0.927	-0.492	-0.248	0.488	-0.237	1.291
Chem	uk	5	-0.171	-0.559	-0.403	-0.618	1.067	-1.056	-0.668
Chem	psm	6	-0.246	-0.375	-0.593	0.158	-0.224	-0.851	0.807
Chem	gra	7	-0.209	-0.375	-0.833	-0.737	-0.221	-0.033	-0.231
Chem	hpc	8	-0.057	-0.743	-0.681	-0.534	0.087	-0.237	0.597
Chem	mtc	9	-0.360	-0.559	-0.416	-0.869	-0.358	0.172	1.331
Chem	acy	10	-0.209	-0.559	-0.593	-1.072	0.132	-0.237	0.668
Chem	cz	11	-0.964	-0.589	-0.757	0.337	-0.402	-0.647	0.655
Chem	ald	12	-1.191	-0.191	-0.176	-0.988	0.354	-0.647	-1.089
Chem	rom	13	-1.002	-0.559	-0.732	-1.215	0.577	-0.647	-0.740
Chem	rci	14	-1.494	-0.559	0.241	-1.943	-1.337	-0.033	-0.190
Heal	hum	15	-0.473	3.672	2.908	2.534	0.666	2.218	-0.135
Heal	hca	16	-0.587	0.361	1.366	1.233	1.067	2.422	-1.981
Heal	nme	17	-0.775	0.913	2.769	1.364	0.265	1.809	-0.841
Heal	ami	18	-0.549	0.729	0.671	1.042	0.755	1.809	-0.839
Heal	ahs	19	0.925	-0.743	-0.100	0.361	0.621	0.786	0.966
Groc	lks	20	1.794	-0.007	-0.189	-0.188	-1.248	-0.442	1.538
Groc	win	21	3.004	-0.927	-0.226	-0.188	-1.204	-0.237	1.372
Groc	sgl	22	-0.209	1.649	-0.909	1.328	-1.471	-0.851	-1.710
Groc	slc	23	-0.209	0.729	0.140	0.254	-1.204	-0.442	-0.696
Groc	kr	24	-0.095	-0.375	-0.530	0.576	-1.515	-0.851	0.370
Groc	sa	25	-0.473	0.545	-0.656	-1.036	-1.560	-0.647	1.506

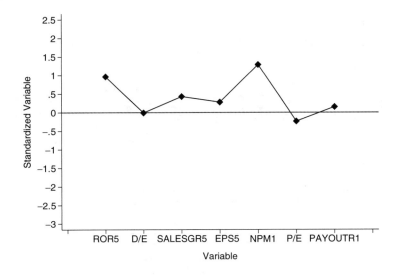

Figure 16.3: *Profile Diagram of a Chemical Company (dia) Using Standardized Financial Performance Data*

points to cross each other excessively, thus making it difficult to identify single company profiles.

Examining Figure 16.4, we note the following.

1. Companies 20 and 21 are very similar.

2. Companies 16 and 18 are similar.

3. Companies 15 and 17 are similar.

4. Company 19 stands alone.

Thus it is possible to identify the above four clusters. It is also conceivable to identify three clusters: (15, 16, 17, 18), (20, 21), and (19). These clusters are consistent with the types of the companies, especially noting that company 19 deals with hospital supplies.

Although this technique's effectiveness is not affected by the number of variables, it fails when the number of observations is large. In Figure 16.4, the impression is clear, because we plotted only seven companies. Plotting all 25 companies would have produced too cluttered a picture.

Distance measures

For a large data set analytical methods such as those described in the next section are necessary. All of these methods require defining some measure of

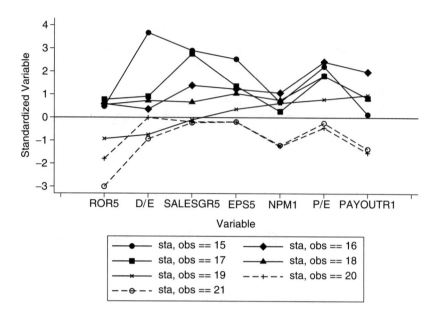

Figure 16.4: *Profile Plot of Health and Supermarket Companies with Standardized Financial Performance Data*

closeness or **similarity** of two observations (see Everitt *et al.*, 2001 or Gan *et al.*, 2007). The converse of similarity is **distance**. Before defining distance measures, though, we warn the investigator that many of the analytical techniques are particularly sensitive to outliers. Some preliminary checking for outliers and blunders is therefore advisable. This check may be facilitated by the graphical methods just described.

The most commonly used distance measurement is the **Euclidian distance**. In two dimensions, suppose that the two points have coordinates (X_{11}, X_{21}) and (X_{12}, X_{22}), respectively. Then the Euclidian distance between the two points is defined as

$$\text{distance} = [(X_{11} - X_{12})^2 + (X_{21} - X_{22})^2]^{1/2}$$

For example, the distance between points 4 and 5 in Figure 16.2 is

$$\text{distance between points } 4,5 = [(5-7)^2 + (8-5)^2]^{1/2} = 13^{1/2} = 3.61$$

For P variables the Euclidian distance is the square root of the sum of the squared differences between the coordinates of each variable for the two observations. Unfortunately, the Euclidian distance is not invariant to changes in scale and the results can change appreciably by simply changing the units of measurement.

In computer program output the distances between all possible pairs of points are usually summarized in the form of a matrix. For example, for our hypothetical data, the Euclidian distances between the five points are as given in Table 16.3. Since the distance between the point and itself is zero, the diagonal elements of this matrix are always zero. Also, since the distances are symmetric, many programs print only the distances above or below the diagonal.

Table 16.3: *Euclidian distance between five hypothetical points*

	1	2	3	4	5
1	0	1.00	5.39	7.21	8.06
2	1.00	0	4.47	6.40	7.62
3	5.39	4.47	0	2.24	5.10
4	7.21	6.40	2.24	0	3.61
5	8.06	7.62	5.10	3.61	0

Since the square root operation does not change the order of how close the points are to each other, some programs use the sum of the **squared differences** instead of the Euclidian distance (i.e., they don't take the square root). Another option available in some programs is to replace the squared differences by another **power** of the absolute differences. For example, if the power of 1 is chosen, the distance is the sum of the absolute differences of the coordinates. The distance is the so-called **city block distance**. In two dimensions it is the distance you must walk to get from one point to another in a city divided into rectangular blocks.

In most situations different distance measures yield different distance matrices, in turn leading to different clusters. When variables have different units, it is advisable to standardize the data before computing distances. Further, when high positive or negative correlations exist among the variables, it may be helpful to consider techniques of standardization that take the covariances as well as the variances into account. An example is given in the SAS ACECLUS write-up using real data where actual cluster membership is known. It is therefore possible to say which observations are misclassified, and the results are striking.

Several rather different types of "distance" measures are available, namely, measures relating to the correlation between variables (if you are clustering variables) or between cases (if you are clustering cases). One measure is $1 - r_{ij}$ where r_{ij} is the Pearson correlation between the ith variable (case) and the jth variable (case). Two variables (cases) that are highly correlated thus are assigned a small distance between them. Other measures of association, not discussed in this book, are available for clustering nominal and interval data (see Kaufman and Rousseeuw, 2005).

In the next section we discuss two of the more commonly used analytical cluster techniques. These techniques make use of the distance functions defined.

16.5   Analytical clustering techniques

The commonly used methods of clustering fall into two general categories: hierarchical and nonhierarchical. First, we discuss the hierarchical techniques.

Hierarchical clustering

Hierarchical methods can be either agglomerative or divisive. In the **agglomerative methods** we begin with N clusters, i.e., each observation constitutes its own cluster. In successive steps we combine the two closest clusters, thus reducing the number of clusters by one in each step. In the final step all observations are grouped into one cluster. In some statistical packages the agglomerative method is called **Join**. In **divisive methods** we begin with one cluster containing all of the observations. In successive steps we split off the cases that are most dissimilar to the remaining ones. Most of the commonly used programs are of the agglomerative type, and we therefore do not discuss divisive methods further.

The centroid procedure is a widely used example of agglomerative methods. In the centroid method the distance between two clusters is defined as the distance between the group centroids (the centroid is the point whose coordinates are the means of all the observations in the cluster). If a cluster has one observation, then the centroid is the observation itself. The process proceeds by combining groups according to the distance between their centroids, the groups with the shortest distance being combined first.

The centroid method is illustrated in Figure 16.5 for our hypothetical data. Initially, the closest two centroids (points) of the five hypothetical observations plotted in Figure 16.2 are points 1 and 2, so they are combined first and their centroid is obtained in step 1. In step 2, centroids (points) 3 and 4 are combined (and their centroid is obtained), since they are the closest now that points 1 and 2 have been replaced by their centroid. At step 3 the centroid of points 3 and 4 and centroid (point) 5 are combined, and the centroid is obtained. Finally, at the last step the centroid of points 1 and 2 and the centroid of points 3, 4, and 5 are combined to form a single group.

Figure 16.6 illustrates the clustering steps based on the standardized hypothetical data. The results are identical to the previous ones, although this is not the case in general.

We could also have used the city-block distance. This distance is available

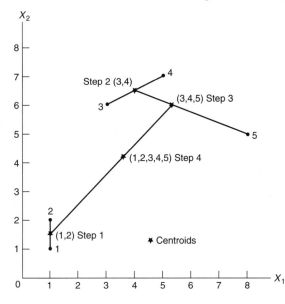

Figure 16.5: *Hierarchical Cluster Analysis Using Unstandardized Hypothetical Data Set*

by that name in many programs. As noted earlier in the discussion of distance measures, this distance can also be obtained from "power" measures.

Several other methods can be used to define the distance between two clusters. These are grouped in the computer programs under the heading of **linkage** methods. In these programs, the linkage distance is the distance between two clusters defined according to one of these methods.

With two variables and a large number of data points, the representation of the steps in a graph similar to Figure 16.5 can get too cluttered to interpret. Also, if the number of variables is more than two, such a graph is not feasible. A clever device called the **dendrogram** or **tree graph** has therefore been incorporated into packaged computer programs to summarize the clustering at successive steps. The dendrogram for the hypothetical data set is illustrated in Figure 16.7. The horizontal axis lists the observations in a particular order. In this example the natural order is convenient. The vertical axis shows the successive steps or the distance between the centers of the clusters. At step 1 points 1 and 2 are combined. Similarly, points 3 and 4 are combined in step 2; point 5 is combined with cluster (3, 4) in step 3; and finally clusters (1, 2) and (3, 4, 5) are combined. In each step two clusters are combined.

A tree graph or dendrogram can be obtained from all six programs. It can be obtained from SAS by calling the TREE procedure, using the output from

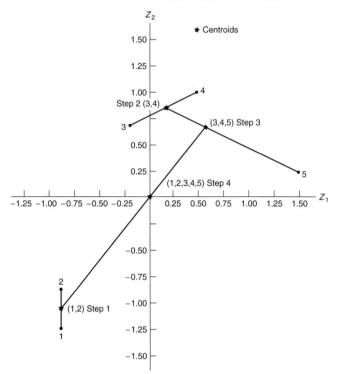

Figure 16.6: *Hierarchical Cluster Analysis Using Standardized Hypothetical Data Set*

the CLUSTER program. In SAS TREE and in R and S-PLUS the number of clusters is printed on the vertical axis instead of the step number shown in Figure 16.7.

SPSS and STATISTICA provide vertical tree plots called icicles. Vertical icicle plots look like a row of icicles hanging from the edge of a roof. The first cases to be joined at the bottom of the graph are listed side by side and will form a thicker icicle. Moving upward, icicles are joined to form successively thicker ones until in the last step all the cases are joined together at the edge of the roof.

The programs offer additional options for combining clusters. The manuals or online Help statements should be read for descriptions of these various options.

Hierarchical procedures are appealing in a taxonomic application. Such procedures can be misleading, however, in certain situations. For example, an undesirable early combination might persist throughout the analysis and may

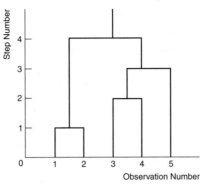

Figure 16.7: *Dendrogram for Hierarchical Cluster Analysis of Hypothetical Data Set*

lead to artificial results. The investigator may wish to perform the analysis several times after deleting certain suspect observations.

For large sample sizes the printed dendrograms become very large and unwieldy to read. One statistician noted that they were more like wallpaper than comprehensible results for large N.

An important problem is how to select the number of clusters. No standard objective procedure exists for making the selection. The distances between clusters at successive steps may serve as a guide. The investigator can stop when the distance exceeds a specified value or when the successive differences in distances between steps make a sudden jump. Also, the underlying situation may suggest a natural number of clusters. If such a number is known, a particularly appropriate technique is the K-means clustering technique.

K-means clustering

K-means clustering is a popular nonhierarchical clustering technique. For a specified number of clusters K the basic algorithm proceeds in the following steps.

1. Divide the data into K initial clusters. The members of these clusters may be specified by the user or may be selected by the program, according to an arbitrary procedure.

2. Calculate the means or **centroids** of the K clusters.

3. For a given case, calculate its distance to each centroid. If the case is closest to the centroid of its own cluster, leave it in that cluster; otherwise, reassign it to the cluster whose centroid is closest to it.

4. Repeat step 3 for each case.

5. Repeat steps 2, 3, and 4 until no cases are reassigned.

Individual programs implement the basic algorithm in different ways. One algorithm is illustrated in Figure 16.8. The first step considers all of the data as one cluster. For the hypothetical data set this step is illustrated in plot (a) of Figure 16.8. The algorithm then searches for the variable with the highest variance, in this case X_1. The original cluster is now split into two clusters, using the midrange of X_1 as the dividing point, as shown in plot (b) of Figure 16.8. If the data are standardized, then each variable has a variance of one. In that case the variable with the smallest range is selected to make the split. The algorithm, in general, proceeds in this manner by further splitting the clusters until the specified number K is achieved. That is, it successively finds that particular variable and the cluster producing the largest variance and splits that cluster accordingly, until K clusters are obtained. At this stage, step 1 of the basic algorithm is completed and it proceeds with the other steps.

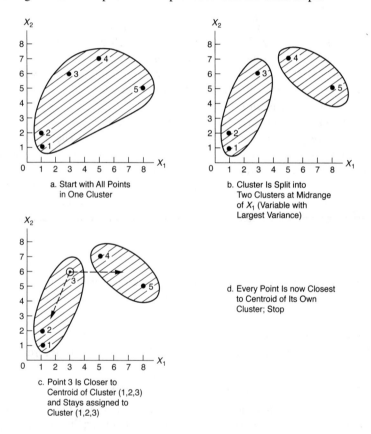

Figure 16.8: *Successive Steps in K-Means Cluster Analysis for $K = 2$, K-Means Clustering with Hypothetical Data Set*

For the hypothetical data example with $K = 2$, it is seen that every case already belongs to the cluster whose centroid is closest to it in plot (c) of Figure 16.8. For example, point 3 is closer to the centroid of cluster $(1, 2, 3)$ than it is to the centroid of cluster $(4, 5)$. Therefore it is *not* reassigned. Similar results hold for the other cases. Thus the algorithm stops, with the two clusters being selected being cluster $(1, 2, 3)$ and cluster $(4, 5)$.

The SAS procedure FASTCLUS is recommended especially for large data sets. The user specifies the maximum number of clusters allowed, and the program starts by first selecting cluster "seeds," which are used as initial guesses of the means of the clusters. The first observation with no missing values in the data set is selected as the first seed. The next complete observation that is separated from the first seed by an optional, specified minimum distance becomes the second seed (the default minimum distance is zero). The program then proceeds to assign each observation to the cluster with the nearest seed. The user can decide whether to update the cluster seeds by cluster means each time an observation is assigned, using the DRIFT option, or only after all observations are assigned. Limiting seed replacement results in the program using less computer time. Note that the initial, and possibly the final, results depend on the order of the observations in the data set. Specifying the initial cluster seeds can lessen this dependence.

The SPSS QUICK CLUSTER program can cluster a large number of cases quickly and with minimal computer resources. Different initial procedures are employed depending upon whether the user knows the cluster center or not. If the user does not specify the number of clusters, the program assumes that two clusters are desired.

In R and S-PLUS, the kmeans option is used. S-PLUS allows the user to specify starting centroids of their own choosing or use the ones that the program provides. It includes other options for centering. Stata also performs K-means clustering but with somewhat fewer options.

As mentioned earlier, the selection of the number of clusters is a troublesome problem. If no natural choice of K is available, it is useful to try various values and compare the results. One aid to this examination is a plot of the points in the clusters, labelled by the cluster they belong to. If the number of variables is more than two, some two-dimensional plots are available from certain programs, such as SAS CLUSTER, ACECLUS, or FASTCLUS. Another possibility is to plot the first two principal components, labelled by cluster membership. A second aid is to examine the F ratio for testing the hypothesis that the cluster means are equal. An alternative possibility is to supply the data with each cluster as a group as input to a K group discriminant analysis program. Such a program would then produce a test statistic for the equality of means of all variables simultaneously, as was discussed in Chapter 11. Then for each variable, separately or as a group, comparison of the P values for various values of K may be a helpful indication of the value of K to be selected. Note, however, that these P values are valid only for comparative purposes and not as significance levels in a hypothesis-testing sense, even if the normality assumption is justified. The individual F statistics for each variable indicate the relative importance of the variables in determining cluster membership (again, they are not valid for hypothesis testing).

Since cluster analysis is an empirical technique, it may be advisable to try several approaches in a given situation. In addition to the hierarchical and K-means approaches discussed above, several other methods are available in the literature cited in this chapter. It should be noted that the K-means approach is gaining acceptability in the literature over the hierarchical approach. In any case, unless the underlying clusters are clearly separated, different methods can produce widely different results. Even with the same program, different options can produce quite different results.

It may also be appropriate to perform other analyses prior to doing a cluster analysis. For example, a principal components analysis may be performed on the entire data set and a small number of independent components selected. A cluster analysis can be run on the values of the selected components. Similarly, factor analysis could be performed first, followed by cluster analysis of factor scores. This procedure has the advantage of reducing the original number of variables and making the interpretations possibly easier, especially if meaningful factors have been selected.

16.6   Cluster analysis for financial data set

In this section we apply some of the standard procedures to the financial performance data set shown in Table 16.1. In all our runs the data are first standardized as shown in Table 16.2. Recall that in cluster analysis the total sample is considered as a single sample. Thus the information on type of company is not used to derive the clusters. However, this information will be used to interpret the results of the various analyses.

Hierarchical clustering

The dendrogram or tree is shown in Figure 16.9. Default options including the centroid method with Euclidian distance were used with the standardized data. The horizontal axis lists the observation numbers in a particular order, which prevents the lines in the dendrogram from crossing each other. One result of this arrangement is that certain subgroups appearing near each other on the horizontal axis constitute clusters at various steps. Note that the distance is shown on the right vertical axis. These distances are measured being the **centers** of the two clusters just joined. On the left vertical axis the number of clusters is listed.

In Figure 16.9, companies 1, 2, and 3 form a single cluster, with the grouping being completed when there are 22 clusters. Similarly, at the opposite end 15, 16, 18, and 17 (all health care companies) form a single cluster at the step in which there are two clusters. Company 22 stays by itself until there are only four clusters.

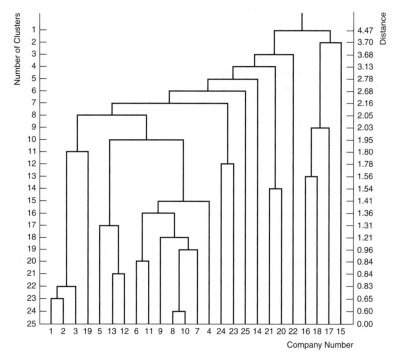

Figure 16.9: *Dendrogram of Standardized Financial Performance Data Set*

The distance axis indicates how disparate the two clusters just joined are. The distances are progressively increasing. As a general rule, large increases in the sequence should be a signal for examining those particular steps. A large distance indicates that at least two very different observations exist, one from each of the two clusters just combined. Note, for example, the large increase in distance occurring when we combine the last two clusters into a single cluster.

It is instructive to look at the industry groups and ask why the cluster analyses, as in Figure 16.9, for example, did not completely differentiate them. It is clear that the clustering is quite effective. First, 13 of the chemical companies, all except no. 14 (rci), are clustered together with only one nonchemical firm, company 19 (ahs), when the number of clusters is eight. This result is impressive when one considers that these are large diversified companies with varied emphasis, ranging from industrial chemicals to textiles to oil and gas production.

At the level of nine clusters three of the four hospital management firms, companies 16, 17, and 18 (hca, nme, and ami), have also been clustered together, and the other, company 15 (hum), is added to that cluster before it is aggregated with any nonhospital management firms. A look at the data in Table 16.1 and Table 16.2 shows that company 15 (hum) has a clearly different D/E value from the others, suggesting a more highly leveraged operation and probably a different management style. The misfit in this health group is company 19 (ahs), clustered with the chemical firms instead of with the hospital management firms. Further examination shows that, in fact, company 19 is a large, established supplies and equipment firm, probably more akin to drug firms than to the fast-growing hospital management business, and so it could be expected to share some financial characteristics with the chemical firms.

The grocery firms do not cluster tightly. In scanning the data of Table 16.1 and Table 16.2 we note that they vary substantially on most variables. In particular, company 22 (sgl) is highly leveraged (high D/E), and company 21 (win) has low leverage (low D/E) relative to the others. Further, if we examine other characteristics in this data set, important disparities show up. Three of the six, companies 21, 24, and 25 (win, ka, sa), are three of the four largest United States grocery supermarket chains. Two others, companies 20 and 22 (kls, sgl), have a diversified mix of grocery, drug, department, and other stores. The remaining firm, company 24 (slc) concentrates on convenience stores (7-Eleven) and has a substantial franchising operation. Thus, the six, while all grocery related, are quite different from each other.

Various K-means clusters

Since these companies do not present a natural application of hierarchical clustering such as taxonomy, the K-means procedure may be more appropriate. The natural variable K is three since there are three types of companies. The middle three columns of Table 16.4 show the results of one run of the SAS FASTCLUS procedure, a run of S-PLUS, and a run of Stata using the default options.

The numbers in each column indicate the cluster to which each company is assigned. One method of summarizing the results from these three programs is given in the last column. This column shows which companies were clustered in a similar fashion and which ones were not. For example, all three runs agreed on assigning companies 1–4 to cluster 1. Similarly, every company assigned unanimously to a cluster is so indicated. The fifth chemical company was assigned to cluster 1 by S-PLUS and Stata but to cluster 3 by SAS FAST-CLUS. The grocery companies appear to be assigned almost equally to 1 or 3. The first four health companies were uniformly assigned to 2 but the last one was assigned to 1 by all three programs. The exception is company 19, a company that deals with hospital supplies. This exception was evident in the hierarchical results as well as in K-means, where company 19 joined the chem-

Table 16.4: *Companies clustered together from K-means standardized analysis of the financial performance data set*

Type of Company	FASTCLUS $K = 3$	S-PLUS $K = 3$	Stata $K = 3$	Summary of three runs
1 Chem	1	1	1	1
2 Chem	1	1	1	1
3 Chem	1	1	1	1
4 Chem	1	1	1	1
5 Chem	3	1	1	1,3
6 Chem	1	1	1	1
7 Chem	3	1	1	1,3
8 Chem	1	1	1	1
9 Chem	1	1	1	1
10 Chem	1	1	1	1
11 Chem	3	1	3	1,3
12 Chem	3	1	1	1,3
13 Chem	3	1	1	1,3
14 Chem	3	1	3	1,3
15 Heal	2	2	2	2
16 Heal	2	2	2	2
17 Heal	2	2	2	2
18 Heal	2	2	2	2
19 Heal	1	1	1	1
20 Groc	1	3	1	1,3
21 Groc	1	3	1	1,3
22 Groc	3	1	3	1,3
23 Groc	3	1	3	1,3
24 Groc	3	1	3	1,3
25 Groc	3	1	3	1,3

ical companies and not the other health companies. With three clusters, twelve companies could not be assigned to the same cluster by the three programs. When this occurs one option is to try a different number of clusters.

Profile plots of means

Graphical output is very useful in assessing the results of cluster analysis. As with most exploratory techniques, it is the graphs that often provide the best understanding of the output. Figure 16.10 shows the cluster profiles based on the means of the standardized variables but here for the sake of illustration we use $K = 4$. Note that the values of ROR5 and PAYOUTR1 are plotted with the signs reversed in order to keep the lines from crossing too much. It is immediately apparent that cluster 2 is very different from the remaining three clusters. Cluster 1 is also quite different from clusters 3 and 4. The latter two clusters are most different on EPS5. Thus the companies forming cluster 2 seem to clearly average higher SALESGR5 and P/E. This cluster mean profile is a particularly appropriate display of the K-means output, because the objective of the analysis is to make these means as widely separated as possible and because there is usually a small number of clusters to be plotted.

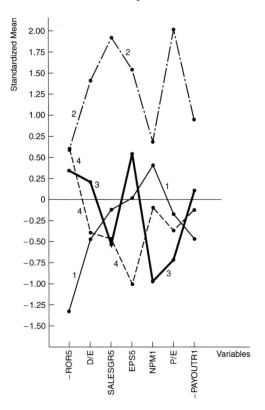

Figure 16.10: *Profile of Cluster Means for Four Clusters (Financial Performance Data Set)*

Use of F tests

For quantification of the relative importance of each variable the univariate F ratio for testing the equality of each variable's means in the K clusters is given in the output of SPSS QUICK CLUSTER, SAS FASTCLUS, and STA-

TISTICA K-Means. A large F value is an indication that the corresponding variable is useful in separating the clusters. For example, the largest F value is for P/E, indicating that the clusters are most different from each other in terms of this variable. Conversely, the F ratio for PAYOUTR1 is very small, indicating that this variable does not play an important role in defining the clusters for the particular industry groups and companies analyzed. Note that these F ratios are used for comparing the variables only, not for hypothesis-testing purposes.

The above example illustrates how these techniques are used in an exploratory manner. The results serve to point out some possible natural groupings as a first step in generating scientific hypotheses for the purpose of further analyses. A great deal of judgment is required for an interpretation of the results since the outputs of different runs may be different and sometimes even contradictory. Indeed, this result points out the desirability of making several runs on the same data in an attempt to reach some consensus.

16.7 Discussion of computer programs

R, S-PLUS, SAS, SPSS, Stata, and STATISTICA all have cluster analysis programs. Since cluster analysis is largely an empirical technique, the options for performing it vary from program to program. It is important to read the manuals or Help statements to determine what the program you use actually does. We have discussed only a small portion of the total options available in the various programs. More complete discussion of the various methods is given in the cluster analysis texts listed in the References. One of the difficulties with cluster analysis is that there is such a wide range of available methods that the choice among them is sometimes difficult. We summarize the options available for hierarchical cluster analysis in Table 16.5. The options that are offered for K-means clustering are listed in Table 16.6. All six software packages will perform hierarchical and K-means clustering but as can be seen by the results in Table 16.4, the results will not necessarily be the same.

In S-PLUS, mclust (called Mclust in R) performs hierarchical clustering and, in both S-PLUS and R, kmeans performs K-means clustering, each with a number of options.

In SAS, CLUSTER is used to obtain hierarchical cluster analysis. This procedure offers a wide variety of options. FASTCLUS is used to perform K-means clustering. VARCLUS clusters the variables instead of the cases and TREE produces dendrograms for the CLUSTER procedure.

SPSS offers two programs: CLUSTER and QUICK CLUSTER. CLUSTER performs hierarchical cluster analysis and QUICK CLUSTER does K-means clustering. CLUSTER can also be used to cluster variables. The SPSS programs are attractive partially because of the rich choice of options and partially because of the readability of the manual.

STATISTICA offers a wide choice of options particularly in its hierarchical

Table 16.5: *Software commands and output for hierarchical cluster analysis*

	S-PLUS/R	SAS	SPSS	Stata	STATISTICA
Standardization					
Not standardized	mclust*,agnes	CLUSTER	CLUSTER	cluster	Cluster Analysis
Standardized	agnes,daisy	CLUSTER	DESCRIPTIVES	egen std;cluster	
Distance					
Euclidian	dist,daisy	CLUSTER	CLUSTER	cluster	Cluster Analysis
Squared Euclidian distance		CLUSTER	CLUSTER	cluster	Cluster Analysis
Sum of pth power of absolute distances			CLUSTER	cluster	Cluster Analysis
Pearson r	cor		CLUSTER	cluster	Cluster Analysis
Other options	dist,daisy	CLUSTER	CLUSTER	cluster	Cluster Analysis
Linkage methods					
Centroid	mclust*	CLUSTER	CLUSTER	cluster	Cluster Analysis
Nearest neighbor (single linkage)	mclust*,agnes	CLUSTER	CLUSTER	cluster	Cluster Analysis
Average linkage	mclust*,agnes	CLUSTER	CLUSTER	cluster	Cluster Analysis
Complete linkage	mclust*,agnes	CLUSTER	CLUSTER	cluster	Cluster Analysis
Ward's method	mclust*,agnes	CLUSTER	CLUSTER	cluster	Cluster Analysis
Other options	mclust*	CLUSTER	CLUSTER	cluster	Cluster Analysis
Printed output					
Distances between combined clusters	agnes	CLUSTER	CLUSTER	cluster,discrim	
Dendrogram (tree)	plclust,plot	TREE	CLUSTER	cluster dendogram	Cluster Analysis
Initial distance between observations	agnes	CLUSTER	CLUSTER	cluster measures;matrix dissim	Cluster Analysis

*mclust in S-PLUS is Mclust in R.

Table 16.6: *Software commands and output for K-means cluster analysis*

	S-PLUS/R	SAS	SPSS	Stata	STATISTICA
Standardization					
Not standardized	kmeans	FASTCLUS	Quick Cluster	cluster	Cluster Analysis
Standardized		STANDARD	DESCRIPTIVES	egen std;cluster	
Distance measured from					
Leader or seed point	kmeans	FASTCLUS	Quick Cluster	cluster	Cluster Analysis
Centroid or center		FASTCLUS	Quick Cluster	cluster	Cluster Analysis
Printed output					
Distances between cluster centers		FASTCLUS	Quick Cluster	cluster;discrim,lda; estat grdistance	Cluster Analysis
Distances between points and center			Quick Cluster	cluster;kmeans;	Cluster Analysis
Cluster means and standard deviations	kmeans	FASTCLUS	Quick Cluster*	tabstat	Cluster Analysis
F ratios		FASTCLUS	Quick Cluster	cluster;kmeans; anova	Cluster Analysis
Plot of cluster membership	plot	CANDISC		cluster; graph matrix	
Save cluster membership	kmeans	FASTCLUS	Quick Cluster	cluster	Cluster Analysis

*Only gives means.

program (called Join). Either Join or K-Means will cluster cases or variables. They also have a program that performs two-way joining of clusters and variables. Being Windows based, all three programs are easy to run.

The Stata cluster performs both types of clustering and includes many useful options.

16.8   What to watch out for

Cluster analysis is a heuristic technique for classifying cases into groups when knowledge of the actual group membership is unknown. There are numerous methods for performing the analysis, without good guidelines for choosing among them.

1. Unless there is considerable separation among the inherent groups, it is not realistic to expect very clear results with cluster analysis. In particular, if the observations are distributed in a nonlinear manner, it may be difficult to achieve distinct groups. For example, if we have two variables and cluster 1 is distributed such that a scatter plot has a U or crescent shape, while cluster 2 is in the empty space in the middle of the crescent, then most cluster analysis techniques would not be successful in separating the two groups.

2. Cluster analysis is quite sensitive to outliers. In fact, it is sometimes used to find outliers. The data should be carefully screened before running cluster programs.

3. Some of the computer programs supply F ratios, but it should be kept in mind that the data have been separated in a particular fashion prior to making the test, so the interpretation of the significance level of the tests is suspect. Usually, the major decision as to whether the cluster analysis has been successful should depend on whether the results make intuitive sense.

4. Most of the methods are not scale invariant, so it makes sense to try several methods of standardization and observe the effect on the results. If the program offers the opportunity of standardizing by the pooled within-cluster covariance matrix, this option should be tried. Note that for large samples, this could increase the computer time considerably.

5. Some investigators split their data and run cluster analysis on both halves to see if they obtain similar results. Such agreement would give further credence to the conclusions.

16.9   Summary

In this chapter we presented a topic still in its evolutionary form. Unlike subjects discussed in previous chapters, cluster analysis has not yet gained a standard methodology. Nonetheless, a number of techniques have been developed for dividing a multivariate sample, the composition of which is not known in advance, into several groups.

In view of the state of the subject, we opted for presenting some of the techniques that have been incorporated into the standard packaged programs. These include a hierarchical and nonhierarchical technique. We also explained the use of profile plots for small data sets. Since the results of any clustering procedure are often not definitive, it is advisable to perform more than one analysis and attempt to collate the results.

16.10   Problems

16.1 For the depression data set, use the last seven variables to perform a cluster analysis producing two groups. Compare the distribution of CESD and cases in the groups. Compare also the distribution of sex in the groups. Try two different programs with different options. Comment on your results.

16.2 For the situation described in Problem 7.7, modify the data for $X1, X2, \ldots, X9$ as follows.

- For the first 25 cases, add 10 to $X1, X2, X3$.
- For the next 25 cases, add 10 to $X4, X5, X6$.
- For the next 25 cases, add 10 to $X7, X8$.
- For the last 25 cases, add 10 to $X9$.

Now perform a cluster analysis to produce four clusters. Use two different programs with different options. Compare the derived clusters with the above groups.

16.3 Perform a cluster analysis on the chemical company data in Table 8.1, using the K-means method for $K = 2, 3, 4$.

16.4 For the accompanying small, hypothetical data set, plot the data by using methods given in this chapter, and perform both hierarchical and K-means clustering with $K = 2$.

Cases	X1	X2
1	11	10
2	8	10
3	9	11
4	5	4
5	3	4
6	8	5
7	11	11
8	10	12

16.5 Describe how you would expect guards, forwards, and centers in basketball to cluster on the basis of size or other variables. Which variables should be measured?

16.6 For the family lung function data described in Appendix A, perform a hierarchical cluster analysis using the variables FEV1 and FVC for mothers, fathers, and oldest children. Compare the distribution of area residence in the resulting groups.

16.7 Repeat Problem 16.6, using AREA as an additional clustering variable. Comment on your results.

16.8 Repeat Problem 16.6, using the K-means method for $K = 4$. Compare the results with the four clusters produced in Problem 16.6.

16.9 Create a data set from the family lung function data described in Appendix A as follows. It will have three times the number of observations that the original data set has — the first third of the observations will be the mothers' measurements, the second third those of the fathers, and the final third those of the oldest children. Perform a cluster analysis, first producing three groups and then two groups, using the variables AGE, HEIGHT, WEIGHT, FEV1, and FVC. Do the mothers, fathers, and oldest children cluster on the basis of any of these variables?

16.10 Repeat Problem 16.1, using the variables age, income, and education instead of the last seven variables.

16.11 For the Parental HIV data consider the subgroup of adolescents who have used marijuana. Perform a hierarchical cluster analysis separately for females on the following variables from the Parental HIV data: the sum of the variables describing the neighborhood (NGHB1–NGHB11), enough money for food (MONFOOD), financial situation (FINSIT), and whether the adolescent likes/liked school (LIKESCH). Graph a dendrogram. If you were to decide to use four groups, how many adolescents would belong to each of the groups?

Chapter 17

Log-linear analysis

17.1 Chapter outline

Log-linear models are useful in analyzing data that are commonly described as categorical, discrete, nominal, or ordinal. We include in this chapter several numerical examples to help clarify the main ideas. Section 17.2 discusses when log-linear analysis is used in practice and Section 17.3 presents numerical examples from the depression data set. The notation and sampling considerations are given in Section 17.4 and the usual model and tests of hypotheses are presented in Section 17.5 for the two-way table. Section 17.6 presents a worked out example for the two-way table. In Section 17.7 models are given when there are more than two variables, and tests of hypotheses are explained for exploratory situations in Section 17.8. Testing when certain preconceived relationships exist among the variables is discussed in Section 17.9, and the issue of small or very large sample sizes is reviewed in Section 17.10. Logit models are defined in Section 17.11 and their relationship to log-linear and logistic models is discussed. The options available in the various computer programs are presented in Section 17.12 while Section 17.13 discusses what to watch out for in applying log-linear models.

17.2 When is log-linear analysis used?

Log-linear analysis helps in understanding patterns of association among categorical variables. The number of categorical variables in one analysis can vary from two up to an arbitrary P. Rarely are values of P as great as six used in actual practice and P equal to three or four is most common. In log-linear analysis the categorical variables are not divided into dependent and independent variables. In this regard, log-linear analysis is similar to principal components or factor analysis. The emphasis is on understanding the interrelationships among the variables rather than on prediction.

If some of the variables are continuous, they must be made categorical by grouping the data into distinct categories prior to performing the log-linear analysis. Log-linear analysis can be used for either nominal or ordinal variables, but we do not discuss models that take advantage of the ordinal nature of the data. Analyses specific to ordinal data are given in Agresti (2002).

Log-linear analysis deals with multiway frequency tables. In Section 12.5 we presented a two-way frequency table where depression (yes or no) was given in the columns and sex (female or male) was listed in the rows of the table. From this table we computed the odds ratio = 2.825 (the odds of a female being depressed are 2.825 times that of a male). For two-way tables most investigators would either compute measures of association such as odds ratios or perform the familiar chi-square test (see Agresti, 2002, Fleiss *et al.*, 2003, or numerous other texts for a discussion of tests of independence and homogeneity using the chi-square test). Log-linear analysis is applicable to two-way tables but it is more commonly used for three-way or larger tables where interpretation of the associations is more difficult.

While the use of odds ratios is mainly restricted to data with only two distinct outcomes (binary data), log-linear analysis can be used when there are two or **more** subgroups or subcategories per variable. We do not recommend that a large number of subcategories be used for the variables as this will lead to very few observations in each cell of the frequency table. The number of subcategories should be chosen to achieve what the investigator believes is meaningful while keeping in mind possible sample size problems.

It is possible to model many different types of relationships in multiway tables. In this chapter we will restrict ourselves to what are called hierarchical models (Section 17.7). For a more complete discussion of log-linear analysis see Agresti (2002), Wickens (1989), Powers and Xie (2008), or Stokes *et al.* (2009).

17.3   Data example

The depression data set described earlier in Section 3.5 will be used in this chapter. Here we will work with categorical data (either nominal or ordinal) or data that have been divided into separate categories by transforming or recoding them. This data set has 294 observations and is part of a single sample taken from Los Angeles County (see Appendix A).

Suppose we wished to study the association between gender and income. Gender is given by the variable SEX and income has been measured in thousands of dollars. We split the variable INCOME into high (greater than 19) and low (less than or equal to 19). The number 19 denotes a yearly income of $19,000. This is an arbitrary choice on our part and does not reflect family size or other factors. This recoding of the data can be done in most programs using the IF THEN transformation statement (see manuals for your program) or by using special grouping options that are present in some programs.

For the depression data set this results in a two-way table (Table 17.1). The totals in the right-hand column and bottom row of Table 17.1 are called

Table 17.1: *Classification of individuals by income and sex*

	INCOME		
SEX	Low	High	Total
Female	125	58	183
Male	54	57	111
Total	179	115	294

marginal totals and the interior of the table is divided into four cells where the frequencies 125, 58, 54, and 57 are located. By simply looking at Table 17.1, one can see that there is a tendency for females to have low income. The entry of 125 females with low income is appreciably higher than the other frequencies in this table. Among the females 125/183 or 68.3% have low income while among the males 54/111 or 48.6% have low income. If we perform a chi-square test of independence, we obtain a chi-square value of 11.21 with one degree of freedom resulting in $P = 0.0008$. We can compute the odds ratio using the results from Section 12.5 as $(125 \times 57)/(58 \times 54) = 2.27$, i.e., the odds of a female having low income are 2.27 times those of a male.

The data in Table 17.1 are assumed to be taken from a cross-sectional survey, and the null hypothesis just tested is that gender is independent of income. Here we assume that the total sample size was determined prior to data collection. In Section 17.4 we discuss another method of sampling.

Now, suppose we are interested in including a third variable. As we shall see, that is when the advantage of the log-linear model comes in. Consider the variable called TREAT (Table 3.3), which was coded yes if the respondent was treated by a medical doctor and no otherwise. If we also categorize the respondents by TREAT as well as SEX and INCOME, then we have a three-way table (Table 17.2). Note that the rows in Table 17.2 have been split into TREAT no and yes (for example, the 125 females with low income have been split into 52 with no treatment and 73 that were treated). In other words, where before we had rows and columns, we now have rows (SEX), columns (INCOME), and layers (TREAT). The largest frequency occurs among the females with low income and medical treatment but it is difficult to describe the results in Table 17.2 without doing some formal statistical analysis. If we proceed one step further and make a four-way frequency table, interpretation becomes very difficult, but let us see what such a table looks like.

We considered adding the variable CASES, which was defined as a score of 16 or more on the CESD (depression) scale (Section 11.3). When we tried this, we got very low frequencies in some of the cells. Low frequencies are a common occurrence when you start splitting the data into numerous variables. (Actually, the problem is one of small theoretical rather than observed

Table 17.2: *Classification by income, sex, and treatment*

		INCOME		
TREAT	SEX	Low	High	Total
No	Female	52	24	76
	Male	34	36	70
Yes	Female	73	34	107
	Male	20	21	41
Total		179	115	294

frequencies; but small theoretical frequencies often result in small observed frequencies.) To avoid this problem, we split the CESD score into low (less than or equal to 10) and high (11 or above). A common choice of the place to split a continuous variable is the median if there is no theoretical reason for choosing other values since this results in an equal or near equal split of the frequencies. If any of our variables had three or more subcategories, our problem of cells with very small frequencies would have been even worse, i.e., the table would have been too **sparse**. If we had a larger sample size, we would have greater freedom in what variables to consider and the number of subcategories for each variable. According to our definition, the high depression score category includes persons who would not be considered depressed by the conventional clinical cutpoint. Thus the need to obtain adequate frequencies in cells has affected the quality of our analysis.

Table 17.3 gives the results when four variables are considered simultaneously. We included in the table several marginal totals. This table is obviously difficult to interpret without the use of a statistical model. Note also how the number of respondents in the various cells have decreased compared to Table 17.2, particularly in the last row for high CESD and medically treated males.

Before discussing the log-linear model, we introduce some notation and discuss two common methods of sampling.

17.4   Notation and sample considerations

The sample results given in Table 17.1 were obtained from a single sample of size $N = 294$ respondents. Table 17.4 presents the usual notation for this case. Here f_{11} (or 125) is the sample frequency in the first row (females) and column (low income) where the first subscript denotes the row and the second the column. The total for the first row is denoted by f_{1+} (or 183 females) where the "+" in place of the column subscripts symbolizes that we added across the columns. The total for the first column is f_{+1} and for the second column f_{+2}. The symbol f_{ij} denotes the observed frequency in an arbitrary row i and

Table 17.3: *Classification by income, sex, treatment, and CESD*

CESD	TREAT	SEX	INCOME Low	INCOME High	Total
Low	No	Female	33	20	53
		Male	23	30	53
		Total	56	50	106
	Yes	Female	48	20	68
		Male	16	16	32
		Total	64	36	100
High	No	Female	19	4	23
		Male	11	6	17
		Total	30	10	40
	Yes	Female	25	14	39
		Male	4	5	9
		Total	29	19	48
			Total		
	TREAT	No	146		
		Yes	148		
	CESD	Low	206		
		High	88		
	SEX	Female	183		
		Male	111		
	INCOME	Low	179		
		High	115		

column j. For a three-way table three subscripts are needed with the third one denoting layer, four-way tables require four subscripts, etc.

If we divide the entire table by N or f_{++}, then the resulting values will be the sample proportions denoted by p_{ij} and the sample table of proportions is shown in Table 17.5. The proportion in the ith row and jth column is denoted by p_{ij}. The sample proportions are estimates of the population proportions that will be denoted by π_{ij}. The total sample proportion for the ith row is denoted by p_{i+} and the total proportion for the jth column is given by p_{+j}. Similar notation is used for the population proportions.

Suppose instead that we took a random sample of respondents from a population that had high income and a second independent random sample of respondents from a population of those with low income. In this case, we will assume that both f_{+1} and f_{+2} can be specified prior to data collection. Here,

Table 17.4: *Notation for two-way frequency table from a single sample*

		Column	
Row	1	2	Total
1	f_{11}	f_{12}	f_{1+}
2	f_{21}	f_{22}	f_{2+}
Total	f_{+1}	f_{+2}	$f_{++} = N$

Table 17.5: *Notation for two-way table of proportions*

		Column	
Row	1	2	Total
1	p_{11}	p_{12}	p_{1+}
2	p_{21}	p_{22}	p_{2+}
Total	p_{+1}	p_{+2}	$p_{++} = 1$

the hypothesis usually tested is that the population proportion of low income people who are female is the same as the population proportion of high income people who are female. This is the so-called null hypothesis of **homogeneity**. The usual way of testing this hypothesis is a chi-square test, the same as for testing the hypothesis of independence.

17.5   Tests and models for two-way tables

In order to perform the chi-square test for a single sample, we need to calculate the expected frequencies under the null hypothesis of independence. The expected frequency is given by

$$\mu_{ij} = N\pi_{ij}$$

where the Greek letters denote population parameters, i denotes the ith row, and j the jth column. Theoretically, if two events are independent, then the probability of both of them occurring is the product of their individual probabilities. For example, the probability of obtaining two heads when you toss two fair coins is $(1/2)(1/2) = 1/4$. If you multiply the probability of an event occurring by the number of trials, then you get the expected number of outcomes of that event. For example, if you toss two fair coins eight times, then the expected number of times in which you obtain two heads is $(8)(1/4) = 2$ times.

Returning to two-way tables, the expected frequency in the ijth cell when the null hypothesis of independence is true is given by

$$\mu_{ij} = N\pi_{i+}\pi_{+j}$$

where π_{i+} denotes the marginal probability in the ith row and π_{+j} denotes the marginal probability in the jth column. In performing the test, sample proportions estimated from the margins of the table are used in place of the population parameters and their values are compared to the actual frequencies in the interior of the table using the chi-square statistic. For example, for Table 17.1, the expected value in the first cell is

$$294(183/294)(179/294) = 111.4$$

which is then compared to the actual frequency of 125. Here, more females have low income than we would have expected if income and gender were independent.

Log-linear models

The **log-linear** model is, as its name implies, a linear model of the logs (*to the base e*) of the population parameters. If we take the log of μ_{ij} when the null hypothesis is true, we have

$$\log \mu_{ij} = \log N + \log \pi_{i+} + \log \pi_{+j}$$

since the log of a product of three quantities is the sum of the logs of the three quantities. This linear model is usually written as

$$\log \mu_{ij} = \lambda + \lambda_{A(i)} + \lambda_{B(j)} \quad \text{(independence model)}$$

where A denotes rows and B columns, and the λ's are used to denote either $\log N$ or $\log \pi$.

There are many possible values of the λ's and so to make them unique two constraints are commonly used, namely, that the sum of the $\lambda_{A(i)} = 0$ and the sum of the $\lambda_{B(j)} = 0$.

Now if the null hypothesis is not true, the log-linear model must be augmented. To allow for association between the row and column variables, a fourth parameter is added to the model to denote the association between the row and column variable:

$$\log \mu_{ij} = \lambda + \lambda_{A(i)} + \lambda_{B(j)} + \lambda_{AB(ij)} \quad \text{(association model)}$$

where $\lambda_{AB(ij)}$ denotes the degree of association between the rows and columns. Note that the association terms are also required to add up to zero across rows and across columns. This latter model is sometimes called the **saturated** model since it contains as many free parameters as there are cells in the interior of the frequency table. The two models only differ by the last term that is added when the null hypothesis is not true. Hence, the null hypothesis of independence can be written

$$H_0: \quad \lambda_{AB(ij)} = 0$$

for all values of i and j.

Note that the log-linear model looks a lot like the two-way analysis of variance model

$$\mu_{ij} = \mu + \alpha_i + \beta_j + (\alpha\beta)_{ij}$$

where constraints are also used (sum of the row means = 0, sum of the column means = 0, and sums of the interaction terms across rows and across columns = 0) (see Wickens (1989), Section 3.10 for further discussion of when and how this analogy can be useful).

In performing the test, two goodness-of-fit statistics can be used. One of these is the usual **Pearson chi-square** goodness-of-fit test. It is the same as the regular chi-square test of independence, namely,

$$\chi^2 = \frac{\Sigma(\text{observed frequency} - \text{expected frequency})^2}{\text{expected frequency}}$$

where the observed frequency is the actual frequency and the expected frequency is computed under the null hypothesis of independence from the margins of the table. This is done for each cell in the interior of the table and then summed for all cells.

For the log-linear model test, the expected frequencies are computed from the results predicted by the log-linear model. To perform the test of no association, the expected frequencies are computed when the $\lambda_{AB(ij)}$ term is not included in the model (independence model above). The estimated parameters of the log-linear model are obtained as solutions to a set of likelihood equations (Agresti (2002), Chapter 8).

If a small chi-square value is obtained (indicating a good fit) then there is no need to include the term measuring association. The simple additive or independence model of the effects of the two variables is sufficient to explain the data. An even simpler model may also be appropriate to explain the data. To check for this possibility, we test the hypothesis that there are no row effects (i.e., all $\lambda_{A(i)} = 0$) or that there are no column effects (i.e., all $\lambda_{B(j)} = 0$). The Pearson chi-square statistic is calculated in the same way, where the expected frequencies are obtained from the model that satisfies the null hypothesis.

The idea is to find the simplest model that will yield a reasonably good fit. In general, if the model does not fit, i.e., chi-square is large, you add more terms to the model. If the model fits all right, you try to find an even simpler model that is almost as good. What you want is the simplest model that fits well.

In this section and the next we are assuming that the sample was from a survey of a single population. For a two-way table, this does not involve looking at many models. But for four- or five-way tables, several models are usually given a close look and you may need to run the log-linear program several times both to find the best fitting model and to get a full description of the selected model and how it fits.

Another chi-square test statistic is reported in some of the computer programs. This is called G^2 or **likelihood ratio chi-square**. Again, the larger the value of G^2, the poorer the fit and the more evidence against the null hypothesis. This test statistic can be written as

$$G^2 = 2\Sigma(\text{observed frequency}) \, \log \left(\frac{\text{observed frequency}}{\text{expected frequency}} \right)$$

where the summation is over all the cells. Note that when the observed and expected frequencies are equal, their ratio is 1 and $\log(1) = 0$, so that the cell will contribute nothing to G^2. The results for this chi-square are usually close to the Pearson chi-square test statistic but not precisely the same. For large sample sizes they are what statisticians call "asymptotically equivalent." That is, for large sample sizes the difference in the results should approach zero. Many statisticians like to use the likelihood ratio test because it is additive in so-called nested or hierarchical models (one model is a subset of the other). However, we would suggest that you use the Pearson chi-square if there is any problem with some of the cells having small frequencies since the Pearson chi-square has been shown to be better in most instances for small sample sizes (see Agresti, 2002, for a more detailed discussion on this point). Otherwise, the likelihood ratio chi-square is recommended.

In order to interpret the size of chi-square, it is necessary to know the degrees of freedom associated with each of the computed chi-squares. The degrees of freedom for the goodness-of-fit chi-square are given by the number of cells in the interior of the table minus the number of free parameters being estimated in the model. The number of free parameters is equal to the number of parameters minus the number of independent constraints. For example, for Table 17.1 we have four cells. For rows, we have one free parameter or $2 - 1$ since the sum of the two row parameters is zero. In general, we would have $A - 1$ free parameters for A rows. Similarly, for column parameters we have one free parameter and, in general, we have $B - 1$ free parameters. For the association term $\lambda_{AB(ij)}$ the sum over i and the sum over j must equal zero. Given these constraints, only $(A - 1)(B - 1)$ of these terms are free. There is also one parameter representing the overall mean. Thus for Table 17.1 the number of free parameters for the independence model is

$$1 + (2 - 1) + (2 - 1) = 3$$

For the association model, there are

$$1 + (2 - 1) + (2 - 1) + (2 - 1)(2 - 1) = 4$$

free parameters. Hence, the number of degrees of freedom is $4 - 3 = 1$ for the independence model and zero for the association or saturated model.

For higher than two-way models computing degrees of freedom can be very complicated. Fortunately, the statistical packages compute this for you and also give the P value for the chi-square with the computed degrees of freedom.

17.6 Example of a two-way table

We analyze Table 17.1 as an example of the two-way models and hypotheses used in the previous section. If we analyze these data using the usual Pearson chi-square test we would reject the null hypothesis of no association with a chi-square value of 11.21 and $P = 0.0008$ with one degree of freedom: (number of rows $-$ 1)$\times$(number of columns $-$ 1) or $(2 - 1)(2 - 1) = 1$.

To understand the log-linear model, we go through some of the steps of fitting the model to the data. The first step is to take the natural logs of the data in the interior of Table 17.1, as shown in Table 17.6. The margins of the table are then computed by taking the means of the appropriate $\log f_{ij}$. Note that this is similar to what one would do for a two-way analysis of variance (for example, 4.444 is the average of 4.828 and 4.060).

The various λ's are then estimated from these data using the maximum likelihood method (Agresti, 2002). Table 17.7 gives the estimates of the λ's for the four cells in Table 17.6. Here, the variable SEX is the row variable with $A = 2$ rows and INCOME is the column variable with $B = 2$ columns.

The parameter λ is estimated by the overall mean from Table 17.6. For cell $(1, 1)$, $\lambda_{A(1)}$ is estimated as $0.214 = 4.444 - 4.230$ (marginal means from Table 17.6) and $\lambda_{B(1)}$ as $0.178 = 4.409 - 4.230$ (minus rounding error of 0.001). For the association term for the first cell, we have $4.828 - (4.230 + 0.214 + 0.178) = 0.206$ (within rounding error of 0.001). The estimates in the remaining cells are computed in a similar fashion. Note that the last three columns all add to zero. The numerical values for $i = 1$ and $i = 2$ sum to zero: $0.214 + (-0.214) = 0$. Likewise for $j = 1$ and $j = 2$, we have $0.178 + (-0.178) = 0$. For the association terms we sum both over i and over j (cells $(1,1) + (1,2) = 0$, cells $(2,1) + (2,2) = 0$, cells $(1,1) + (2,1) = 0$, and cells $(1,2) + (2,2) = 0$). This is analogous to the two-way analysis of variance where the main effects and interaction terms sum to zero.

In this example with only two rows and two columns, each pair of estimates of the association terms (± 0.205) and the row (± 0.214) and column (± 0.178) effects are equal except for their signs. If we had chosen an example with three or more rows and/or columns, pairs of numbers would not all be the same, but row, column, and interaction effects would still sum to zero. Note that $\lambda_{\text{SEX}(1)}$

Table 17.6: *Natural logs of entries in Table 17.1*

SEX	INCOME		Mean
	Low	High	
Female	4.828	4.060	4.444
Male	3.989	4.043	4.016
Mean	4.409	4.052	4.230

Table 17.7: *Estimates of parameters of log-linear model for Table 17.6*

Cell (i, j)	λ	$\lambda_{A(i)}$	$\lambda_{B(j)}$	$\lambda_{AB(ij)}$
1, 1	4.230	0.214	0.178	0.205
1, 2	4.230	0.214	−0.178	−0.205
2, 1	4.230	−0.214	0.178	−0.205
2, 2	4.230	−0.214	−0.178	0.205

is positive, indicating that there are more females than males in the sample. Likewise, $\lambda_{INC(1)}$ is positive, indicating more low income respondents than high. Also, $\lambda_{SEX \times INC(11)}$ and $\lambda_{SEX \times INC(22)}$ are positive, indicating that there are more respondents in the first cell $(1, 1)$ and the last $(2, 2)$ than would occur if a strictly additive model held with no association (there is an excess of females with low income and males with high income).

Statistical packages provide estimates of the parameters of the log-linear model along with their standard errors or the ratio of the estimate to its standard error to assist in identifying which parameters are significantly different from zero.

The programs also present results for testing that the association between the row and column variables is zero. Here they are computing goodness of fit chi-square statistics for the model with the association term included as well as for the model without it. The chi-square test statistic and its degrees of freedom are obtained by taking the **difference** in these chi-squares and their degrees of freedom. A large test statistic would indicate that the association is significant and should be included in the model. The Pearson chi-square for this test is 11.21 with one degree of freedom, giving $P = 0.0008$, or the same as was given for the regular chi-square test mentioned earlier. The real advantage of using the log-linear model will become evident when we increase the number of rows and columns in the table and when we analyze more than two variables. The G^2 or likelihood ratio chi-square test statistic was calculated as 11.15 with $P = 0.00084$, very similar to the Pearson chi-square in this case. We reject the null hypothesis of no association and would say that income and gender are associated.

There is also a significant chi-square value for the row and for the column effects. Thus, we would reject the null hypotheses of there being equal numbers of males and females and low and high income respondents. These last two tests are often of little interest to the investigator, who usually wants to test for associations. In this example, the fact that we had more females than males is a reflection of an unequal response rate for males and females in the original sample and may be a matter of concern for some analyses.

The odds ratio is related to the association parameters in the log-linear model. For the two-way table with only two rows and columns, the relationship is very straightforward, namely,

$$\log_e(\text{odds ratio}) = 4\lambda_{AB(11)}$$

In the SEX by INCOME example the odds ratio is 2.275 and the computed value from the association parameter using the above equation is given by 2.270 (in this case they are not precisely equal since only three decimal places were reported for the association parameter).

17.7   Models for multiway tables

The log-linear model for the three-way table can be used to illustrate the different types of independence or lack of association that may exist in multiway tables. The simplest model is given by the complete independence (mutual independence) model. In order to demonstrate the models we will first redo Table 17.2 in symbols so the correspondence between the symbols and the contents of the table will be clear. We have written Table 17.8 in terms of the sample proportions that can be computed from a table such as Table 17.2 by dividing the entries in the entire table by N. The marginal table given at the bottom is the sum of the top two-way tables (for layer $C = 1$ plus for layer $C = 2$). If we knew the true probabilities in the population, then the p_{ijk}'s would be replaced by π_{ijk}'s. Here i denotes the ith row, j the jth column, and k the kth layer and a "+" signifies a sum over that variable. Another marginal table of variable A versus C could be made using the column margins (column labelled Total) of the top two partial layer tables. A third marginal table for variables B and C can be made of the row marginals for the top two partial layer-tables; some programs can produce all of these tables by request.

Next we consider some special cases with and without associations.

Table 17.8: *Notation for three-way table of proportions*

Layer C		Column B		
1	Row A	1	2	Total
	1	p_{111}	p_{121}	p_{1+1}
	2	p_{211}	p_{221}	p_{2+1}
Partial layer 1		p_{+11}	p_{+21}	p_{++1}
2	Row A	1	2	Total
	1	p_{112}	p_{122}	p_{1+2}
	2	p_{212}	p_{222}	p_{2+2}
Partial layer 2		p_{+12}	p_{+22}	p_{++2}
		Column B		
Marginal table	Row A	1	2	Total
	1	p_{11+}	p_{12+}	p_{1++}
	2	p_{21+}	p_{22+}	p_{2++}
	Total	p_{+1+}	p_{+2+}	p_{+++}

Complete independence model

In this case, all variables are unassociated with each other. This is also called mutual independence. The probability in any cell (i, j, k) is then given by

$$\pi_{ijk} = \pi_{i++}\pi_{+j+}\pi_{++k}$$

where π_{i++} is the marginal probability for the ith row (the lower marginal table row total), π_{+j+} is the marginal probability for the jth column (the lower marginal table column total), and π_{++k} is the marginal probability for the kth layer (the total for layer 1 and layer 2 tables). The sample estimates for the right side of the equation are p_{i++}, p_{+j+}, and p_{++k}, which can be found in Table 17.8.

When complete independence holds, the three-way log-linear model is given by

$$\log \mu_{ijk} = \lambda + \lambda_{A(i)} + \lambda_{B(j)} + \lambda_{C(k)}$$

The model includes no parameters that signify an association.

One variable independent of the other two

If variable C is unrelated to (jointly independent of) both variable A and variable B then the only variables that can be associated are A and B. Hence, only parameters denoting the association of A and B need be included in the model. The three-way log-linear model then becomes

$$\log \mu_{ijk} = \lambda + \lambda_{A(i)} + \lambda_{B(j)} + \lambda_{C(k)} + \lambda_{AB(ij)}$$

Saying that C is jointly independent of A and B is analogous to saying that the **simple** correlations between variables C and A and variables C and B are both zero in regression analysis. Note that, similarly, A could be jointly independent of B and C or B could be jointly independent of A and C.

Conditional independence model

Suppose A and B are independent of each other for the cases in layer 1 of C in Table 17.8 and also independent of each other for the cases in layer 2 of C in Table 17.8. Then A and B are said to be conditionally independent in layer k of C. This model includes associations between variables A and C and variables B and C but not the association between variables A and B. The model can be written as

$$\log \mu_{ijk} = \lambda + \lambda_{A(i)} + \lambda_{B(j)} + \lambda_{C(k)} + \lambda_{AC(ik)} + \lambda_{BC(jk)}$$

If we say that A and B are independent of each other conditional on C, this is analogous to the **partial** correlation of A and B given C being zero (Wickens, 1989, Sections 3.3–3.6).

Note that, similarly, the A and C association term or the B and C association term could be omitted to produce other conditional independence models.

Models with all two-way associations

It is also possible to have a model with all three two-way associations present. This is called the homogeneous association model and would be written as

$$\log \mu_{ijk} = \lambda + \lambda_{A(i)} + \lambda_{B(j)} + \lambda_{C(k)} + \lambda_{AB(ij)} + \lambda_{AC(ik)} + \lambda_{BC(jk)}$$

In this case all three two-way association terms are included. This would be analogous to an analysis of variance model with the main effects and all two-way interactions included.

Saturated model

The final model is the saturated model that also includes a three-way association term, $\lambda_{ABC(ijk)}$, in addition to the parameters in the last model. In this model there are as many free parameters as cells in the table and the fit of the model to the log of the cell frequencies is perfect. Here we are assuming that there is a three-way association among variables A, B, and C that is above and beyond the two-way interactions plus the row, column, and layer effects. These higher order interaction terms are not common in real-life data but significant ones sometimes occur.

Associations in partial and marginal tables

Some anomalies can occur in multiway tables. It is possible, for example, to have what are called positive associations in the partial layers 1 and 2 of Table 17.8 while having a negative association in the marginal table at the bottom of Table 17.8. A positive association is one where the odds ratio is greater than one and a negative association has an odds ratio that lies between zero and one. Thus, the associations found in the partial layer tables can be quite different from the associations among the marginal sums.

It is important to carefully examine the partial and marginal frequency tables computed by the statistical packages in an attempt to determine both from the data and from the test results how to interpret the outcome.

Hierarchical log-linear models

When one is examining log-linear models from multiway tables, there is a large number of possible models. One restriction on the models examined that is often used is to look at **hierarchical models**. Hierarchical models obey the following rule: whenever the model contains higher order λ's (higher order associations or interactions) with certain variables, it also contains the lower order effects with these same variables. For instance, if $\lambda_{AB(ij)}$ is included in the model, then λ, $\lambda_{A(i)}$, and $\lambda_{B(j)}$ must be included. (An example of a non-hierarchical model would be a model only including λ, $\lambda_{A(i)}$, and $\lambda_{BC(jk)}$.) A hierarchical model is analogous to keeping the main effects in analysis of variance if an interaction is present. Most investigators find hierarchical models easier to interpret than nonhierarchical ones and restrict their search to this type. Note that when you get up to even four variables there is a very large number of possible models, so although this restriction may seem limiting, actually it is a help in reducing the number of models and restricting the choice to ones that are easier to explain.

Four-way and higher dimensional models

As stated in Section 17.2, the use of log-linear analysis for more than four variables is rare since, in those cases, the interpretation becomes difficult and insufficient sample sizes are often a problem. Investigators who desire to either use log-linear analyses a great deal or analyze higher order models should read texts such as Agresti (2002), Wickens (1989), or Bishop *et al.* (2007).

For a four-way model, the log-linear model can be written as follows. (Here we have omitted the $ijkl$ subscripts to shorten and simplify the appearance of the equation.)

$$\log \mu = \lambda + \lambda_A + \lambda_B + \lambda_C + \lambda_D + \lambda_{AB} + \lambda_{AC} + \lambda_{AD} + \lambda_{BC} + \lambda_{BD} +$$
$$\lambda_{CD} + \lambda_{ABC} + \lambda_{ABD} + \lambda_{ACD} + \lambda_{BCD} + \lambda_{ABCD}$$

In all, there are over a hundred possible hierarchical models to explore for a four-way table. Obviously, it becomes necessary to restrict the ones examined. In the next section, we describe how to go about navigating through this sea of possible models.

17.8   Exploratory model building

Since there are so many possible tests that can be performed in three- or higher-way tables, most investigators try to follow some procedure to restrict their search. Here we will describe some of the strategies used and discuss tests to assist in finding a suitable model. In this section we consider a single sample with no restrictions on the role of any variable. In Section 17.9 strategies will be discussed for finding models when some restrictions apply.

There are two general strategies. One is to start with just the main effects and add more interaction terms until a good-fitting model is obtained. The other is to start with a saturated or near-saturated model and remove the terms that appear to be least needed. The analogy can be made to forward selection or backward elimination in regression analysis. Exploratory model building for multiway tables is often more difficult to do than for linear regression analysis. The output from the computer programs requires more attention and the user may have to run the program several times before a suitable model is found, especially if several of the variables have significant interaction terms. In other cases, perhaps only two of the variables have a significant interaction, so the results are quite obvious.

Adding terms

For the first strategy, we could start with the main effects (single order factors) only. In Table 17.2, we presented data for a three-way classification using the variables INCOME (i), SEX (s), and TREAT (t). We ran log-linear analyses of these data. When we included just the main effects i, s, and t in the model, we got a likelihood ratio (goodness-of-fit) chi-square of 24.08 with four degrees of freedom and $P = 0.0001$, indicating a very poor fit. So we tried adding the three second order interaction terms (si, it, and st). The likelihood ratio chi-square dropped to < 0.001 with one degree of freedom and $P = 0.9728$, indicating an excellent fit.

But we may not need all three second order interaction terms, so we tried just including si. When we specify a model with si, the program automatically includes i and s, since we are looking at hierarchical models. However, in this case we did need to specify t in order to include the three main effects. Here we got a likelihood ratio chi-square of 12.93 with three degrees of freedom and $P = 0.0048$. Thus, although this was an improvement over just the main effect terms, it was not a good fit. Finally, we added st to the model and got a very good fit with a likelihood ratio chi-square of < 0.001 with two degrees of freedom and $P = 0.9994$. Therefore, our choice of a model would be one with all three main effects and the INCOME and SEX interaction as well as the SEX and TREAT interaction.

Eliminating terms

The second strategy used when the investigator has a single sample (cross-sectional survey) and there is no prior knowledge about how the variables should be interrelated is to examine the chi-square statistics starting with fairly high order interaction term(s) included. This will usually have a small value of the goodness-of-fit chi-square, indicating an acceptable model. Note that for the saturated model, the value of the goodness-of-fit chi-square will be zero. But a simpler model may also provide a good starting point. After eliminating nonsignificant terms, less complex models are examined to find one that still gives a reasonably good fit. Significant fourth or higher order interaction terms are not common and so, regardless of the number of variables, the investigator rarely has to start with a model with more than a four-way interaction. Three-way interactions are usually sufficient.

We instructed the program to run a model that includes the three factor association term (λ_{ABC}) so that we would see the results for the model with all possible terms in it. Note that if we specify that the third order interaction term be included then the program will automatically include all second order terms, since, at this point, we are assuming a hierarchical model.

First, we obtain results for three tests. The first tests that all main effects (λ_A, λ_B, and λ_C) are zero. The second tests that all two-factor associations ($\lambda_{AB}, \lambda_{AC}$, and λ_{BC}) are zero. The third tests that the third order association term (λ_{ABC}) is zero. The results using SAS were as follows:

```
Simultaneous Test that All K-Factor Interactions Are Zero
```

K-Factor	DF	LR Chi-Sq	P	Pearson Chi-Sq	P
1	3	31.87	0.00000	36.67	0.00000
2	3	24.08	0.00002	24.65	0.00002
3	1	0.00	0.97262	0.00	0.97262

These test results were obtained from the goodness-of-fit chi-squares for the following models:

1. all terms included;

2. all two factor association and first order (main effect) terms included;

3. all first order terms included;

4. only the overall mean included.

Note that we are assuming hierarchical models and, therefore, the overall mean is included in all models. The chi-squares listed above were then computed by subtraction as follows. The LR Chi-sq G^2 of 31.87 was computed by subtracting the goodness of fit chi-square for model 3 from the goodness-of-fit chi-square for model 4. The chi-square of 24.08 was computed by subtracting the chi-square for model 2 from the chi-square for model 3. The 0.00 chi-square was computed by subtracting the chi-square for model 1 from the chi-square for model 2. Similar subtractions were performed for the Pearson chi-square.

We start our exploration at the bottom of the table. The third order interaction term was not even near significant by either the likelihood ratio or the Pearson test, being 0.00 (first two decimal points were zero). Now we figure we only have to be concerned with the two factor association and main effect terms. Since we have a large chi-square when testing that all K-factors for $K = 2$ are zero, we decide that we have to include some measures of association between two variables. But we do not know which two-way associations to include.

The next step is to try to determine which second order association terms to include. There are two common ways of doing this. One is to examine the ratio of the estimates of each second order interaction term to its estimated standard error and keep the terms that have a ratio of, say, two or more (Agresti, 2002). The ratios were greater than three for both the SEX by INCOME interaction and the SEX by TREAT interaction. The third one was not greater than two and thus should not be kept.

The other common method includes examining output for individual tests for the three possible two factor associations. Using the SAS software we obtained two tests, one called partial association and one marginal association as follows:

Effect	DF	Partial Association Chi-Sq	P	Marginal Association Chi-Sq	P
is	1	10.67	0.0011	11.15	0.0008
it	1	0.00	0.9934	0.48	0.4895
st	1	12.45	0.0004	12.93	0.0003

In the partial association test, the program computes the difference in chi-square when all three second order association terms are included, and when one is deleted. This is done for each association term separately. A large value of chi-square (small P value) indicates that deleting that association term results in a poor fit and the term should be kept.

The marginal association test is done on the margins. For example, to test the SEX by INCOME association term, the program sums across TREAT to get a marginal table and obtains the chi-square for that table. This test may or may not give the same results as the partial test. If both tests result in a large chi-square, then that term should be kept. Also, if both result in a small chi-square value then the term should be deleted. It is less certain what to do if the two tests disagree. One strategy would be to keep the interaction and report the disagreement.

The results in our example indicate that we should retain the INCOME by SEX and the SEX by TREAT association terms, but there is no need for the INCOME by TREAT association term in the model. Here the partial and marginal association tests agree even though they do not result in precisely the same numerical values. These results also agree with our previous analysis.

To check our results we ran a model with si and st (which also includes i, s, and t). The fitted model has a very small goodness of fit chi-square, indicating a good fit. In summary our analysis indicates that there is a significant association between gender and income as well as between gender and getting medical treatment but there was no significant association between income and getting medical treatment and no significant third order association. Thus we obtained the conditional independence model described in Section 17.7. INCOME and TREAT are conditionally independent given SEX.

Stepwise selection

Some programs offer stepwise options to assist the user in the choice of an appropriate model. In SPSS both forward and backward elimination are available, but backward elimination or backward stepwise is the preferred method (Benedetti and Brown, 1978; Oler, 1985). The criterion used for testing is the P value from the likelihood ratio chi-square test. The program keeps eliminating terms that have the least effect on the model until the test for lack of fit is significant. The procedure continues in a comparable fashion to stepwise procedures in regression or discriminant function analysis. Again, it should be noted that with all this multiple testing, that levels of P should not be taken seriously.

We recommend trying the backward elimination more as a check to make sure that there is not a better model than the one that you have previously chosen. The process should not be relied on to automatically find the "best" model. Nevertheless, it does serve as a useful check on previous results and should be tried. Sometimes it is difficult to anticipate what associations will occur and trying the backward elimination method may lead to an interesting result.

For example, the data in Table 17.3, which include four variables (SEX(s), INCOME(i), TREAT (t), and CESD (c)), were analyzed by SPSS using backward elimination. We started with a model with all three-way interaction terms included (*isc, itc, stc, ist*). The program eliminates at each step the term with the largest P value for the likelihood ratio chi-square test (least effect on the model). In this case, we instructed the program to stop when the largest term to be eliminated had a P value less than 0.05. The program deleted each three-way interaction term one at a time and computed the chi-square, degrees of freedom, and the P value.

The three-way interaction term with the largest P value was *ist*, so it was eliminated in the first step, leaving *isc, itc*, and *stc*. At the second step, the largest P value occurred when *isc* was tested, so it was eliminated, keeping *itc* and *stc* and also *si*. Note that we are testing only three-way interaction terms at this step and thus the program keeps all second order interactions.

At step 3, the interaction *stc* has the largest P value, so it is removed and the resulting model is *itc, si, sc*, and *st*. The terms *it, tc*, and *ic* are contained in *itc* and are thus also included since we have a hierarchical model. At step four, *itc* was **not** removed and the term with the largest P value was *sc*, a second order interaction term. So the model at the end of step three is *itc, si* and *st*.

At the next step the largest P value was 0.0360 for *itc*, which is less than 0.05, so it stays in and the process stops since no term should be removed. So the program selected the final model as *itc, si*, and *st*. Also, the main effects *i, t, c*, and *s* must be included since they occur in at least one of the interaction terms.

These results can be compared with the likelihood ratio chi-square tests for partial associations obtained when all three-way interaction terms were included, as given below in the listing that follows.

Here also we would include the *itc* three-way interaction term, and the *si* and *st* two-way interaction terms. We also need to include the *it, ic*, and *tc* and *t* terms even though they are not significant if we include *itc* in order to keep the model hierarchical. The model is written *itc, si* and *st* since *ti, ic* and *tc* are automatically included according to the hierarchical principle.

In this example we retained a three-way interaction term among INCOME, TREAT, and CESD, although no pair of the terms involved in the three-way interaction had a significant two-way interaction. This illustrates some of the complexity that can occur when interpreting multiway tables.

Term	Degrees Freedom	LR Chi-Sq	P
i	1	14.04	0.0002
s	1	17.81	0.0000
t	1	0.01	0.9071
c	1	48.72	0.0000
si	1	9.91	0.0016
it	1	0.00	0.9653
ic	1	1.17	0.2799
st	1	11.98	0.0005
sc	1	2.34	0.1259
tc	1	0.32	0.5733
ist	1	0.01	0.9294
isc	1	0.01	0.9225
itc	1	4.74	0.0294
stc	1	1.23	0.2681

There was also a significant interaction between SEX and INCOME, as we had found earlier, and also between SEX and TREAT, with women being more apt to be treated by a physician.

Model fit

The fit of the model can be examined in detail for each cell in the table. This is often a useful procedure to follow particularly when two alternative models are almost equally good since it allows the user to see where the lack of model fit may occur. For example, there may be small differences throughout the table or perhaps larger differences in one or two cells. The latter would point to either a poor model fit or possibly errors in the data. This procedure is somewhat comparable to residual analysis after fitting a regression model.

The expected values from the model are computed for each cell. Given this information, several methods of examining the frequencies between the expected values and the actual frequencies are possible. The simplest is to just examine the difference, observed minus expected. But many find it easier to interpret the components of chi-square for each cell. For the Pearson chi-square this is simply (observed − expected)^2/(expected) computed for each cell. Large values of these components indicate where the model does not fit well. Standardized deviates are also available and many users prefer to examine them for much the same reasons that standardized residuals are used in regression analysis. These are computed as (observed − expected)/(square root of expected). The standardized deviates are asymptotically normally distributed but are somewhat less variable than normal deviates (Agresti, 2002). Thus, standardized deviates greater than 1.96 or less than -1.96 will be at least two standard deviations away from the expected value and are candidates for inspection.

17.9   Assessing specific models

The log-linear model is sometimes used when the sample is more complicated than a single random sample from the population. For example, suppose a sample of males and a separate sample of females are taken. This is called a **stratified** sample. Here we assume that the sample size in each subgroup or stratum (males and females) is fixed. This implies that the marginal totals for the variable SEX are fixed (unlike in the case of a single random sample where the marginal totals for sex are random). When we collect a stratified sample, we have two independent samples, one for each gender. In this case we should always include the main effect for SEX, λ_{SEX}, in the model whether or not it is significant according to the chi-square test. This is done since we want to test the other effects when the main effects of SEX are accounted for.

If we stratify by both income and gender, and take independent samples with fixed sample sizes among low income males, high income males, low income females, and high income females, then we should include $\lambda_{SEX} + \lambda_{INC} + \lambda_{SEX \times INC}$ terms in the model.

Furthermore, in Section 8.2, in connection with regression analysis, we indicated that often investigators have prior justification for including some variables and they wish to check if other variables improve prediction. Similarly, in log-linear modelling, it often makes the interpretation of the model conceptually simpler if certain variables are included, even if this increases the number of variables in the model. Since the results obtained in log-linear models depend on the particular variables included, care should be taken to keep in variables that make logical sense. If, theoretically, a main effect and/or certain association terms belong in a model to make the tests for other variables reasonable, then they should be kept.

As mentioned earlier, the goodness of fit chi-square for any specified model measures how well the model fits the data; the smaller the chi-square, the better the fit. Also, it is possible to test a specific model against another model obtained by adding or deleting some terms. Suppose that Model 2 is obtained from Model 1 by deleting certain interaction and/or main effect terms. Then we can test whether Model 2 fits the data as well as Model 1 as follows:

test chi-square $=$ (goodness-of-fit chi-square for Model 2)

$\qquad -$(goodness-of-fit chi-square for Model 1)

test degrees of freedom $=$ (degrees of freedom for Model 2)

$\qquad -$(degrees of freedom for Model 1)

If the test chi-square value is large (or the P value is small) then we conclude that Model 2 is not as good as Model 1, i.e., the deleted terms do improve the fit.

This test is similar to the general F test discussed in Section 8.5 for regression analysis. It can be used to test two theoretical models against each other or in exploratory situations as was done in the previous section. For example, in the SEX (s), INCOME (i), TREAT (t) data suppose we believe that the ist term should not be included. Then our base model is st, is, it. Thus Model 1 includes these interactions and all main effects. We could then begin by testing the hypothesis that INCOME has no effect by formulating Model 2 to include only the st interaction (as well as s and t). If the P value of the test chi-square is small, we would keep INCOME in the model and proceed with the various methods described in Section 17.8 to further refine the model.

In some cases, the investigator may conceive of some variables as explanatory and some as response variables. The log-linear model was not derived originally for this situation but it can be used if care is taken with the model. If this is the case, it is important to include the highest order association term among the explanatory variables in the model (Agresti, 2002; Wickens, 1989, Section 7.1). For example, if A, B, and C are explanatory variables and D is considered a response variable, then the model should include the λ_{ABC} third order association term and other terms introduced by the hierarchical model, along with $\lambda_{AD}, \lambda_{BD}$, and λ_{CD} terms.

17.10 Sample size issues

The sample size used in a multiway frequency table can result in unexpected answers either because it is too small or too large. As with any statistical analysis, when the sample is small there is a tendency not to detect effects unless these effects are large. This is true for all statistical hypothesis testing. The problem of low power due to small samples is common, especially in fields where it is difficult or expensive to obtain large samples. On the other hand, if the sample is very large, then very small and unimportant effects can be statistically significant.

There is an additional difficulty in the common chi-square test. Here, we are using a chi-square statistic (which is a continuous number) to approximate a distribution of frequencies (which are discrete numbers). Any statistics computed from the frequencies can only take on a limited (finite) number of values. Theoretically, the approximation has been shown to be accurate as the sample size approaches infinity. The critical criteria are the sizes of the expected values that are used in the chi-square test. For small tables (only four or so cells) the minimum expected frequency in each cell is thought to be two or three. For larger tables, the minimum expected frequency is supposed to be around one, although for tables with a lot of cells, some could be somewhat less than one.

Clogg and Eliason (1987) point out three possible problems that can occur with sparse tables. The first is essentially the same one given above. The goodness-of-fit chi-square may not have the theoretical null distribution. Second, sparse tables can result in difficulties in estimating the various λ parameters. Third, the estimate of λ divided by its standard error may be far from normal, so inferences based on examining the size of this ratio may be misleading.

For the log-linear model, we are looking at differences in chi-squares. Fortunately, when the sample size is adequate for the chi-square tests, it is also adequate for the difference tests (Wickens, 1989, Section 5.7; Haberman, 1977).

One of the reasons that it is difficult to make up a single rule is that, in some tables, the proportions in the subcategories in the margins are almost equal, whereas in other tables they are very unequal. Unequal proportions in the margins mean that the total sample size has to be much larger to get a reasonably sized expected frequency in some of the cells.

It is recommended that the components of chi-square for the individual cells be examined if a model is fit to a table that has many sparse cells to make sure that most of the goodness-of-fit chi-square is not due to one or two cells with small expected values(s).

In some cases, a particular cell might even have a zero observed frequency. This can occur due to sampling variation (called **sampling zero**) or because it is impossible to get a frequency in the cell (no pregnant males), which is called **structural zeros**. In this latter case, some of the computer programs will exclude cells that have been identified to have structural zeros from the computation.

On the other hand, if the sample size is very large as it sometimes is when large samples from the census or other governmental institutions are analyzed, the resulting log-linear analysis will tend to fit a very complicated model, perhaps even a saturated model. The size of some of the higher order association terms may be small, but they will still be significantly different from zero when tested due to the large sample size. In this case, it is suggested that a model be fitted simply with the main effects terms included and G^2 (the likelihood ratio chi-square) be computed for this baseline model. Then alternative models are compared to this baseline model to see the proportional improvement in G^2. This is done using the following formula:

$$\frac{G^2_{\text{baseline}} - G^2_{\text{alternative}}}{G^2_{\text{baseline}}}$$

As more terms are added, it is suggested that you stop when the numerical value of the above formula is above 0.90 and the effects of additional terms are not worth the added complexity in the model.

17.11 The logit model

Log-linear models treat all the variables equally and attempt to clarify the inter-relationships among all of them. However, in many situations the investigator may consider one of the variables as outcome or dependent and the rest of the variables as explanatory or independent.

For example, in the depression data set, we may wish to understand the relationship of TREAT (outcome) and the explanatory variables SEX and IN-COME. We can begin by considering the log-linear model of the three variables with the following values:

$$I(= \text{INCOME}) = \begin{cases} 0 & \text{if low (19 or less)} \\ 1 & \text{if high (20 or more)} \end{cases}$$

$$S(= \text{SEX}) = \begin{cases} 0 & \text{if male} \\ 1 & \text{if female} \end{cases}$$

$$T(= \text{TREAT}) = \begin{cases} 0 & \text{if treatment is recommended} \\ 1 & \text{if no treatment is recommended} \end{cases}$$

Analysis of this model shows that the three-way interaction is unnecessary ($\chi^2 = 0.0012$ and $P = 0.97$). Therefore we consider the model which includes only the two-way interactions. The model equation is

$$\log \mu_{ijk} = \lambda + \lambda_{I(i)} + \lambda_{S(j)} + \lambda_{T(k)} + \lambda_{IS(ij)} + \lambda_{IT(ik)} + \lambda_{ST(jk)}$$

The estimates of the parameters are obtained as follows:

$$\hat{\lambda}_{I(0)} = 0.178 = -\hat{\lambda}_{I(1)}$$
$$\hat{\lambda}_{S(0)} = -0.225 = -\hat{\lambda}_{S(1)}$$
$$\hat{\lambda}_{T(0)} = -0.048 = -\hat{\lambda}_{T(1)}$$
$$\hat{\lambda}_{IS(0,0)} = -0.206 = \hat{\lambda}_{IS(1,1)} = -\hat{\lambda}_{IS(1,0)} = -\hat{\lambda}_{IS(0,1)}$$
$$\hat{\lambda}_{IT(0,0)} = -0.001 = \hat{\lambda}_{IT(1,1)} = -\hat{\lambda}_{IT(1,0)} = -\hat{\lambda}_{IT(0,1)}$$
$$\hat{\lambda}_{ST(0,0)} = -0.219 = \hat{\lambda}_{ST(1,1)} = -\hat{\lambda}_{ST(1,0)} = -\hat{\lambda}_{ST(0,1)}$$

These numbers do not readily shed light on treatment as a function of sex and income.

One method for better understanding this relationship is the **logit model**. First, we consider the odds of **no** treatment, i.e., the probability of no recommended treatment divided by the probability of recommended treatment. Thus,

$$\text{odds} = \frac{\pi_{ij1}}{\pi_{ij0}}$$

for a person for whom SEX $= i$ and INCOME $= j$. This can equivalently be written as

$$\text{odds} = \frac{\mu_{ij1}}{\mu_{ij0}}$$

since $\mu_{ijk} = N\pi_{ijk}$.

Although this is an easily interpreted quantity, statisticians have found it easier to deal mathematically with the **logit** or the logarithm of the odds. (Note that this is also what we did in logistic regression.) Thus,

$$\text{logit} = \log(\text{odds})$$
$$= \log\left(\frac{\mu_{ij1}}{\mu_{ij0}}\right)$$

This equation may be equivalently written as

$$\text{logit} = \log\mu_{ij1} - \log\mu_{ij0}$$

which, after substituting the log-linear model, becomes

$$\lambda + \lambda_{I(i)} + \lambda_{S(j)} + \lambda_{T(1)} + \lambda_{IS(ij)} + \lambda_{IT(i1)} + \lambda_{ST(j1)}$$
$$-\lambda - \lambda_{I(i)} - \lambda_{S(j)} - \lambda_{T(0)} - \lambda_{IS(ij)} - \lambda_{IT(i0)} - \lambda_{ST(j0)}$$

which reduces to

$$(\lambda_{T(1)} - \lambda_{T(0)}) + (\lambda_{IT(i1)} - \lambda_{IT(i0)}) + (\lambda_{ST(j1)} - \lambda_{ST(j0)})$$

But, recall that the λ parameters sum to zero across treatment categories, i.e., $\lambda_{T(0)} = -\lambda_{T(1)}$ and hence

$$\lambda_{T(1)} - \lambda_{T(0)} = 2\lambda_{T(1)}$$

Similar manipulations for the other parameters reduce the expression for the logit to

$$\text{logit} = 2\lambda_{T(1)} + 2\lambda_{IT(i1)} + 2\lambda_{ST(j1)}$$

Note that this is a linear equation of the form

$$\text{logit} = \text{constant}$$
$$+ \text{effect of sex on treatment}$$
$$+ \text{effect of income on treatment}$$

For example, to compute the logit or log odds of no recommended treatment for a high income ($i = 1$) female ($j = 1$), we identify

$$\lambda_{T(1)} = 0.048, \quad \lambda_{IT(1,1)} = -0.001, \quad \lambda_{ST(1,1)} = -0.219$$

and substitute in the logit equation to obtain:

$$\text{logit} = 0.096 - 0.002 - 0.438 = -0.344$$

By exponentiating, we can compute the odds of no recommended treatment as $\exp(-0.344) = 0.71$. Conversely, the odds of recommended treatment is $\exp(0.344) = 1.411$. Similarly, for a low income male, we compute the logit of no recommended treatment as 0.536 or odds of 1.71.

Similar manipulations can be performed to produce a logit model for any log-linear model in which one of the variables represents a binary outcome. Demaris (1992) and Agresti (2002) discuss this subject in further detail, including situations in which the outcome and/or the explanatory variables may have more than two categories.

In order to understand the relationship between the logit and logistic models, we define the following variables:

$$Y = \text{TREAT} (= 0 \text{ for yes and 1 for no})$$
$$X1 = \text{SEX} (= 0 \text{ for male and 1 for female})$$
$$X2 = \text{INCOME} (= 0 \text{ for low and 1 for high})$$

Some numerical calculations show that the logit equation for the above example can be written as

$$\log\left(\frac{\text{prob } Y = 1}{\text{prob } Y = 0}\right) = 0.536 - 0.004X_1 - 0.876X_2$$

which is precisely in the form of the logistic regression model presented in Chapter 12. In fact, these coefficients agree within round-off error with the results of logistic regression programs which we ran on the same data.

In general, it can be shown that any logit model is equivalent to a logistic regression model. However, whereas logit models include only categorical (discrete) variables, logistic regression models can handle discrete **and** continuous variables. On the other hand, logit models including discrete and continuous variables are also used and are called **generalized logit models** (Agresti, 2002). For these reasons, the terms "logit models" and "logistic regression models" are used interchangeably in the literature.

17.12 Discussion of computer programs

All six packages described in this book fit the log–linear model. However, programs that examine several models and allow you to get the output of the results in one run appear to be the simplest to use. The features of the six programs are summarized in Table 17.9. Note that SPSS has a useful selection in its HILOG-LINEAR program when combined with its CROSSTAB program. The CAT-MOD and GENMOD procedures in SAS include log-linear modelling along with several other methods. In Stata, log-linear analysis can be performed by using several of the procedures indicated in Table 17.9. Printing the output is recommended for log-linear analysis because it often takes some study to decide which model is desired.

Several of the programs also have options for performing "stepwise" operations. SPSS provides the backward elimination method. STATISTICA starts off with a forward addition mode until it includes all significant k-way interactions and then it does backward elimination. The forward selection method has been shown to have low power unless the sample size is large. When backward elimination is used, the user specifies the starting model. Terms are then deleted in either a *simple* or *multiple effect*. Simple signifies that one term is removed at a time and multiple means that a combination of terms is removed at each step. We recommend the simple option for deletion. SPSS and STATISTICA use the simple option.

The delta option allows the user to add an arbitrary constant to the frequency in each cell. Some investigators who prefer Yates's correction in Pearson chi-square tests add 0.5, particularly when they suspect they have a table that has several cells with low or zero frequencies. This addition tends to result in simpler models and can be overly conservative. It is not recommended that it be uniformly used in log-linear analysis. If a constant is added, the user should try various values and consider constants that are considerably smaller than 0.5. In most programs the default option is delta = 0 but in STATISTICA it is set at 0.5.

17.13   What to watch out for

Log-linear analysis requires special care and the results are often difficult for investigators to interpret. In particular the following should be noted.

1. The choice of variables for inclusion is as important here as it is in multiple regression. If a major variable is omitted you are in essence working with a marginal table (summed over the omitted variables). This can lead to the wrong conclusions since the results for partial and marginal tables can be quite different (as noted in Section 17.7). If the best variables are not considered, no method of log-linear modelling will make up for this lack.

2. It is necessary to carefully consider how many and what subcategories to choose for each of the categorical variables. The results can change dramatically depending on which subcategories are combined or kept separate. We recommend that this be done based on the purpose of the analysis and what you think makes sense. Usually, the results are easier to explain if fewer subcategories are used.

3. Since numerous test results are often examined, the significance levels should be viewed with caution.

Table 17.9: *Software commands and output for log-linear analysis*

	S-PLUS/R	SAS	SPSS	Stata	STATISTICA
Multiway table	table	FREQ	CROSSTAB	table	
Pearson's chi-square	loglin	GENMOD	HILOGLINEAR	glm	Log-Linear Analysis
Likelihood ratio chi-square	loglin	CATMOD*	HILOGLINEAR	glm	Log-Linear Analysis
Test of K-factor interactions		CATMOD*	HILOGLINEAR	lrtest,glm	Log-Linear Analysis
Differences in chi-squares			HILOGLINEAR	lrtest,glm	
Stepwise			HILOGLINEAR	sw:glm	Log-Linear Analysis
Partial association			HILOGLINEAR	poisson;lrtest	Log-Linear Analysis
Structural zeros	loglin	CATMOD*	HILOGLINEAR	zip	Log-Linear Analysis
Delta			HILOGLINEAR		Log-Linear Analysis
Expected values	loglin	CATMOD*	HILOGLINEAR	tabulate	Log-Linear Analysis
Residuals		CATMOD*	HILOGLINEAR	predict	Log-Linear Analysis
Standardized residuals			HILOGLINEAR	predict	Log-Linear Analysis
Lambda estimates	loglin	CATMOD*	HILOGLINEAR	lambda[a]	Log-Linear Analysis

[a] User written command
*Can also use GENMOD.

4. In general, it is easier to interpret the results if the user has a limited number of models to check. The investigator should consider how the sample was taken and any other factors that could assist in specifying the model.

5. If the model that you expect to fit well does not, then you should look at the cell residuals to determine where it fails. This may yield some insight into the reason for the lack of fit. Unless you get a very good fit, residual analysis is an important step.

6. Collapsing tables by eliminating variables should not be done without first considering how this will affect the conclusions. If there is a causal order to the variables (for example gender preceding education preceding income) it is best to eliminate a variable such as income that comes later in the causal order (see Wickens, 1989, for further information on collapsing tables).

7. Clogg and Eliason (1987) provide specific tables for checking whether or not the correct degrees of freedom are used in computer programs. Some programs use incorrect degrees of freedom for models that have particular patterns of structural zeros or sampling zeros.

17.14   Summary

Log-linear analysis is used when the investigator has several categorical variables (nominal or ordinal variables) and wishes to understand the interrelationships among them. It is a useful method for either exploratory analysis or confirmatory analysis. It provides an effective method of analyzing categorical data. The results from log-linear (or logit) analysis can also be compared to what you obtain from logistic regression analysis (if categorical or dummy variables are used for the independent variables).

Log-linear analysis can be used to model special sampling schemes in addition to the ones mentioned in this chapter. Bishop *et al.* (2007) provide material on these additional models.

For log-linear analysis there is not the wealth of experience and study done on model building as there is with regression analysis. This forces users to employ more of their own judgment in interpreting the results. In this chapter we provided a comprehensive introduction to the topic of exploratory model building. We also described the general method for testing the fit of specific models. We recommend that the reader who is approaching the subject for the first time analyze various subsets of the data presented in this book and also some data from a familiar field of application. Experience gained from examining and refining several log-linear models is the best way to become comfortable with this complex topic.

17.15 Problems

17.1 Using the variables AGE and CESD from the depression data, make two new categorical variables called CAGE and CCESD. For CAGE, group together all respondents that are 35 years old or less, 36 up to and including 55, and over 55 into three groups. Also group the CESD into two groups; those with a CESD of ten or less and those with a CESD greater than ten. Compute a two-way table and obtain the usual Pearson chi-square statistic for test of independence. Also, run a log-liner analysis and compare the results.

17.2 Run a log-linear analysis using the variables DRINK, CESD (split at ten or less), and TREAT from the depression data set to see if there is any significant association among these variables.

17.3 Split the variable EDUCAT in the depression data set into two groups, those who completed high school and those who did not. Also split INCOME at 18 or less and greater than 18 (this is thousands of dollars). Run a log-linear analysis of SEX versus grouped EDUCAT versus grouped INCOME and describe what associations you find. Does the causal ordering of these variables suggested in paragraph 6 of Section 17.13 help in understanding the results?

17.4 Perform a log-linear analysis using data from the lung cancer data set described in Section 13.3 and Appendix A. Check if there are any significant associations among Staget, Stagen and Hist. Which variable is resulting in a small number of observations in the cells?

17.5 Run a four-way log-linear analysis using the variables Stagen, Hist, Smokfu and Death in the lung cancer data set. Report on the significant associations that you find.

17.6 Rerun the analyses given in Problem 17.4 adding a delta of 0.01, 0.1, and 0.5 to see if it changes any of the results.

17.7 In problem 17.5, eliminate the four-way interaction, then compare the two models: the one with and the one without Smokfu. Interpret the results.

17.8 Look for significant associations among the variables gender, currently living with (LIVWITH), the financial situation (FINSIT), and having enough money for food (MONFOOD) in the Parental HIV data using log-linear analyses. For the final model assess goodness-of-fit. Interpret the findings.

17.9 This problem uses the Northridge earthquake data set. Look for significant associations in homeowner status (V449), home evacuation status (V173), and status of home damage (V127) using a log-linear analysis.

17.10 (Problem 17.9 continued) Compare the results you get from problem 17.9 and problem 12.35.

Chapter 18

Correlated outcomes regression

18.1 Chapter outline

In Chapter 6 we discussed the correlation coefficient and its interpretation in the context of simple linear regression. In Chapter 7 this concept was extended to multiple linear regression. In addition, Chapter 10 was devoted to canonical correlation analysis. In those chapters the correlation that was examined was between two or more factors. In the example in Chapter 6 the correlation between respiratory function (measured as forced expiratory volume in 1 second) and father's height was examined. The underlying observations (for fathers in this example) were assumed to be independent. If the underlying observations are not independent, regression approaches presented in previous chapters will usually not be appropriate. In this chapter we present statistical methods that can be used when observations cannot be assumed to be independent.

Section 18.2 defines correlated outcome data and describes two fundamental ways in which such data can arise, namely, clustered and longitudinal data. Data examples for this chapter are introduced in Section 18.3. In Section 18.4 the basic concepts associated with correlated outcome data are presented. Sections 18.5 and 18.6 present regression approaches for clustered and longitudinal data. Additional analytical approaches for correlated outcome data are discussed in Section 18.7. The output of the statistical packages is described in Section 18.8 and what to watch out for is given in Section 18.9.

18.2 When is correlated outcomes regression used?

When outcome data cannot be assumed to be independent special regression techniques are needed to account for the dependence. For example, if different clinics have different approaches and procedures for treating patients, we might expect outcome data for patients from the same clinic to be more similar than outcome data from patients from different clinics. When outcome data are expected to be more similar within subgroups than across subgroups the data are typically called **clustered**.

Recall that Chapter 16 focuses on cluster analysis. The difference in the definition of clusters in Chapter 16 and in this chapter is that clusters as defined in Chapter 16 are *unknown* before the analysis and determining which observations belong to which cluster is of central interest. In this chapter membership in a cluster is *known* before data are collected and is typically not of specific interest. Rather, it is known or suspected that membership in a cluster will make it more likely that outcomes for participants or units of the same cluster are more similar compared to outcomes for participants or units from different clusters. In other words, outcomes are expected to be correlated within clusters. We need to ensure that analyses of the outcome take into account such correlation; otherwise conclusions might not be valid.

Correlations that can arise through clusters are often not limited to one level of subgroups. Different groups of clinics might belong to different health care organizations and procedures for patient care might be more similar for clinics belonging to the same health care organization compared to clinics belonging to different health care organizations. In this case, the different levels of clusters form a hierarchy. Models that account for correlated outcome data arising from multiple levels of clusters are often called **hierarchical models** or **contextual models**.

A different setting where outcome data can be expected to be correlated is when the outcome is observed **longitudinally**, i.e., at two or more time points for the same individual or unit. Consider body weight of adult participants as an example. We would not expect body weight to be substantially different when comparing it from one day to the next or even from one week to the next (with the exception of intermittent surgery or giving birth). As a result, we would expect a high correlation of body weights on the same individual across short time periods. When making comparisons across longer time periods, e.g., when comparing body weight at age 20 to age 60 it is unclear how strong such correlation would be (see, e.g., Serdula *et al.,* 1993 or Whitaker *et al.,* 1997). Nevertheless, we might still expect that body weight at age 60 is correlated with body weight of the same person at age 20. But we would not expect body weight at age 60 of any specific individual to be correlated with body weight at age 60 (or any other age for that matter) of a randomly selected different individual. To take this one step further and introduce another level of correlation, we might expect body weight at age 60 to be correlated among siblings. If we were to follow body weight of siblings over time, correlation between different outcome measurements might be expected for multiple measurements over time as well as for clusters that represent families. Also in this case, the correlation structure could be thought of as a hierarchical structure, because observations for the same individuals are nested within observations for family members.

An important characteristic that both settings, hierarchical clusters and **longitudinal data**, have in common is the fact that outcomes are assessed at the individual level, but factors that might impact the individual level outcomes might arise from any or all of the different levels.

18.3 Data examples

The school data set was introduced in Chapter 1 and is described in Appendix A. We will use this data set to illustrate some of the characteristics of clustered outcome data. In the data set, eighth grade math scores are reported for 519 students from 23 schools. We might be interested in identifying factors which are associated with math performance. These factors might be student specific such as gender or how much time per week the student spends on completing the homework. The factors could also be school specific, such as whether the school is public or private, or whether the school is part of an urban, suburban, or rural area. A list of school and student level variables is provided in Table 18.1. Because students are considered nested within schools, any student specific variables are also called *lower level* or *micro level* variables. School specific variables are also called *higher level* or *macro level* or *contextual* variables. In the school example one might imagine a number of additional levels, e.g., students within classes, classes within schools, schools within school districts, and school districts within states. In this case, the lowest level of nesting (here again the student specific variables) is called the micro level and the highest level (state specific variables in the hypothetical hierarchy or school specific variables in the school data set) is called the macro level (see Figure 18.1).

Table 18.1: *School and student level variables in the school data set*

School (macro) level	Variable
	School type
	Class structure
	School size
	Urbanity
	Geographic region
	Percent minority
	Student-teacher ratio
Student (micro) level	Variable
	Gender
	Race
	Time spent on math homework
	Socioeconomic status (SES)
	Parental education
	Math score

For these data the math scores might be more similar within schools than

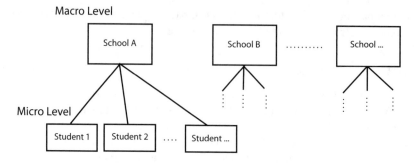

Figure 18.1: *Graphical Representation of the School (Hierarchical) Data*

across schools. One way of exploring this possibility graphically is to present summary statistics for math scores by school. In Figure 18.2 boxplots for math scores are plotted separately for each school. When ordering the schools by their median scores, as in this graph, it is easy to see that there is quite a difference between the lowest and highest median math scores and the variability in scores. However, it is difficult to tell from this graph whether and to what extent the clustering could affect estimation of the effect of other factors.

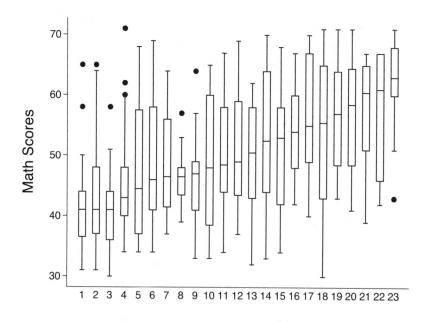

Figure 18.2: *Boxplots of Math Scores by Schools*

Assume for now that the main question we would like to answer is whether or not average math scores differ between students who attend different school types: public schools vs. private schools. For purposes of comparison only we will initially present results from three linear regression models that naïvely ignore the hierarchical structure of the school data. Results from these analyses are presented in Table 18.2. Model 1 includes as the only covariate an indicator variable with a value of one for students who attend public schools and zero for students who attend private schools. The estimated coefficient has a value of -7.8. This might (naïvely) be interpreted as follows: students who attend public schools have an eighth grade math score that is on average 7.8 points lower than students who attend private schools. With a p value that is less than 0.001 this difference is statistically significant. In Model 2, the effect of school type is adjusted for the student's socioeconomic status (SES). After adjusting for SES, the effect of school type is notably reduced to -2.7, but this difference remains statistically significant. Model 3 not only adjusts for SES, but also for the number of hours the student spends on math homework per week. Note that we are using dummy variables for the number of hours spent on completing homework as coded in the original data set (with no time spent on homework as a referent, e.g., the value of 1 represents "less than one hour" and the value of 6 represents 7 to 9 hours, etc.). Because the coded values do not represent the actual number of hours it is unclear whether it would be appropriate to assume a linear relationship between the coded values and average math scores. Because there are only very few students who report spending 7 to 9 hours or 10 or more hours on their homework (6 and 3, respectively), we combine the three categories with the highest number of hours. After adjusting for SES and the number of hours spent on math homework per week, the estimate for school type is further reduced to -1.6 and is not statistically significant at an alpha level of 0.05. Note that it is debatable whether the number of hours spent on homework should be adjusted for or not. On the one hand, it might be argued that we want to compare math performance between public and private schools between students who spend the same amount of time on homework. In this case we would want to adjust for this factor. On the other hand, it might be argued that one of the factors through which math performance might be improved is through spending more time on homework. If students at private schools (or public schools for that matter) were more successful in achieving high math scores because students do more homework, adding this factor to the model would "remove" the effect of the homework from the effect the school type has. To decide whether or not this factor should be included one might ask the questions "Do we want to know what the difference is with respect to math scores between public and private schools if the same average amount of homework time was spent in both settings?" or "Do we want to know what the difference is with respect to math scores between public and private schools even if this in part might be due to differences between the amount of time spent on homework?". In the first case we would want to adjust for homework time and in the second we would not.

Thus far, we have ignored the hierarchical structure of the data and the fact that students from the same school might be expected to perform more similarly than students from different schools. We will discuss the potential impact of this aspect and how to account for it in the analyses in the next sections.

Table 18.2: *Estimated coefficients from three naïve linear regression models ignoring the hierarchical structure of the school data*

Factor	Coefficient	Standard error	P value
Model 1			
Public vs Private School	−7.8	0.9	< 0.001
Constant	56.5	0.7	< 0.001
Model 2			
Public vs Private School	−2.7	1.0	0.007
SES (per one unit increase)	5.1	0.6	< 0.001
Constant	53.4	0.7	< 0.001
Model 3			
Public vs Private School	−1.6	1.0	0.09
SES (per one unit increase)	4.3	0.5	< 0.001
Time per week spent			
on Math Homework			
Referent: None			
Less than one hour	−1.4	1.5	0.34
One hour	0.2	1.6	0.89
Two hours	5.2	1.9	0.005
Three hours	7.6	1.9	< 0.001
Four or more hours	7.6	1.9	< 0.001
Constant	51.4	1.5	< 0.001

18.4   Basic concepts

Model 2 in Table 18.2 can be written as

$$\text{MathScore}_i = \beta_0 + \beta_1 \text{SchoolType}_i + \beta_2 \text{SES}_i + e_i$$

where MathScore$_i$ represents the eighth grade math test score for individual i,

SchoolType$_i$ represents an indicator variable with a value of one if student i attends a public school and a value of zero if student i attends a private school, SES$_i$ represents the socioeconomic status of student i, and e_i represents the residual. This residual is defined in a way similar to the definition used in Chapter 6 in that it is the difference between the individual outcome values (here the math scores) and the ones that are predicted by the model. Of note, SES is a derived continuous, centered variable which was created from parents' education levels, both parents' occupations, and family income (the values range from -2.41 to 1.85; for more information consult the web site mentioned in Appendix A). In more general terms, the model can be written as:

$$Y_i = \beta_0 + \beta_1 X_{1i} + \beta_2 X_{2i} + e_i$$

Note that in the above formulation of the model the subscript i runs from 1 to 519 for the school data set, which contains 519 students.

Again, the above model does not take into account the hierarchical nature of the data, i.e., the fact that students are nested within schools. One might consider adding indicator variables for each school to the model above. Such a model could be written as

$$\begin{aligned} \text{MathScore}_i \;=\; & \beta_0 + \beta_1 \text{SchoolType}_i + \beta_2 \text{SES}_i \\ & + \beta_3 (\text{School2})_i + \ldots + \beta_{24} (\text{School23})_i + e_i \end{aligned}$$

where School2 up to School23 represent indicator variables which take on the value of one if student i belongs to the school indicated.

Or again in more general terms

$$Y_i = \beta_0 + \beta_1 X_{1i} + \beta_2 X_{2i} + \beta_3 X_{3i} + \ldots + \beta_{24} X_{24i} + e_i$$

Note that this model contains 22 dummy variables representing schools 2 through 23 with the first school serving as the reference category. In addition, there are only a few students (5 and 8) who are contributing math scores for two of the schools.

Fixed and random effects

The above model treats differences between schools as **fixed effects**. When considering school as a fixed effect each school (or in general each level of the factor) is thought to have its own specific effect on the math scores. This effect is then estimated in relation to the referent group mean (β_0). Some factors are fixed effects by their nature. Examples are school type, where we are interested in the average difference in math scores between public and private schools and gender, where we are interested in the average difference in math scores between males and females. If schools are chosen at random from all schools that might be of interest, then the schools that are sampled might be thought of as representative of all the schools that could have been sampled. In this case, the effect that we observe for any particular school would be viewed as a random draw from all possible schools that we could have sampled. When we view a factor in this way, we consider the effect of the factor as a **random effect**. As such we are more interested in the distribution of the effects of a number of different schools than the effect of any individual school.

For simplicity, we first consider a model which assesses only two effects, the effect of SES (fixed) and the effect of school (random). The model can be described at two levels, the individual (micro) level and the school (macro) level. For the individual level we are considering

$$\text{MathScore}_{ij} = \beta_{0j} + \beta_1 \text{SES}_{ij} + e_{ij}$$

where i now runs from 1 to the number of students who contribute math scores from their specific school and j is a subscript for schools which runs from 1 to 23. At the school level we are considering

$$\beta_{0j} = \gamma_{00} + U_{0j}$$

where γ_{00} represents an average intercept across all schools and U_{0j} represents the value of the random effect (random intercept) for school j (which is expressed as the difference between the slope for school j and the average slope across all schools). Since this model is expressed in terms of two equations, one at the student level and one at the school level, it is called a **multilevel model**. It is also known as a **hierarchical model** because it incorporates the hierarchical nature of the data. This model can also be written as a single equation by replacing β_{0j} in the first equation with the right side expression in the second equation:

$$\text{MathScore}_{ij} = \gamma_{00} + \beta_1 \text{SES}_{ij} + U_{0j} + e_{ij}$$

This single equation is called the **linear mixed model** since it combines the fixed effect of SES (slope) with a random intercept effect. From a pure notational perspective, the term relating to SES has not changed compared to the model introduced in the beginning of Section 18.4 (except for the numbering, which initially ran from 1 to 519 for a single subscript and now has a double subscript but includes the same 519 individuals). However, we now have an intercept term called γ_{00}, which represents an overall average intercept across all individual school intercepts, and an additional term U_{0j}, which does not have a coefficient and represents the difference between the overall average school intercept and the school specific intercept for school j. From a conceptual perspective the estimation and interpretation of the slope coefficient for SES have not changed, but the estimation and interpretation of the intercept and the effect of school have changed. Instead of assuming that there is an overall average of math scores (the former intercept coefficient β_0) from which each school deviates by a fixed amount, the above model assumes that the adjusted average school math scores are normally distributed and are centered at γ_{00}. Because we assume that there is a distribution of intercepts, this model is also called a **random intercept model**. This might seem like a subtle difference, but it represents a fundamental conceptual departure. In addition to changes in estimation as well as interpretation of some of the estimated coefficients, this model can readily incorporate effects that act at the school level.

Consider now the effect of school type (public versus private). Because this effect is school (not student) specific, it is incorporated into the school level equation.

$$
\begin{aligned}
\text{MathScore}_{ij} &= \beta_{0j} + \beta_1 \text{SES}_{ij} + e_{ij} \\
\beta_{0j} &= \gamma_{00} + \gamma_{01} \text{SchoolType}_j + U_{0j}
\end{aligned}
$$

Now consider adding the effect of the number of hours spent on homework to this model. Because this effect is student specific, it is incorporated into the student level equation.

$$
\begin{aligned}
\text{MathScore}_{ij} &= \beta_{0j} + \beta_1 \text{SES}_{ij} + \beta_2 \text{Homework1}_{ij} \\
&\quad + \beta_3 \text{Homework2}_{ij} + \cdots + \beta_6 \text{Homework5}_{ij} + e_{ij} \\
\beta_{0j} &= \gamma_{00} + \gamma_{01} \text{SchoolType}_j + U_{0j}
\end{aligned}
$$

Note, in the above equation, Homework1 is an indicator variable for students who spend less than one hour on homework and, similarly, Homework2, Homework3, Homework4 and Homework5 represent one hour, two hours, three hours, and four or more hours of homework, respectively. Again, both of the above equations could be written in a single line (equation), but we will continue to write them in a hierarchical manner as it is easier to distinguish which factors are individual or school level factors and how they are conceptualized and interpreted.

Components of variance

Results for the above random intercept models are presented in Table 18.3. The first part of Table 18.3 contains results regarding the fixed effects. When comparing the Model 3 estimates from Table 18.3 with the Model 3 estimates from Table 18.2 we notice that the coefficient for school type (private versus public) is almost exactly the same (-1.6 in Table 18.2 and -1.7 in Table 18.3). However, the estimate for the standard deviation is different (1.0 in Table 18.2 and

Table 18.3: *Estimated coefficients from three random slope regression models accounting for the hierarchical structure of the school data*

Fixed Terms	Coefficient	Standard error	P value
Model 1			
Public vs Private School	−5.9	2.1	0.006
Constant	54.7	01.7	< 0.001
Model 2			
Public vs Private School	−2.5	1.8	0.18
SES (per one unit increase)	4.1	0.6	< 0.001
Constant	52.8	1.5	< 0.001
Model 3			
Public vs Private School	−1.7	1.8	0.36
SES (per one unit increase)	3.5	0.6	< 0.001
Time per week spent			
on Math Homework			
Referent: None			
Less than one hour	−1.3	1.8	0.37
One hour	0.7	1.5	0.64
Two hours	5.3	1.8	0.003
Three hours	7.6	1.8	< 0.001
Four or more hours	7.7	1.9	< 0.001
Constant	51.0	1.9	< 0.001

Random Terms Standard Deviation	Estimate
Model 1	
School	4.4
Residual Error	9.0
Model 2	
School	3.5
Residual Error	8.7
Model 3	
School	3.5
Residual Error	8.7

1.8 in Table 18.3). This leads to a change in p value from 0.09 (Table 18.2) to 0.36 (Table 18.3). It is often, but not always, the case that the estimates of variability increase after the hierarchical nature of the data is taken into account. As a result, p values typically (but not always) are larger when compared to models which do not account for the hierarchical nature of the data.

The second part of Table 18.3 contains results regarding the random effects. Two items are presented for each of the three models, one for school and one for residual error. For Models 2 and 3 the standard deviation of the school random effect is 3.5 and the standard deviation of the residual error is 8.7. This can be interpreted as follows. After adjusting for other factors, the standard deviation of the average school math scores random intercepts U_{0j} is 3.5 and the standard deviation of the individual math scores e_{ij} is 8.7. These two standard deviations, σ_U and σ_e (after dropping the subscripts), represent the variability that remains after adjusting for the fixed effects in the model. Their squared values are called the **components of variance**. The ratio

$$\rho = \sigma_U^2/(\sigma_U^2 + \sigma_e^2)$$

is called the **intraclass correlation coefficient**. It represents the proportion of total variance that is explained by differences between schools. In our example (Table 18.3) the intraclass correlation coefficient for Model 1 is approximately

$$\rho = 3.6^2/(3.6^2 + 8.7^2) = 0.15$$

i.e., about 15% of the total variability is explained by the clusters. The intraclass correlation coefficient can also be interpreted as the expected correlation between two students who are randomly drawn from the same cluster.

We will discuss extensions of the above concepts in subsequent sections.

18.5 Regression of clustered data

Random slopes

More complex models are possible. For example, we might think that the effect of SES on math scores is different for different schools. We can choose a model that allows each school to have its own SES slope. Because the effect of schools is considered a random effect, the SES slope would be a random effect as well as the intercept. In this case not only the intercept, but also the slope for SES is modeled at the school level and represents a second random effect called a **random slope**. The mathematical representation of this random intercept and random slope model is

$$
\begin{aligned}
\text{MathScore}_{ij} &= \beta_{0j} + \beta_{1j}\text{SES}_{ij} + e_{ij} \\
\beta_{0j} &= \gamma_{00} + \gamma_{01}\text{SchoolType}_j + U_{0j} \\
\beta_{1j} &= \gamma_{10} + U_{1j}
\end{aligned}
$$

where γ_{10} represents an average slope across all schools and U_{1j} represents the value of the random effect (random slope) for school j (which is expressed as the difference between the slope for school j and the average slope across all schools). For this model, the combined equation becomes:

$$\text{MathScore}_{ij} = \gamma_{00} + \gamma_{01}\text{SchoolType}_j + \gamma_{10}\text{SES}_{ij} + U_{1j}\text{SES}_{ij} \\ + U_{0j} + e_{ij}$$

The term U_{1j} in the combined equation represents the random slope for SES.

If the effect of gender is added to the model and considered to be different for different schools, then the coefficient for gender could be conceptualized as a random effect as well and another equation added for such an effect β_{2j}.

$$\begin{aligned} \text{MathScore}_{ij} &= \beta_{0j} + \beta_{1j}\text{SES}_{ij} + \beta_{2j}\text{gender}_{ij} + e_{ij} \\ \beta_{0j} &= \gamma_{00} + \gamma_{01}\text{SchoolType}_j + U_{0j} \\ \beta_{1j} &= \gamma_{10} + U_{1j} \\ \beta_{2j} &= \gamma_{20} + U_{2j} \end{aligned}$$

The above model includes a random intercept as well as two random slopes. If there are two or more random effects included in the model, then these random effects potentially could be correlated. This correlation would be separate from the one that is introduced by the hierarchical nature of the data, but statistical programs provide an option to model any correlation between random effects. Unless there is a reason to believe that two (or more) random effects are correlated, the correlation between different random effects is typically assumed to be zero and this is the default for many statistical software packages. Nevertheless, other correlation structures can be important. Because specific correlation structures are essential for longitudinal data they are described in Section 18.6.

To graphically illustrate the differences between the naïve, the random intercept, and the random intercept and slope models, we fit each of these models using SES as the only variable. For the naïve model, an indicator variable is created for each of the schools. Note again that there are several concerns regarding the naïve model and we recommend to not use this approach and only include it here for illustrative purposes. For the random intercept and the random intercept and slope models we predict the random effects for each of the schools. Because the graph becomes very cluttered if all schools are depicted, only six schools are shown in Figure 18.3. For these data there is not much difference between the random intercept and the random intercept and slope models. This would remain true if all schools were shown. Another observation that can be made from this graph relates to the location of the lines in the graph. When focusing on the lower half of the graph, the predicted lines from the mixed effects models are higher than the ones predicted from the naïve model. The opposite is true for the lines in the upper half of the graph. For the upper half of the graph, the predicted lines from the mixed effects models are lower than the ones predicted from the naïve model. That is, the predicted lines are closer to what would be considered the center of the lines for the mixed effects models. This characteristic is also called shrinkage (towards the average) and describes a mathematical property of mixed effects models. A detailed description of this property is beyond the scope of this book, but further information regarding prediction and shrinkage in mixed effects models is given in Fitzmaurice, Laird and Ware (2004) or Weiss (2005).

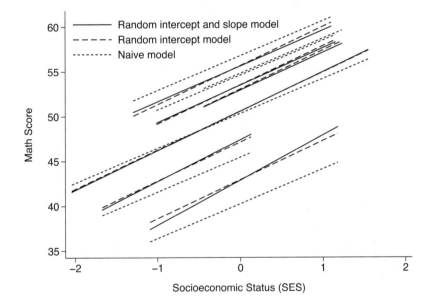

Figure 18.3: *Predicted Slopes for Naïve, Random Intercept and Random Intercept and Slope Models for Six Schools*

Interactions

If school type was thought to influence the effect SES has on math scores (it might, e.g., be hypothesized that the effect of SES is different for private and public schools), school type could be incorporated into the equation for β_{1j} just as it was incorporated into the equation for β_{0j} (the equation that characterizes the random intercept). Suppose we model math scores as a function of SES and school type using a random intercept model (with fixed effects/slopes for SES and school type, Model 2 in Table 18.3). To incorporate an interaction between SES and school type, we write the model as follows:

Micro level:

$$\text{MathScore}_{ij} = \beta_{0j} + \beta_{1j}\text{SES}_{ij} + e_{ij}$$

Macro level:

$$\beta_{0j} = \gamma_{00} + \gamma_{01}\text{SchoolType}_j + U_{0j}$$
$$\beta_{1j} = \gamma_{10} + \gamma_{11}\text{SchoolType}_j$$

The combined equation (linear mixed model) becomes:

$$\text{MathScore}_{ij} = \gamma_{00} + \gamma_{01}\text{SchoolType}_j + \gamma_{10}\text{SES}_{ij} \\ + \gamma_{11}\text{SchoolType}_j\text{SES}_{ij} + U_{0j} + e_{ij}$$

In this equation, the term $\text{SchoolType}_j\text{SES}_{ij}$ represents an interaction of SES and school type, while γ_{11} denotes its regression coefficient. This interaction is called a cross-level interaction since SES is a student level variable and school type is a school level variable.

As one might suspect, a hierarchical model can easily become very complicated. It is always helpful, if not essential, to write down the model similarly to what has been done above. This process clarifies which factors are thought to act at which level and which interactions will be considered. Only those interactions that seem sensible in the research context should be considered. With respect to the interpretation of effects that are involved in interaction terms, caution is in order, as we mentioned regarding interpretation of interaction terms in other chapters of this book. That is, if an interaction effect is included in the model the main effects should be included as well and the presentation and interpretation of effects of the factors that are involved in an interaction need to take into account that the effect of one factor depends on the level/value of the other factor.

Centering within clusters

An additional consideration for hierarchical models is that it might be better to measure the effect of a specific level of a factor relative to the average cluster rather than to the overall average. An example of such a factor in the educational setting is the student's intelligence quotient (IQ). As pointed out by many authors (the "frog pond effect", see, e.g., Bachman and O'Malley, 1986), a student with an average IQ might be more confident and excel in a group of students with less than average IQ. He or she might, on the other hand, be discouraged and not perform to his or her potential in a group of students with higher than average IQ.

In general, if the effect of a specific level of a factor is dependent on where the level is in reference to other cluster members more so than where the level is in reference to all other participants, a few simple steps are necessary to model the covariate accordingly. Instead of using the actual value (e.g., IQ score of the individual) in the regression model, one first calculates cluster specific (in this case school) averages and calculates the difference between the individual values and their specific cluster (school) average. Both the school average and the difference (between individuals' IQ and their school's average) are then included in the model. A term for the difference between individual value and cluster average is then modeled at the individual level and an additional term for the cluster (school) average is modeled at the school level.

Options for model fitting

Some practical challenges arise when fitting hierarchical models (such as models 1, 2 and 3 in Table 18.3) in contrast to fitting linear models that assume independent observations (such as models 1, 2 and 3 in Table 18.2). One of them is the fact that estimates cannot be calculated as a straightforward solution to a number of equations, but that estimates need to be calculated in an iterative manner. This is similar to what was described, e.g., for logistic regression and survival analysis. As a result, the estimates that are obtained represent approximations. The **maximum likelihood (ML)** approach can be used in this setting. ML was mentioned previously in this book as an approach to estimating parameters. A more detailed discussion of this approach is beyond the scope of this book, but in a nutshell, maximum likelihood can be described as a method of finding those parameter estimates that best fit the observed data. To apply the ML method, some assumptions are necessary. The most commonly made assumptions are that the random e and U terms in the model equations are independent and normally distributed. In the context of hierarchical models, there is another method which is often used for obtaining estimates that is closely related to maximum likelihood estimation. This other method is called **restricted (or residual) maximum likelihood (REML)**. REML attempts to account for the number of fixed effects that are estimated whereas ML does not (see, e.g., Verbeke and Molenberghs, 2000).

It is important to know which estimation procedure is used by the specific statistical software, because some of the typical testing procedures (e.g., likelihood ratio test) might not be valid depending on what is tested. When the likelihood ratio test is used for nested models based on REML estimates, the test does not yield valid results when the models compared include different fixed effects. Note that REML is the default estimation approach for most software packages. If we want to compare two models with different fixed effects using a likelihood ratio test, we need to use the maximum likelihood estimation method rather than the restricted maximum likelihood estimation method. This comment applies only to the likelihood ratio test. We could use the Wald test statistic with either the REML or ML method and obtain valid results for both.

Consider again the models presented in Table 18.3. These estimates were obtained using the REML method. The test statistic presented for the effect of school type in model 3, e.g., represents the Wald test statistic and as such is valid whether the MLE or the REML approach is used. For the amount of time spent on homework, on the other hand, each p value represents valid Wald test statistics comparing each of the levels to the referent group (of no time spent on homework). To test the overall effect of the amount of time spent on homework, a likelihood ratio test statistic can be performed comparing a model with all homework indicators included to a model with all homework indicators excluded. For this likelihood ratio test to be valid, we would need to use the ML estimation approach and explicitly specify this in the statistical software.

If there is interest in testing whether a random effect has variability of zero (which would represent, for example, the question of whether it is necessary to include a random effect for schools), one has to be cautious as well regarding the likelihood ratio test. In this case the test statistic does not follow a chi-square distribution with the typical degrees of freedom, and an adjustment has to be made to the degrees of freedom. Some implementations of the likelihood ratio test make this adjustment automatically and others do not. It is important to check the specific software manuals.

Another potential issue that can occasionally arise when fitting hierarchical models is that the iterative process of obtaining estimates might never arrive at a solution. Because of the iterative process of obtaining estimates, the statistical program typically has to perform estimation steps (as part of a programming loop) over and over until the change in the estimates from one iteration to the next is extremely small. Once that happens the program stops the iterative process and reports the last iteration. Occasionally, the change in estimates from one iteration to the next never reaches a level where it is small enough to meet the criteria to stop the loop. In that case, the software typically informs the user that the model did not converge after performing the loop a larger number of times. This can also happen for other iterative estimation procedures such as multiple logistic regression and survival analysis, but it very rarely does. Depending on the complexity of the model and the specific data this can be an issue for fitting hierarchical models. Cheng *et al.* (2010) report that in their experience scaling, centering and avoiding collinearity can alleviate problems with convergence.

Despite the challenges described above, other conceptual aspects of model fitting are very similar for hierarchical models as compared to linear, logistic and survival analysis regression. One of the similar tasks is to identify predictors that are to be included in the model. When **nested models** are compared, the Wald test and the likelihood ratio test can be used with the caveats described above. Models are called nested if the set of factors considered for one of the models is a subset of the factors considered for the other model and both models have the same parameterization for the common factors. When models are compared that are not nested, neither the Wald test nor the likelihood ratio test is valid. For non-nested comparisons the Akaike Information Criterion (AIC) and the Bayes Information Criterion (BIC) can be used as a guideline. Similar to the likelihood ratio test, AIC and BIC should only be compared for models that use the ML estimation approach. The AIC is described in Section 8.4. The BIC is defined in a similar manner and relies on Bayesian concepts (see, e.g., Fitzmaurice, Laird and Ware, 2004 or Weiss, 2005). If the REML estimation approach is used, the fixed effects of the compared models need to be the same to ensure that the comparisons are valid.

Again, AIC and BIC should be used only as a guideline, because they cannot be interpreted as statistical tests of a specific hypothesis. Higher values for the AIC or BIC values are desirable. Note, though, that some programs report the AIC and BIC values multiplied by a negative factor, in which case smaller values would be desired. Both criteria make use of an overall likelihood for a fitted model. With more covariates in the model the likelihood value will typically be higher. As a result, if based mainly on the likelihood value, a model which includes more covariates would be considered preferable. Both criteria (AIC and BIC) attempt to penalize models with more covariates.

One might want to find the best model with exactly two (or three or four, etc.) individual level covariates. When such models are compared this is often called subset regression (see Section 8.7). A comparison of the AIC and BIC criteria for all possible combinations of two and three individual level covariates (Student or micro level variables from Table 18.1) is provided in Table 18.4.

For the models in Tables 18.4 a random intercept was used for schools and race was modeled using three categories (black, other and white), homework was modeled using six categories (none, less than one hour, one hour, two hours, three hours, four or more hours), and parental education was modeled using six categories (no high school degree, high school or General Educational Development (GED) exam, high school and less than four years of college, college graduate, master's degree, doctorate degree).

Overall, the values of AIC and BIC seem relatively close and range from

Table 18.4: *AIC and BIC criteria for all possible combinations of two individual level factors (with random intercept for schools)*

Factor 1	Factor 2	Factor 3	AIC	BIC
Two factors				
Sex	Race		3792.3	3799.1
Sex	Homework		3733.1	3743.4
Sex	SES		3758.3	3764.0
Sex	Parental Education		3745.7	3756.0
Race	Homework		3719.7	3731.1
Race	SES		3751.3	3758.1
Race	Parental Education		3734.6	3746.0
Homework	SES		3692.6	3702.8
Homework	Parental Education		3681.4	3696.2
SES	Parental Education		3744.6	3754.8
Three factors				
Sex	Race	Homework	3721.0	3733.5
Sex	Race	SES	3753.2	3761.2
Sex	Race	Parental Education	3736.6	3749.1
Sex	Homework	SES	3694.1	3705.5
Sex	Homework	Parental Education	3683.3	3699.2
Sex	SES	Parental Education	3746.6	3757.9
Race	Homework	SES	3688.7	3701.2
Race	Homework	Parental Education	3674.5	3691.5
Race	SES	Parental Education	3736.6	3749.1
Homework	SES	Parental Education	3683.2	3699.1

3681.4 to 3799.1 for models with two covariates and from 3674.5 to 3761.2 for models with three covariates. Of note, BIC values are always slightly higher than AIC values. Because SAS was used for these analysis and SAS reports the AIC and BIC after multiplying them with a negative factor, smaller values for AIC and BIC are desirable. The best model with two factors includes homework and parental education and the best model with three factors includes homework, parental education and race. Based on the values of AIC and BIC, the model with three variables is preferable to the one with two. Again, the AIC and BIC statistics should be used as a guideline and cannot be used to test whether one model represents a statistically significant improvement over another.

Another modeling task that is conceptually similar for hierarchical models and other regression approaches is the need for assessing model fit and iden-tification of highly influential individual observations or clusters. For details the reader is referred to, e.g., Weiss (2005), McCulloch and Searle (2001), and McCullagh and Nelder (1989).

18.6 Regression of longitudinal data

As mentioned in Section 18.2, a special case of clustered outcome data occurs when the same person or unit is observed at multiple time points. These data are called **longitudinal data**. Statistical approaches that are used for clustered data are very similar to those used for longitudinal measurements. The main difference between the correlation arising from the types of data is that there is an order according to chronological time for the longitudinal data, whereas there is typically no order regarding the clustered data that have no longitudinal component.

Consider again the example of measuring body weight mentioned in Sec-tion 18.2. If body weight is measured in adults over a number of weeks the strength of the correlation is likely stronger than if body weight is measured over a number of years or decades. For example, if weight is measured at base-line, after 3 months, 2 years and 20 years, we would expect the correlation to be strongest for the measurements made at baseline and 3 months. We would expect it to be weak between the measurements made at baseline and 20 years or between 3 months and 20 years.

These differences in strength of the correlation can be modeled by assum-ing a specific structure of a correlation matrix that represents the correlation between the different time points for the observations. Such a correlation ma-trix is arranged similarly to a covariance matrix (see, e.g., Section 7.5) with the difference that the entries in the correlation matrix represent correlation coeffi-cients rather than covariances. As a result, the diagonal terms in the correlation matrix are all ones. All non-diagonal entries are between zero and one (inclu-sive) and represent correlations between successive measurements that are one time period apart, two time periods apart, and so forth. Like covariance matri-ces, all correlation matrices are symmetric in the sense that the entry in column i and row j is the same as in column j and row i. If the rows and columns repre-sent time points that are equally spaced, e.g., four follow-up time points every three months, then a potential structure and representation of the correlation induced by such a design might be:

	y_1	y_2	y_3	y_4
y_1	1	a	b	c
y_2	a	1	a	b
y_3	b	a	1	a
y_4	c	b	a	1

In the correlation matrix above outcome measures obtained at the first visit are labeled y_1, outcome measures obtained at the second visit are labeled y_2, etc. For the structure of this correlation matrix it is assumed that the correlation is the same between visits that are the same number of months apart: the correlation between visits 1 and 2, visits 2 and 3 and visits 3 and 4 is the same (a) because each pair is separated by 3 months. A different correlation (b) is assumed between visits 1 and 3 and visits 2 and 4. Each of these are 6 months apart and we might expect that b is less than a. Visits 1 and 4 are 9 months apart and the correlation between them is labeled c.

As a result, there are three correlation coefficients which would be estimated as part of the estimation process. A special case of the above matrix which is often used for longitudinal studies is called **autoregressive of first order [AR(1)]**. Under AR(1), the above matrix has the form:

	y_1	y_2	y_3	y_4
y_1	1	ρ	ρ^2	ρ^3
y_2	ρ	1	ρ	ρ^2
y_3	ρ^2	ρ	1	ρ
y_4	ρ^3	ρ^2	ρ	1

As an example, if $\rho = 0.5$, then the correlation matrix becomes:

	y_1	y_2	y_3	y_4
y_1	1	0.5	0.25	0.125
y_2	0.5	1	0.5	0.25
y_3	0.25	0.5	1	0.5
y_4	0.125	0.25	0.5	1

Whether or not the above correlation structure provides a reasonable representation of the data will depend on the specific outcome and the spacing between the visits. If follow-up time points are not equally spaced in a substantial way (such as the example of baseline, 3 months, 2 years and 20 years follow-up), then neither of the above matrices is likely to provide an adequate correlation structure. In this case it might be necessary to estimate all of the non-diagonal elements of the correlation matrix separately. Such a correlation matrix is called **unstructured** and can be represented as

	y_1	y_2	y_3	y_4
y_1	1	a	b	c
y_2	a	1	d	e
y_3	b	d	1	f
y_4	c	e	f	1

With four follow-up visits six parameters need to be estimated for this correlation matrix. With additional follow-up visits the number of parameters that need to be estimated can increase quickly though.

Another correlation structure that is available in most statistical software is called **exchangeable** or **compound symmetric**. In this structure it is assumed that all correlation coefficients have the same value (with the exception of the diagonal). A representation of such a correlation matrix is provided below.

	y_1	y_2	y_3	y_4
y_1	1	a	a	a
y_2	a	1	a	a
y_3	a	a	1	a
y_4	a	a	a	1

On the one hand, this correlation structure seems to rely on a strong assumption in the sense that the correlation between any of the follow-up visits is assumed to be the same regardless whether it is, e.g., between the first and second or the first and last visit. On the other hand, if the correlation is not of primary interest, this representation might be sufficient in terms of estimating the effects of other factors.

Data example

To illustrate longitudinal data analyses we use the publicly available data on the growth of mice (see, e.g., Izenman and Williams, 1989, for a more detailed description). The original data consist of four groups of mice. The weights of the mice were measured on a rotating basis for three of the groups and the fourth group was measured every day. For illustrative purposes, we use the outcome measures for groups 3 and 4 for only those days where group 3 was measured. Figure 18.4 provides a graphical representation of the weight measures over time for groups 3 and 4.

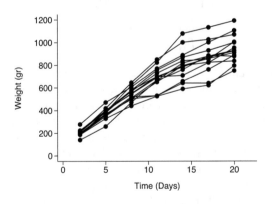

Figure 18.4: *Weight over Time for 14 Mice*

Based on Figure 18.4, one might model the growth of the mice linearly. In the hierarchical model notation using subscript i for observations made on different days and subscript j for different mice and using a model with only a random intercept, this can be described as

$$
\begin{aligned}
\text{Weight}_{ij} &= \beta_{0j} + \beta_{1j}\text{Day}_i + e_{ij} \\
\beta_{0j} &= \gamma_{00} + U_{0j} \\
\beta_{1j} &= \gamma_{10}
\end{aligned}
$$

For the above model it is typically assumed that the e_{ij} are normally distributed with mean zero. The weights of different mice (on the same or different days) are assumed to not be correlated, i.e., the correlation between e_{ij} and e_{i*j*} (where $j \neq j^*$) is zero. But the correlation between e_{ij} and e_{i*j} (i.e., the correlation between different times of observation for the same mouse) is modeled according to one of the correlation matrices described above [AR(1), unstructured or exchangeable]. In other words, observations that lie on the same line in Figure 18.4 are assumed to be correlated, but any observations from different lines are not.

As with the hierarchical models described in previous sections, the hierarchical nature of the data can easily become very complicated. In the case of the mice, data potentially could be correlated on a different level, e.g., if the mice came from different litters.

If we want to model not only the intercept, but also the slope, as random, then this can be formulated as:

$$
\begin{aligned}
\text{Weight}_{ij} &= \beta_{0j} + \beta_{1j}\text{Day} + e_{ij} \\
\beta_{0j} &= \gamma_{00} + U_{0j} \\
\beta_{1j} &= \gamma_{10} + U_{1j}
\end{aligned}
$$

If there are multiple random effects (in the above model there are two, the intercept and the slope), then there might be a correlation between those random effects. In some situations, the outcome measure might have a natural maximum level. In the case of the mice, there is a natural limit in maximum weight and if the mice had been observed longer, we would have expected that the weight measures for each mouse might plateau after some time. The value that the slope could achieve might then depend on the level of the intercept. An initial high value of weight might not leave as much room for growth as a lower initial value of weight. This might therefore induce a correlation between the random slope and intercept parameters. Such a correlation is separate from the correlation induced by observing the same mouse multiple times, but potential correlation structures are the same as described above. However, because there is no natural ordering between the components of multiple random effects, the AR(1) structure is ordinarily not appropriate for this type of correlation. Also,

Table 18.5: *Estimates for random intercept and random slope models with different correlation structures*

	Un, AR(1)	Un, CS
Intercept Estimate	156.8	180.4
Intercept, Standard Error	22.0	11.3
Intercept, p value	<0.0001	<0.0001
Day, Estimate	41.1	41.1
Day, Standard Error	2.0	2.2
Day, p value	<0.0001	<0.0001
AIC	1093.7	1128.3
BIC	1096.9	1131.5

with only two random effects, the exchangeable and unstructured correlation structures are the same.

Note also that in the case of a plateau effect of weight (if the mice are observed sufficiently long), it would likely be unreasonable to model the weight as a linear function. In fact, examining Figure 18.4 more closely, one might suspect that a quadratic function might fit the data better than a linear function. Using a quadratic function for days and evaluating different models is part of the exercises at the end of this chapter.

Estimates for the average intercept and slope are provided in Table 18.5 for two models. Both models use an unstructured correlation matrix for the correlation between the random effects. One model uses the AR(1) structure for the correlation over time, whereas the other model uses a compound symmetric structure (also called exchangeable). Of note, when attempting to use an unstructured correlation matrix for the correlation over time for the mice data, the model fails to converge, because there are 28 parameters to be estimated for correlation matrix alone.

Recall that lower values for AIC and BIC are preferable. Because of the lower values for AIC and BIC for the first model and because the AR(1) model should conceptually be a better fit for these data, we would prefer the first model (with AR(1)).

18.7   Other analyses of correlated outcomes

All of the approaches described thus far in this chapter represent what are called hierarchical or mixed effects models. Another regression approach that can be used for correlated outcome data is called **generalized estimating equations (GEE)**. One of the key features of this approach is the fact that the correlation structure is not specified, but estimated from the data. For data that represent a continuous outcome (such as the math scores in the school data) which can be assumed to be normally distributed, GEE and hierarchical or mixed effects models give virtually the same results. These methods can yield quite different results with different interpretations for outcomes which are not continuous (e.g., a binary outcome). Because we do not address binary or other non-continuous outcome data in this chapter, we do not describe the GEE approach in more detail. The interested reader is referred to Verbeke and Molenberghs (2000), Hardin and Hilbe (2002), Fitzmaurice, Laird and Ware (2004), or Vittinghoff *et al.* (2005).

18.8   Discussion of computer programs

Table 18.6 summarizes the options available in the six statistical packages that perform the clustered outcome regression analyses presented in this chapter. It is important to note that different software might have very different ways of specifying options and also might use different defaults for some of the options (e.g., which correlation structure is assumed in case none is explicitly specified).

Some areas of research rely heavily on hierarchical modeling techniques. As a consequence some specialized programs have been developed which primarily serve specific areas of research. Zhou, Perkins and Hui (1999) provide a comparison of a number of software packages and features for the analysis of hierarchical data.

A potentially confusing difference between packages and statistical text books is the fact that different areas of research might use different labels and nomenclature. For example, some packages use level 1 to describe what we call the macro level and others use level 1 to describe what we call the micro level here. Users are encouraged to carefully read the software manual to ensure that options and descriptions fit what is intended.

Before entering the data for regression analysis of correlated data, the user should check what the options are for entering dates in the program. Most programs have been designed so that the user can simply use one record for each observation and indicate which observations belong to the same cluster. This format is called long format by some programs. An alternative representation is the wide format, in which each record represents a cluster. Most programs expect the format for the analysis to be the long format, but some tasks might be easier to perform in the wide format (e.g., some descriptive statistics or graphs).

18.9   What to watch out for

In addition to the remarks made in the regression analysis chapters, several potential problems exist for correlated regression analyses.

1. If a trial is randomized at the cluster level the required sample size will

Table 18.6: *Software commands and output for correlated outcomes regression*

	S-PLUS/R	SAS	SPSS	Stata	STATISTICA
Mixed effects linear regression	lme	MIXED	mixed	xtmixed, xtreg	see note*
Variance structure					
Independent	corCompSymm	MIXED	mixed	xtmixed, xtreg	
Exchangeable	corCompSymm	MIXED	mixed	xtmixed, xtreg	
Unstructured	CorSymm	MIXED	mixed	xtmixed, xtreg	
AR(k)	corAR1	MIXED	mixed	xtreg or xtmixed	
Estimation Method					
MLE	lme	MIXED	mixed	xtmixed, xtreg	
REML	lme	MIXED	mixed	xtmixed	
Predicting random effects	predict.lme	MIXED	mixed	predict	
Criteria					
AIC	AIC	MIXED	mixed	estat	
BIC	BIC	MIXED	mixed	estat	

*Does not have a general mixed model module.
STATISTICA does offer Variance Components and Mixed Model ANOVA/ANCOVA, but not mixed effects linear regression.

depend more heavily on the number of clusters rather than the number of subjects (see, e.g., Donner and Klar, 2000).

2. Missing values and informative drop out can present substantial challenges for the analysis and interpretation of longitudinal studies. Almost any longitudinal study which involves humans will have some data missing. Depending on the degree and nature of the missingness, appropriate approaches might require complex analysis techniques. A general introduction to missing data issues is provided in Section 9.2. Approaches specific to hierarchical/longitudinal data are described, e.g., by Schafer (1997), Molenberghs and Kenward (2007) and Daniels and Hogan (2008).

3. As a result, the estimates that are obtained represent approximations and different software packages might have different solutions; better packages might have fewer problems with convergence.

4. Other statistical textbooks might have correlated outcomes data under longitudinal data analysis and some might have it under hierarchical or multilevel modeling.

5. For likelihood ratio tests to be valid when comparing models with different fixed effects, the maximum likelihood method of estimation, and not REML, must be used. The maximum likelihood method might not represent the default in the statistical software used and might need to be explicitly requested.

18.10 Summary

In this chapter, we presented analysis techniques appropriate for correlated outcome data. Many authors make a clear distinction between hierarchical and longitudinal data. They are described here in the same chapter because the techniques that are used to analyze them are closely related. One even could think of longitudinal data as a special case of hierarchical data. The goal of the analyses of hierarchical data and outcome data that are not correlated are often very similar (e.g., elucidating differences between treatment or exposure groups). Nevertheless, there are conceptual as well as technical differences.

If outcome data are correlated, the analysis needs to accommodate such correlation. We described hierarchical data in conceptual as well as statistical terms. A distinction is made between fixed and random effects. Random intercept and random slope models were introduced. We provided an example of longitudinal data and the distinguishing features between hierarchical and longitudinal data and analysis.

18.11 Problems

Problems involving data sets refer to the school or the small mice data described in Section 1.1 and/or Appendix A.

18.1 Three student have the following scores on their math, english and social science classes. Student 1: 87.3, 91.2 and 86.0, Student 2: 75.4, 81.3 and 79.6, Student 3: 98.8, 95.0 and 94.7.
(a) Using a spread sheet program or a calculator, calculate the intra-class correlation coefficient.
(b) Change three of the values such that the intra-class correlation coefficient is between 0 and 0.5.

18.2 For the school data set generate box plots of SES by schools (similar to Figure 18.2). Interpret the graph with respect to potential differences in mean SES and differences in variability of SES across schools.

18.3 For the school data set using math score as the outcome variable, fit the following models:
(a) A fixed effects model with schools and SES as fixed effects.
(b) The same as (a), but with each SES centered around its respective school mean.
Compare the models in (a) and (b). What effect did centering SES have on the results? Why do you think this happened? (One or two sentences are sufficient.)

18.4 For the school data set (again, using math score as the outcome variable) fit a random intercept model with random intercepts for schools and SES centered around its respective school mean as a fixed effect. What is the estimated within school slope for SES? Compute the estimated between school slopes. How are this model and model (b) from Problem 18.3 the same and how are they different? (Again, one or two sentences are sufficient.)

18.5 For the school data set (again, using math score as the outcome variable) fit the following models:
(a) A random intercept model with random intercepts for schools and both the raw SES scores and the respective school means as fixed effects.
(b) The same model as in (a) except using SES centered around its respective school mean (instead of the raw SES scores) as well as school mean SES as fixed effects.
Consider model (a). Does school mean SES contribute significantly to explaining the child's math score? Would your conclusions change if you based them on model (b) instead? Why or why not?

18.6 For the school data set (again, using math score as the outcome variable) start with fitting a model with the following fixed effects: SES centered around its respective school mean, school mean SES, public school (vs private), hours spent on homework, race, gender, and the following random effects: SES centered around its respective school mean and hours spent on homework. Select a final model explain how you arrived at the final model, and justify any decisions made in the process. Also graph the between and within schools regressions.

18.7 There might be siblings represented in the school data set. The factor parental education could potentially be considered another level in the hierarchical model. Give reasons for when and why this might or should (not) be addressed.

18.8 For the mice data, instead of modeling weight linearly over time, model weight as a quadratic function of time by including a term for *days* as well as for $days^2$. Which model (random intercept, random slope) and correlation structure(s) seem most appropriate for these data? Describe your reasoning for the model you choose.

Appendix A

A.1 Data sets and how to obtain them

In the following we give a brief description of the data sets used in this book and where to find the data sets. All nine data sets and codebooks described in this Appendix can be obtained by going to the CRC Press web site

http://www.crcpress.com/product/isbn/9781439816806

and clicking on the Downloads & Updates tab. The data sets are also available at the following UCLA web site:

http://statistics.ats.ucla.edu/stat/examples/pma5/default.htm

At this web site each of the data sets is provided in five formats (R/S-PLUS, SAS, SPSS, Stata, and Statistica). In addition, the UCLA web site shows how to do the examples from each of the chapters using these statistical packages. Thus, it includes the data sets as well as the code necessary to produce the output included in the book.

A.2 Chemical companies financial data

The chemical companies data set includes seven measures of a company's financial performance. The data set is given in Table 8.1 and the measures are described in Section 8.3. The data set includes 30 companies that were the top performers listed in the January 5, 1981, volume 127 issue of *Forbes* magazine. The price earnings ratio was used as the dependent or outcome variable and the other six indicators were used as independent or predictor variables to see which of them best predicted the price earnings ratio.

Data set on CRC web site: Chemical.txt
Codebook: ChemicalCodebook.txt

A.3 Depression study data

The depression data set is from the first set of interviews of a prospective study of depression in the adult residents of Los Angeles County and includes 294 observations. See Table 3.3 for a codebook that describes the variables. The study is described in Section 1.2.

The authors wish to thank Dr. Ralph Frerichs and Dr. Carol Aneshensel at UCLA for their kind permission to use this data set.

Data set on CRC web site: Depress.txt
Codebook: DepressCodebook.txt

A.4   Financial performance cluster analysis data

The financial performance data set is given in Table 16.1 of Section 16.3. This data set includes the same seven financial performance measures used in Table 8.1. The 25 companies have been chosen from three types of companies: chemical companies, health related companies, and grocery store companies. This was done so that the *true* industry corresponding to each company can be compared to the results from the cluster analysis. Again these data were taken from the January 5, 1981, volume 127 issue of *Forbes* magazine.

Data set on CRC web site: Cluster.txt
Codebook: ClusterCodebook.txt

A.5   Lung cancer survival data

The lung function data set is from a multicenter clinical trial of an experimental treatment called BCG. It was carried out to see if BCG would lengthen the life of patients with lung cancer. All patients received the standard treatment and in addition they received either a saline control or BCG. This data set includes a sample of size 410 from the original data set. For further description see Sections 13.3 and 13.4 and the codebook given in Table 13.1.

The authors wish to thank Dr. E. Carmack Holmes, Professor of Surgical Oncology at the UCLA School of Medicine, for his kind permission to use the data.

Data set on CRC web site: Surv.txt
Codebook: SurvCodebook.txt

A.6   Lung function data

The lung function data set includes information on nonsmoking families from the UCLA study of chronic obstructive respiratory disease (CORD). In the CORD study persons 7 years old and older from four areas (Burbank, Lancaster, Long Beach, and Glendora) were sampled, and information was obtained from them at two time periods. The data in this book are a subset including 150 families with a mother and a father, and one, two, or three children between the ages of 7 and 17 who answered the questionnaire and took the lung function tests at the first time period. The purpose of the CORD study was to determine the effects of different types of air pollutants on respiratory function, but numerous other types of studies have been performed on this data set. Data on age, sex, height, weight, FVC, and FEV1 are included for the members of each family. Some families have only one or two children and if there is only one child it is listed as the oldest child. Since many families have only one child (considered the oldest), there are many missing values in the data for the middle and youngest child.

The authors wish to thank Dr. Roger Detels, the principal investigator of the CORD project, for the use of this data set and Miss Cathy Reems for assembling it.

Data set on CRC web site: Lung.txt
Codebook: LungCodebook.txt

A.7 Parental HIV data

The parental HIV data are part of a clinical trial to evaluate a behavioral intervention for families with a parent with HIV. The data include information on a subset of 252 adolescent children of parents with HIV. The codebook describes the variables and gives a brief description of their meaning.

The authors wish to thank Dr. Mary Jane Rotheram-Borus, Professor of Psychiatry and Biobehavioral Sciences, Director of the Center for Community Health, Neuropsychiatric Institute, UCLA for her kind permission to use the data.

Data set on CRC web site: Parhiv.txt
Codebook: ParhivCodebook.txt

A.8 Northridge earthquake data

The Northridge earthquake data set comes from a set of telephone surveys on the experiences of Los Angeles, CA county residents following the 1994 Northridge earthquake. Subjects were asked for their demographic information and about damage to their home, sustaining injury, and psychological responses to the earthquake. The data used here include 506 randomly selected subjects with relevant variables for the problems. The authors wish to thank Professor Linda Borque of the UCLA School of Public Health, Department of Community Health Sciences for her permission to use the data. We include it on the web site both as an ASCII data file and as a SAS data file.

Data set on CRC web site: Earthq.txt (ASCII data file)
Codebook: EarthqCodebook.txt
SAS data set: Earthq.sas7bdat

A.9   School data

The school data set is a publicly available data set that is provided by the National Center for Educational Statistics. The data come from the National Education Longitudinal Study of 1988 (called NELS:88). Extensive documentation of all aspects of the study is available at the following web site: `http://nces.ed.gov/surveys/NELS88/`.

Data set on CRC web site: School23.txt
Codebook: School23Codebook.txt

A.10   Mice data

The mice data set is a publicly available data set. The original data consist of four groups of mice. The weights of the mice were measured on a rotating basis for three of the groups and the fourth group was measured every day. We use the outcome measures for groups 3 and 4 for only and only the days where group three was measured.

Data set on CRC web site: Mice.txt
Codebook: MiceCodebook.txt

References

For all books cited, we use the copyright date.

Abelson, R.P. and Tukey, J.W. (1963). Efficient utilization of non-numerical information in quantitative analysis: General theory and the case of simple order. *Annals of Mathematical Statistics,* **34,** 1347–69.

Abu-Bader, S.H. (2010). *Advanced and Multivariate Statistical Methods for Social Science Research with a Complete SPSS Guide,* 3rd ed. Lyceum Books, Chicago.

Afifi, A.A. and Azen, S.P. (1979). *Statistical Analysis: A Computer Oriented Approach,* 2nd ed. Academic Press, New York.

Afifi, A.A. and Elashoff, R.M. (1966). Missing observations in multivariate statistics. I: Review of the literature. *Journal of the American Statistical Association,* **61,** 595–604.

Afifi, A.A. and Elashoff, R.M. (1969a). Missing observations in multivariate statistics. III: Large sample analysis of simple linear regression. *Journal of the American Statistical Association,* **64,** 337–58.

Afifi, A.A. and Elashoff, R.M. (1969b). Missing observations in multivariate statistics. IV: A note on simple linear regression. *Journal of the American Statistical Association,* **64,** 359–65.

Afifi, A.A., Kotlerman, J., Ettner, S. and Cowan, M. (2007). Methods for improving regression analysis for skewed or counted responses. *Annual Review of Public Health,* **28,** 95–111.

Agresti, A. (2002). *Categorical Data Analysis,* 2nd ed. Wiley, New York.

Allison, P.D. (2000). Multiple imputation for missing data: A cautionary tale. *Sociological Methods and Research,* **28,** 301–309.

Allison, P.D. (2001). *Missing Data,* Sage Publications, Thousand Oaks, CA.

Allison, P.D. (2010). *Survival Analysis Using the SAS System: A Practical Guide,* 2nd ed. SAS Institute Inc., Cary, NC.

Alwan, L.C. (2000). *Statistical Process Analysis,* Irwin, McGraw-Hill, New York.

Andersen, P.K., Borgan, Ø., Gill, R.D. and Keiding, N. (1993). *Statistical Models Based on Counting Processes,* Springer-Verlag, New York.

Andreasen, N.C. and Grove, W.M. (1982). The classification of depression:

Traditional versus mathematical approaches. *American Journal of Psychiatry,* **139,** 45–52.

Andrews, F., Klem, L., O'Malley, P., Rodgers, W., Welch, K. and Davidson, T. (1998). *Selecting Statistical Techniques for Social Science Data: A Guide for SAS Users,* SAS Institute Inc., Cary, NC.

Aneshensel, C.S. and Frerichs, R.R. (1982). Stress, support, and depression: A longitudinal causal model. *Journal of Community Psychology,* **10,** 363–76.

Atkinson, A.C. (1980). A note on the generalized information criterion for a choice of a model. *Biometrika,* **67,** 413–18.

Bachman, J.G. and O'Malley, P.M. (1986). Self-concepts, self-esteem, and educational experiences: The frog pond revisited (again). *Journal of Personal and Social Psychology,* **50(1),** 35–46.

Barnett, V. and Lewis, T. (2000). *Outliers in Statistical Data,* 3rd ed. Wiley, New York.

Bartholomew, D.J. and Knott, M. (1999). *Latent Variable Models and Factor Analysis,* 2nd ed. Charles Griffin and Company, London.

Bartlett, M.S. (1941). The statistical significance of canonical correlations. *Biometrika,* **32,** 29–38.

Beale, E.M.L. and Little, R.J.A. (1975). Missing values in multivariate analysis. *Journal of the Royal Statistical Society,* **37,** 129–45.

Belsley, D.A., Kuh, E. and Welsch, R.E. (1980). *Regression Diagnostics: Identifying Influential Data and Sources of Collinearity,* Wiley, New York.

Bendel, R.B. and Afifi, A.A. (1977). Comparison of stopping rules in forward stepwise regression. *Journal of the American Statistical Association,* **72,** 46–53.

Benedetti, J.K. and Brown, M.B. (1978). Strategies for the selection of log-linear models. *Biometrics,* **34,** 680–86.

Bickel, P.J. and Doksum, K.A. (1981). An analysis of transformations revisited. *Journal of the American Statistical Association,* **76,** 296–311.

Bishop, Y.M., Fienberg, S.E., Holland, P.W. and Light, R.J. (2007). *Discrete Multivariate Analysis: Theory and Practice,* MIT Press, Cambridge, MA.

Box, G.E.P. (1966). Use and abuse of regression. *Technometrics,* **8,** 625–29.

Box, G.E.P. and Cox, D.R. (1964). An analysis of transformations. *Journal of the Royal Statistical Society,* Series B, **26,** 211–52.

Bozdogan, H. (1987). Model selection and Akaike's Information Criterion (AIC): The general theory and its analytical extensions. *Psychometrika,* **52,** 345–70.

Braun, W. J. and Murdoch, D.J. (2007). *A First Course in Statistical Programming with R,* Cambridge University Press, New York.

Breiman, L., Friedman, J.H., Olshen, R.A. and Stone, C.J. (1984). *Classification and Regression Trees,* Chapman & Hall/CRC, Boca Raton, FL.

Breslow, N.E. (1996). Statistics in epidemiology: The case-control study. *Journal of the American Statistical Association,* **91,** 14–28.

Breslow, N.E. and Day, N.E. (1993). *Statistical Methods in Cancer Research: Volume I: The Analysis of Case-Control Studies,* International Agency for Research in Cancer, Scientific Publications No. 32. World Health Organization, Lyons, France.

Breslow, N.E. and Day, N.E. (1994). *Statistical Methods in Cancer Research: Volume II: The Design and Analysis of Cohort Studies,* International Agency for Research in Cancer, Scientific Publications No. 32. World Health Organization, Lyons, France.

Brigham, E.F. and Houston, J.F. (2009). *Fundamentals of Financial Management,* 12th ed. Western Cengage Learning, Mason, OH.

Bring, I. (1994). How to standardize regression coefficients. *The American Statistician,* **48,** 209–13.

Brownlee, K.A. (1965). *Statistical Theory and Methodology in Science and Engineering,* 2nd ed. Wiley, New York.

Cameron, A. C. and Trivedi, P. K. (1998). *Regression Analysis of Count Data,* Cambridge University Press, Cambridge, UK.

Chambers, J.M. (1983). *Graphical Methods for Data Analysis,* Wadsworth, Belmont, CA.

Chatfield, C. (2004). *The Analysis of Time Series: An Introduction,* 6th ed. Chapman & Hall, London.

Chatterjee, S. and Hadi, A.S. (1988). *Sensitivity Analysis in Linear Regression,* Wiley, New York.

Chatterjee, S. and Hadi, A.S. (2006). *Regression Analysis by Example,* 4th ed., Wiley, New York.

Cheng, J., Edwards, L.J., Maldonado-Molina, M.M. *et al.* (2010). Real longitudinal data analysis for real people: Building a good enough mixed model. *Statistics in Medicine,* **29,** 510–520.

Claeskens, G. and Hjort, N. (2008). *Model Selection and Model Averaging,* Cambridge University Press, New York.

Clark, V.A., Aneshensel, C.S., Frerichs, R.R. and Morgan, T.M. (1981). Analysis of effects of sex and age in response to items on the CES-D scale. *Psychiatry Research,* **5,** 171–81.

Cleveland, W.S. (1993). *Visualizing Data,* Hobart Press, Summit, NJ.

Cleves, M., Gould, W.W. and Gutierrez, R. (2008). *An Introduction to Survival Analysis Using Stata,* 2nd ed. Stata Press, College Station, TX.

Clogg, C.C. and Eliason, S.R. (1987). Some common problems in log-linear analysis. *Sociological Methods and Research,* **16,** 8–44.

Cohen, J., Cohen, P., West, S.G. and Aiken, L.S. (2002). *Applied Multiple Regression/Correlation Analysis for the Behavioral Sciences,* 3rd ed. Routledge Academic, Mawah, NJ.

Comstock, G.W. and Helsing, K.J. (1976). Symptoms of depression in two communities. *Psychological Medicine*, **6**, 551–63.

Conover, W.J. (1999). *Practical Nonparametric Statistics*, 3rd ed. Wiley, New York.

Cook, R.D. (1977). Detection of influential observations in linear regression. *Technometrics*, **19**, 15–18.

Cook, R.D. and Weisberg, S. (1982). *Residuals and Influence in Regression*, Chapman & Hall, London.

Cook, R.D. and Weisberg, S. (1994). *An Introduction to Regression Graphics*, Wiley, New York.

Costanza, M.C. and Afifi, A.A. (1979). Comparison of stopping rules for forward stepwise discriminant analysis. *Journal of the American Statistical Association*, **74**, 777–85.

Crawley, M. (2002). *Statistical Computing: An Introduction to Data Analysis Using S-PLUS*, Wiley, New York.

Crowley, J. and Hu, M. (1977). Covariance analysis of heart transplant data. *Journal of the American Statistical Association*, **72**, 27–36.

D'Agostino, R.B., Belanger, A. and D'Agostino, Jr., R.B. (1990). A suggestion for using powerful and informative tests of normality. *The American Statistician*, **44**, 316–21.

Dalgaard, P. (2008). *Introductory Statistics with R*, Springer, New York.

Daniels, M.J. and Hogan, J.W. (2008). *Missing Data in Longitudinal Studies, Strategies for Bayesian Modeling and Sensitivity Analysis*, Chapman and Hall/CRC Press, Boca Raton, FL.

Das-Munshi, J., Goldberg, D, Bebbington P.E., Bhugra D.K., Brugha T.S., Dewey, M.E., Jenkins, R., Stewart, R. and Prince, M. (2008). Public health significance of mixed anxiety and depression: beyond current classification. *British Journal of Psychiatry*, **192**, 171–7.

David, M., Little, R.J.A., Samuhel, M.E. and Triest, R.K. (1986). Alternative methods for CPS income imputation. *Journal of the American Statistical Association*, **81**, 29–41.

Davison, A.C. and Hinkley, D.V. (1997). *Bootstrap Methods and Their Application*, Cambridge University Press, New York.

Dean, A. and Voss, D. (1999). *Design and Analysis of Experiments*, Springer, New York.

Demaris, A. (1990). Interpreting logistic regression results: A critical commentary. *Journal of the Marriage and the Family*, **52**, 271–77.

Demaris, A. (1992). *Logit Modeling: Practical Applications*, Sage, Newbury Park, CA.

Der, G. and Everitt, B.S. (2008) *A Handbook of Statistical Analyses Using SAS Software*, 3rd ed. CRC Press, Boca Raton, FL.

Detels, R., Coulson, A., Tashkin, D. and Rokaw, S. (1975). Reliability of plethysmography, the single breath test, and spirometry in population studies. *Bulletin de Physiopathologie Respiratoire,* **11,** 9–30.

Diggle, P.J. (1996). *Time Series: A Biostatistical Introduction,* Oxford University Press, New York.

Donner, A. and Klar N. (2000). *Design and Analysis of Cluster Randomization Trials in Health Research,* Arnold, New York.

Draper, N.R. and Hunter, W.G. (1969). Transformations: Some examples revisited. *Technometrics,* **11,** 23–40.

Draper, N.R. and Smith, H. (1998). *Applied Regression Analysis,* 3rd ed. Wiley, New York.

Dunteman, G.H. (1989). *Principal Components Analysis,* Sage University Papers. Sage, Newbury Park, CA.

Eastment, H.T. and Krzanowski, W.J. (1982). Cross-validory choice of the number of components from a principal component analysis. *Technometrics,* **24,** 73–77.

Efron, B. and Tibshirani, R.J. (1994). *An Introduction to the Bootstrap,* Vol. 57, CRC Press, Boca Raton, FL.

Enders, C.K. (2001). A primer on maximum likelihood algoriths available for use with missing data. *Structural Equation Modeling,* **8,** 128–141.

Enders, C.K. (2010). *Applied Missing Data Analysis.* The Guilford Press, New York.

Everitt, B.S. (2007). *An R and S-PLUS Companion to Multivariate Analysis,* Springer, London.

Everitt, B.S. and Dunn, G. (2001). *Applied Multivariate Data Analysis,* 2nd ed. Wiley, New York.

Everitt, B.S., Landau, S. and Leese, M. (2001). *Cluster Analysis,* 4th ed. Wiley-Blackwell.

Fisher, R.A. (1936). The use of multiple measurements in taxonomic problems. *Annals of Eugenics,* **7,** 179–88.

Fisher, L.D. and Lin, D.Y. (1999). Time-dependent covariates in the Cox proportional hazards regression model. *Annual Review of Public Health,* **20,** 145–157.

Fitzmaurice, G.M., Laird, N.M. and Ware J.H. (2004). *Applied Longitudinal Analysis.* Wiley, Hoboken.

Flack, V.F. and Chang, P.C. (1987). Frequency of selecting noise variables in a subset regression analysis: A simulation study. *The American Statistician,* **41,** 84–86.

Fleiss, J.L., Levin, B. and Paik, N.C. (2003). *Statistical Methods for Rates and Proportions,* 3rd ed. Wiley, Hoboken, NJ.

Fox, J. (1991). *Regression Diagnostics,* Sage, Newbury Park, CA.

Fox, J. and Long, J.S. (Eds.) (1990). *Modern Methods of Data Analysis,* Sage, Newbury Park, CA.

500 REFERENCES

Frerichs, R.R., Aneshensel, C.S. and Clark, V.A. (1981). Prevalence of depression in Los Angeles County. *American Journal of Epidemiology,* **113,** 691–99.

Frerichs, R.R., Aneshensel, C.S., Clark, V.A. and Yokopenic, P. (1981). Smoking and depression: A community survey. *American Journal of Public Health,* **71,** 637–40.

Freund, R.J. and Littell, R.C. (2006). *SAS System for Regression,* 3rd ed. Wiley, New York.

Frigge, M., Hoaglin, D.C. and Iglewicz, B. (1989). Some implementations of the boxplot. *The American Statistician,* **43,** 50–4.

Fuller, W.A. (1987). *Measurement Error Models,* Wiley, New York.

Gail, M.H., Eagan, R.T., Feld, R., Ginsberg, R., Goodell, B., *et al.* (1984). Prognostic factors in patients with resected stage 1 non-small-cell lung cancer. *Cancer,* **9,** 1802–13.

Gan, F.F., Koehler, K.J. and Thompson, J.C. (1991). Probability plots and distribution curves for assessing the fit of probability models. *The American Statistician,* **45,** 14–21.

Gan, G., Ma, C. and Wu, J. (2007). *Data clustering: Theory, Algorithms, and Applications,* SIAM, Society for Industrial and Applied Mathematics, Philadelphia, PA.

Gerber, S.B. and Finn, K.V. (2005). *Using SPSS for Windows: Data Analysis and Graphics,* 2nd ed. Springer-Verlag, New York.

Gorsuch, R.L. (1983). *Factor Analysis,* 2nd ed. Psychology Press, New York, London.

Green, M.S. and Symons, M.J. (1983). A comparison of the logistic risk function and the proportional hazards model in prospective epidemiologic studies. *Journal of Chronic Disease,* **36,** 715–24.

Greene, W.H. (2008). *Econometric Analysis,* 6th ed. Prentice Hall, Upper Saddle River, NJ.

Grønnesby, J. and Borgan, Ø. (1996). A method for checking regression models in survival analysis based on the risk score. *Lifetime Data Analysis,* **2,** 315–328.

Groves, R.M., Dillman, D.A., Eltinge, J.L. and Little, R.J.A. (Eds.) (2002). *Survey Nonresponse,* Wiley, New York, 329-341.

Gruber, M. (2010). *Regression Estimators: A Comparative Study,* 2nd ed. Johns Hopkins University Press, Baltimore.

Gunst, R.F. and Mason, R.L. (1980). *Regression Analysis and Its Application,* Marcel Dekker, New York.

Haberman, S.J. (1977). Log-linear models and frequency tables with small expected cell counts. *Annals of Statistics,* **5,** 1148–69.

Hamilton, L.C. (2009). *Statistics with Stata: Version 10,* 7th ed., Duxbury Press, Toronto, Canada.

Hardin, J.W. and Hilbe, J.M. (2002) *Generalized Estimating Equations*, Chapman & Hall/CRC, Boca Raton, FL.

Harman, H.H. (1976). *Modern Factor Analysis,* University of Chicago Press, Chicago.

Harris, E.K. and Albert, A. (1991). *Survivorship Analysis for Clinical Studies,* Marcel Dekker, New York.

Hartley, H.O. and Hocking, R.R. (1971). The analysis of incomplete data. *Biometrics,* **27,** 783–824.

Hatcher, L. (1994). *A Step-by-Step Approach to Using the SAS System for Factor Analysis and Structural Equation Modeling,* SAS Institute Inc., Cary, NC.

Heiberger, R.M. and Neuwirth, E. (2009). *R through Excel: A Spreadsheet Interface for Statistics, Data Analysis, and Graphics,* Dordrecht: Springer, New York.

Hills, M. and De Stavola, B.L. (2009). *A Short Introduction to Stata for Biostatistics,* Timberlake Consultants Press, London, UK.

Hines, W.G.S. and O'Hara Hines, R.J. (1987). Quick graphical power-law transformation selection. *The American Statistician,* **41,** 21–4.

Hirji, K.F., Mehta, C.R. and Patel, N.R. (1987). Computing distributions for exact logistic regression. *Journal of the American Statistical Association,* **82,** 1110–17.

Hoaglin, D.C., Mosteller, F. and Tukey, J.W. (Eds.) (1985). *Understanding Robust and Exploratory Data Analysis,* Wiley, New York.

Hocking, R.R. (1976). The analysis and selection of variables in linear regression. *Biometrics,* **32,** 1–50.

Hoerl, A.E. and Kennard, R.W. (1970). Ridge regression: Application to nonorthogonal problems. *Technometrics,* **12,** 55–82.

Holford, T.R., White, C. and Kelsey, J.L. (1978). Multivariate analyses for matched case-control studies. *American Journal of Epidemiology,* **107,** 245–56.

Horton, N.J. and Lipsitz, S.R. (2001). Multiple imputation in practice: Comparison of software packages for regression models with missing variables. *The American Statistician,* **55,** 244–254.

Hosmer, D.W. and Lemeshow, S. (1980). Goodness-of-fit tests for the multiple logistic regression model. *Communications in Statistics: Theory and Methods,* **A10,** 1043–1069.

Hosmer, D.W. and Lemeshow, S. (2000). *Applied Logistic Regression,* 2nd ed., Wiley, New York.

Hosmer, D.W., Lemeshow, S. and May, S. (2008). *Applied Survival Analysis: Regression Modeling of Time-to-Event Data*, 2nd ed., Wiley, Hoboken, NJ.

Hotelling, H. (1933). Analysis of a complex of statistical variables into principal components. *Journal of Educational Psychology,* **24,** 417–41.

Ingram, D.D. and Kleinman, J.C. (1989). Empirical comparison of proportional hazards and logistic regression models. *Statistics in Medicine,* **8,** 525–38.

Izenman, A. and Williams, J. (1989). A class of linear spectral models and analyses for the study of longitudinal data. *Biometrics,* **45,** 831–849.

Jaccard, J. (2001). *Interaction Effects in Logistic Regression,* Sage Publicatons, Thousand Oaks, CA.

Jaccard, J., Turrisi, R. and Wan, C.K. (1990). *LISREL Approaches to Interaction Effects in Multiple Regression,* Sage, Newbury Park, CA.

Jackson, J.E. (2003). *A User's Guide to Principal Components,* Wiley, New York.

Jackson, J.E. and Hearne, F.T. (1973). Relationships among coefficients of vectors in principal components. *Technometrics,* **15,** 601–10.

Jewell, N.P. (2004). *Statistics for Epidemiology,* Chapman & Hall/CRC, Boca Raton, FL.

Johnson, R.A. and Wichern, D.W. (2007). *Applied Multivariate Analysis,* 5th ed. Prentice Hall, Upper Saddle River, NJ.

Jolliffe, I.T. (2010). *Principal Components Analysis.* 2nd Ed., Springer-Verlag, New York.

Kachigan, S.K. (1991). *Multivariate Statistical Analysis: A Conceptual Introduction,* 2nd ed. Radius Press, New York.

Kalbfleisch, J.D. and Prentice, R.L. (2002). *The Statistical Analysis of Failure Time Data*, 2nd ed. Wiley, New York.

Kalton, G. and Kasprzyk, D. (1986). The treatment of missing survey data. *Survey Methodology,* **12,** 1–16.

Kaufman, L. and Rousseeuw, P.J. (2005). *Finding Groups in Data: An Introduction to Cluster Analysis,* Wiley, Hoboken, NJ.

Kedem, B. and Fokianos, K. (2002). *Regression Models for Time Series,* Wiley, New York.

Khattree, R. and Naik, D. (2003). *Applied Multivariate Statistics with SAS Software,* 2nd ed. SAS Institute, Cary, NC.

King, E.N. and Ryan, T.R. (2002). A preliminary investigation of maximum likelihood logistic regression versus exact logistic regression. *The American Statistician,* **56,** 163–170.

Klein, J.P. and Moeschberger, M.L. (2003). *Survival Analysis: Techniques for Censored and Truncated Data*, 2nd ed. Springer, New York.

Kleinbaum, D.G. and Klein, M. (2005). *Survival Analysis: A Self-Learning Text,* 2nd ed. Springer, New York.

Kleinbaum, D.G. and Klein, M. (2010). *Logistic Regression: A Self-learning Text,* 3rd ed. Springer, New York.

Kleinbaum, D.G., Kupper, L.L., Nizam, A. and Muller, K.E. (2007). *Applied Regression Analysis and Multivariable Methods,* 4th ed. Duxbury Press, Pacific Grove, CA.

Kraemer, H.C. (1988). Assessment of 2×2 associations: Generalization of signal-detection methodology. *The American Statistician,* **42,** 37–49.

Lawley, D.N. (1959). Tests of significance in canonical analysis. *Biometrika,* **46,** 59–66.

Lee, E.T. and Wang, J.W. (2003). *Statistical Methods for Survival Data Analysis,* 3rd ed. Lifetime Learning, Belmont, CA.

Lee, J.J., Hess, K.R. and Dubin, J.A. (2000). Extensions and applications of event charts. *The American Statistician,* **54,** 63–70.

Lemeshow, S. and Hosmer, D.W. (1982). A review of goodness-of-fit statistics for use in the development of logistic regression models. *American Journal of Epidemiology,* **115,** 92–106.

Little, R.J.A. (1982). Models for nonresponse in sample surveys. *Journal of the American Statistical Association,* **77,** 237–50.

Little, R.J.A. (1992). Regression with missing X's: A review. *Journal of the American Statistical Association,* **87,** 1227–37.

Little, R.J.A. and Rubin, D.B. (1990). The analysis of social science data with missing values, in *Modern Methods of Data Analysis* (Eds. J. Fox and J.S. Long), Sage, Newbury Park, CA.

Little, R.J.A. and Rubin, D.B. (2002). *Statistical Analysis with Missing Data,* Wiley, New York.

Long, J.S. (1983). *Confirmatory Factor Analysis: A Preface to LISREL,* Sage, Newbury Park, CA.

Luce, R.D. and Narens, L. (1987). Measurement scales on the continuum. *Science,* **236,** 1527–1532.

Lumley, T., Diehr, P., Emerson, S. and Chen, L. (2002). The importance of the normality assumption in large public health data sets. *Annual Review of Public Health,* **23,** 151–69.

Mallows, C.L. (1973). Some comments on C_p. *Technometrics,* **15,** 661–76.

Mallows, C.L. (1986). Augmented partial residuals. *Technometrics,* **28,** 313–19.

Manly, B.F.J. (2004.) *Multivariate Statistical Methods: A Primer,* 3rd ed. CRC Press, Boca Raton, FL.

Marks, S. and Dunn, O.J. (1974). Discriminant functions when covariance matrices are unequal. *Journal of the American Statistical Association,* **69,** 555–559.

Marquardt, D.W. and Snee, R.D. (1975). Ridge regression in practice. *American Statistician,* **29,** 3–20.

Marsh, L. and Cormier, D.R. (2002). *Spline Regression Models,* Sage Publications, Thousand Oaks, CA.

May, S. and Hosmer, D.W. (1998). A simplified method of calculating an overall goodness-of-fit test for the Cox proportional hazards model. *Lifetime Data Analysis,* **4,** 109–120.

McCulloch, C.E., Searle, S.,R. and Neuhaus, J.M. (2008). *Generalized, Linear, and Mixed Model*, 2nd Ed. Springer, New York.

McCullagh, P. and Nelder, J.A. (1989). *Generalized Linear Models*, 2nd Ed. Chapman & Hall/CRC, Boca Raton, FL.

McKnight, P.E., McKnight, K.M., Sidani, S. and Figueredo, A.J. (2007). *Missing Data: A Gentle Introduction*, Guilford Press, New York.

McLachlan, G.J. (2004). *Discriminant Analysis and Statistical Pattern Recognition,* Wiley, New York.

McQuarrie, A.D. and Tsai, C.L. (1998). *Regression and Time Series Model Selection,* World Scientific Publishing Company, River Edge, NJ.

Mehta, C.R. and Patel, N.R. (1995). Exact logistic regression: Theory and examples. *Statistics in Medicine, **14,*** 2143–2160.

Menard, S. (2001). *Applied Logistic Regression Analysis,* 2nd ed. Sage Publications, Thousand Oaks, CA.

Meredith, W. (1964). Canonical correlation with fallible data. *Psychometrika* **29,** 55–65.

Metz, C.E. (1978). Basic principles of ROC analysis. *Seminars in Nuclear Medicine,* **8,** 283–98.

Mickey, R.M., Dunn, O.J. and Clark, V.A. (2009). *Applied Statistics: Analysis of Variance and Regression,* 3rd ed. Wiley, NY.

Miller, A.J. (2002). *Subset Selection in Regression,* CRC Press, Boca Raton, FL.

Miller, R.G. (1981). *Survival Analysis*, Wiley, New York.

Molenberghs, G. (2007). Editorial: What to do with missing data? *Journal of the Royal Statistical Society Series A*, **170,** Part 4, 861–863.

Molenberghs, G. and Kenward, M.G. (2007). *Missing Data in Clinical Studies*, Wiley, Chichester, UK.

Mosteller, F. and Tukey, J.W. (1977). *Data Analysis and Regression: A Second Course in Statistics,* Addison-Welsley, Reading, MA.

Muller, K.E. (1981). Relationships between redundancy analysis, canonical correlation, and multivariate regression. *Psychometrika,* **46,** 139–42.

National Center for Educational Statistics. (1988). *National Education Longitudinal Study of 1988*, U.S. Department of Education, Washington, DC.

Nguyen, L.H., Shen, H., Ershoff, D., Afifi, A.A., and Bourque, L.B. (2006). Exploring the causal relationship between exposure to the 1994 Northridge earthquake and pre- and post-earthquake preparedness activities. *Earthquake Spectra,* **22,** 569–87.

Nishii, R. (1984). Asymptotic properties of criteria for selection of variables in multiple regression. *The Annals of Statistics*, **12,** 758–65.

O'Gorman, T.W. and Woolson, R.F. (1991). Variable selection to discriminate between two groups: Stepwise logistic regression or stepwise discriminant analysis? *The American Statistician,* **45,** 187–93.

Okunade, A.A., Chang, C.F. and Evans, R.D. (1993). Comparative analysis of regression output summary statistics in common statistical packages. *The American Statistician*, **47**, 298–303.

Oler, J. (1985). Noncentrality parameters in chi-squared goodness-of-fit analyses with an application to log-linear procedures. *Journal of the American Statistical Association*, **80**, 181–89.

Parzen, M. and Lipsitz, R. (1999). A Global goodness-of-fit statistic for Cox regression models. *Biometrics*, **55**, 580–584.

Pett, M.A., Lackey, N.R. and Sullivan, J.J. (2003). *Making Sense of Factor Analysis: The Use of Factor Analysis for Instrument Development in Health Care Research,* Sage Publications, Inc., Thousand Oaks, CA.

Pough, F.H. (1996). *Field Guide to Rocks and Minerals,* 5th ed. Houghton Mifflin, Boston, MA.

Powers, D.A. and Xie, Y. (2008). *Statistical Methods for Categorical Data Analysis,* 2nd ed. Elsevier Science & Technology Books, New York.

Pugesek, B., Tomer, A. and Von Eye, A. (Eds.) (2003). *Structural Equation Modeling: Applications in Ecological and Evolutionary Biology,* Cambridge University Press, New York.

Quandt, R.E. (1972). New approaches to estimating switching regression. *Journal of the American Statistical Association*, **67**, 306–310.

Rabe-Hesketh, S. and Everitt, B. (2007). *A Handbook of Statistical Analyses Using Stata,* 4th ed. CRC Press, Boca Raton, FL.

Radloff, L.S. (1977). The CES-D scale: A self-report depression scale for research in the general population. *Applied Psychological Measurement*, **1**, 385–401.

Rao, C.R. (1965). Covariance adjustment and related problems in multivariate analysis, in Dayton Symposium on Multivariate Analysis, *Multivariate Analysis 1*, pp. 87–103, Academic Press, New York.

Rao, C.R. (1973). *Linear Statistical Inference and Its Applications,* 2nd ed. Wiley Interscience, Hoboken, NJ.

Rencher, A.C. (2002). *Methods of Multivariate Analysis,* 2nd ed. Wiley, New York.

Rencher, A.C. and Larson, S.F. (1980). Bias in Wilks' Λ in stepwise discriminant analysis. *Technometrics*, **22**, 349–56.

Reyment, R.A. and Jőreskog, K.G. (1996). *Applied Factor Analysis in the Natural Sciences,* 2nd ed. Cambridge University Press, New York.

Rotheram-Borus, M.J., Lee, M.B., Gwadz, M. and Draimin, B. (2001). An intervention for parents with AIDS and their adolescent children. *American Journal of Public Health*, **91**, 1294–1302.

Rothman, K.J., Greenland, S. and Lash, T.L. (2008). *Modern Epidemiology,* 3rd ed., Wolters Kluwer Health/Lippincott Williams & Wilkins, Philadelphia, PA.

Royston, P. (2004). Multiple imputation of missing values. *Stata Journal*, **4**, 227–241.

Royston, P., Carlin, J.B. and White I.R. (2009). Multiple imputation of missing values: New features for mim. *Stata Journal*, **9**, 252-264.

Rubin, D.B. (1987). *Multiple Imputation for Nonresponse in Surveys*, Wiley, New York.

Rubin, D.B. (2004). *Multiple Imputation for Nonresponse Surveys*, Wiley, New York.

Ryan, T.P. (2009). *Modern Regression Methods*, 2nd ed., Wiley, New York.

Schafer, J.L. (1997). *Analysis of Incomplete Multivariate Data*, Chapman & Hall, New York.

Schafer, J.L. (1999). Multiple imputation: A primer. *Statistical Methods in Medical Research*, **8**, 3–15.

Schlesselman, J.J. (1982). *Case-Control Studies: Design, Conduct, Analysis*, Oxford University Press, New York.

Schwaiger, M. and Opitz, O. (2003). *Exploratory Data Analysis in Empirical Research: Proceedings of the 25th Annual Conference of the Gesellschaft fr Klassifikation e.V., University of Munich, March 14–16, 2001*, Springer-Verlag, New York.

Selvin, S. (1998). *Modern Applied Biostatistical Methods Using S-PLUS*, Oxford University Press, New York.

Serdula, M.K., Ivery, D., Coates, R.J. *et al.* (1993) Do obese children become obese adults? A review of the literature. *Preventive Medicine*, **22**, 167–177.

Smith, P.J. (2002). *Analysis of Failure and Survival Data*, Chapman and Hall/CRC, Boca Raton, FL.

Snijders, T.A. and Bosker, R.J. (1999). *Multilevel Analysis: An Introduction to Basic and Advanced Multilevel Modeling*, Sage Publications, Thousand Oaks, CA.

Sprent, P. and Smeeton, N.C. (2007). *Applied Nonparametric Statistical Methods*, 4th ed. Chapman & Hall/CRC, Boca Raton, FL.

Stevens, S.S. (1966). Mathematics, measurement and psychophysics, in *Handbook of Experimental Psychology* (Ed. S.S. Stevens), Wiley, New York, 1–49.

Stokes, M.E., Davis, C.S. and Koch, G.G. (2009). *Categorical Data Analysis Using the SAS System*, 2nd ed. Wiley, New York.

Stuart, A. and Ord, J.K. (1994). *Kendall's Advanced Theory of Statistics, Vol. 1: Distribution Theory*, 6th ed., Wiley, New York.

Swets, J.A. (1973). The receiver operating characteristic in psychology. *Science*, **182**, 990–1000.

Tashkin, D.P., Clark, V.A., Simmons, M., Reems, C., Coulson, A.H., Bourque, L.B., Sayre, J.W., Detels, R. and Rokaw, S. (1984). The UCLA population studies of chronic obstructive respiratory disease. VII. Relationship between parents smoking and children's lung function. *American Review of Respiratory Disease*, **129**, 891–97.

Tchira, A.A. (1973). Stepwise regression applied to a residential income valuation system. *Assessors Journal*, **8**, 23–35.

Timm, N.H. (2002). *Applied Multivariate Analysis*, Springer-Verlag, Inc., New York.

Tobin, J. (1958). Estimation of relationship for limited dependent variables. *Econometrica*, **26**, 24–36.

Tufte, E.R. (1997). *Visual Explanations: Images and Quantities, Evidence and Narrative*, Graphics Press, Cheshire, CT.

Tufte, E.R. (2001). *The Visual Display of Quantitative Information*, 2nd ed., Graphics Press, Cheshire, CT.

Tukey, J.W. (1977). *Exploratory Data Analysis*, Addison-Wesley, Reading, MA.

US Bureau of Census (1998). Available at http://www.census.gov/prod/3/98pubs/98statab/sasec30.pdf

van Belle, G., Fisher, L.D., Heagerty, P.J. and Limley, T. (2004). *Biostatistics: A Methodology for the Health Sciences*, 2nd ed., Wiley, New York.

Velleman, P.F. and Wilkinson, L. (1993). Nominal, ordinal, interval, and ratio typologies are misleading. *The American Statistician*, **47**, 65–72.

Verbeke, G. and Molenberghs, G. (2000). *Linear Mixed Models for Longitudinal Data*, Springer, New York.

Vittinghoff, E., Glidden, D., Shiboski, S. *et al.* (2005) *Regression Methods in Biostatistics*, Springer, New York.

Waller, N.G. (1993). Seven confirmatory factor analysis programs – EQS, EZ-PATH, LINCS, LISCOMP, LISREL-7, SIMPLIS, and CALIS. *Applied Psychological Measurement*. **17**, 73–100.

Waugh, F.V. (1942). Regressions between sets of variables. *Econometrica*, **10**, 290–310.

Weiss, R. (2005). *Modeling Longitudinal Data*, Springer, New York.

Whitaker, R.C., Wright, J.A., Pepe, M.S. *et al.* (1997). Predicting obesity in young adulthood from childhood and parental obesity. *N Engl J Med.*, **337**, 869–873.

Wickens, T.D. (1989). *Multiway Contingency Tables Analysis for the Social Sciences*, Lawrence Erlbaum Associates, Hillsdale, NJ.

Wilson, E.O., Eisner, T., Dickerson, R.E., Metzenberg, R.L., O'Brien, R.D., Susman, M. and Boggs, W.E. (1973). *Life on Earth*, Sinauer Associates, Inc., Publisher, Stamford, CT.

Winkelmann, R. (2008). *Econometric Analysis of Count Data*, 5th ed. Springer, New York.

Wishart, J. (1931). The mean and second moment coefficient of the multiple correlation coefficient, in samples from a normal population. *Biometrika*, **22**, 353–367.

Woodbury, M.A., Manton, K.G. and Stallard, E. (1981). Longitudinal models for chronic disease risk: An evaluation of logistic multiple regression and alternatives. *International Journal of Epidemiology,* **10,** 187–97.

Woolson, R.F. and Lachenbruch, P.A. (1982). Regression analysis of matched case-control data. *American Journal of Epidemiology,* **115,** 442–52.

Yaffee, R. and McGee, M. (2000). *Introduction to Time Series and Forecasting: With Applications of SAS and SPSS,* Academic Press, New York.

Yamaguchi, K. (1991). *Event History Analysis,* Sage, Newbury Park, CA.

Zhang, H. and Singer, B. (1999). *Recursive Partitioning in the Health Sciences,* Springer-Verlag, New York.

Zhou X.H., Perkins A.J. and Hui, S.L. (1999). Comparisons of software packages for generalized linear multilevel models. *American Statistician,* **53(3),** 282–90.

Index

514

INDEX

logit model, 455
models, 442
notation, 435, 442
sample size issues, 453
three-way, 433

Nominal logistic regression, 298
Nominal variables, 17, 68
Normal probability plots, 55
Normal quantile plots, 56
Normality, tests for, 59
Northridge earthquake study, 5

Oblique rotations, 393
Odds, 272, 274
Odds ratio, 274, 433
Ordinal logistic regression, 302
Ordinal variables, 17, 70
Orthogonal rotations, 391
Outcome variables, 19, 81
Outliers, 37, 100, 138

Paired sample, 111
Pairwise deletion, 192
Parental HIV study, 4
Partial correlation, 129, 137
Partial regression coefficient, 123
Partial regression plots, 181
Partial residual plots, 182
Poisson distribution, 309
Poisson regression
computer programs, 312
counted data, 309
generalized linear model, 311
link function, 311
maximum likelihood, 312
outcome variable, 309
rate ratios, 313
Polynomial regression, 140
Posterior probabilities, 254
Prediction interval, 88
Prediction intervals, 124
Predictor variables, 19, 81
Principal components

analysis of variance, 373
CESD scale, 367
characteristic root, 361
coefficients, 360
communalities, 382
computer programs, 373
concentration ellipse, 361
data example, 358
definition, 360
eigenvalues, 361
Kaiser normalization, 392
last eigenvalue, 371
multicollinearity, 371
number of
components, 363, 367
outliers, 372
rotation, 360
rules for number used, 363
specificity, 382
standardized variables, 364
uncorrelated, 358
use in regression, 371
variances, 360
what to watch for, 374, 400
when used, 357
Profile diagrams, 409
Proportional hazards model, 338
Proportional odds model, 302
Pseudo R squared, 295

Quantiles, 52

Random effects, 470
Random intercept model, 471
Random slope, 473
Rate ratios, 313
Ratio variables, 18, 71
Receiver operating curve, 296
Recursive partitioning, 206
Redundancy analysis, 232
Referent group, 199
Reflected, 43
Regression trees, 206